# Physical Metallurgy

# Physical Metallurgy

Edited by **Darren Wang**

**C**WILLFORD PRESS

New York

Published by Willford Press,
118-35 Queens Blvd., Suite 400,
Forest Hills, NY 11375, USA
www.willfordpress.com

**Physical Metallurgy**
Edited by Darren Wang

International Standard Book Number: 978-1-68285-018-3 (Hardback)

Printed in the United States of America.

# Contents

# Preface

Physical metallurgy is a dynamic field that studies the process of forming metals with varying properties. This book is a valuable compilation of topics, ranging from the basic to the most complex advancements in the field of alloy design, processing of ores, and metallurgical extraction, etc. As this field is emerging at a rapid pace, the contents of this book along with the contributions made by eminent scholars will help the readers understand the modern concepts and applications of the subject.

After months of intensive research and writing, this book is the end result of all who devoted their time and efforts in the initiation and progress of this book. It will surely be a source of reference in enhancing the required knowledge of the new developments in the area. During the course of developing this book, certain measures such as accuracy, authenticity and research focused analytical studies were given preference in order to produce a comprehensive book in the area of study.

This book would not have been possible without the efforts of the authors and the publisher. I extend my sincere thanks to them. Secondly, I express my gratitude to my family and well-wishers. And most importantly, I thank my students for constantly expressing their willingness and curiosity in enhancing their knowledge in the field, which encourages me to take up further research projects for the advancement of the area.

**Editor**

# Synthesis of perovskite CaTiO$_3$ nanopowders with different morphologies by mechanical alloying without heat treatment

**Sahebali Manafi and Mojtaba Jafarian**

Department of Engineering, Shahrood Branch, Islamic Azad University, Shahrood, Iran.

**Mechanical alloying (MA) method is one of the methods used for large scale production of different nanopowders. In this study, calcium titanate (CaTiO$_3$: CTO) nanoparticles have been synthesized via mechanical alloying (MA) without using heat treatment. The milled powders and CTO were characterized by XRD, SEM, and zetasizer. It is found that the CTO has a diameter of 30 - 70 nm with different morphologies. The results showed the minimum time of calcium titanate synthesis via mechanical alloying without heat treatment is 70 h that formed and the range of grain size (apparent size) using Williamson-Hall equation is 69 nm.**

**Key words:** Mechanical alloying/activation, morphology, perovskite, CaTiO$_3$.

## INTRODUCTION

Calcium titanate (CaTiO$_3$: CTO) belongs to the important group of compounds with a perovskite structure. Its most important features are high dielectric constant, large positive temperature of the resonance frequency, but also high dielectric loss that could be decreased by substitution of the A-site with trivalent ions (Evans et al., 2003). It is promising material for microwave tunable devices and is also used for modification of ferroelectric perovskites, such as PbTiO$_3$ or BaTiO$_3$, for various applications (Kim, 2000; Ganesh and Goo, 1997). Calcium titanate is mostly prepared by a solid state reaction between CaCO$_3$ or CaO and TiO$_2$ at 1350°C, but also by some other methods such as sol–gel processing, thermal decomposition of peroxo-salts, and mechano-chemical synthesis from different precursors, such as CaCO$_3$, Ca(OH)$_2$ or CaO, with TiO$_2$ (Vukotic et al., 2004; Mi et al., 1998). Up until now, various methods have been reported in the literatures for the syntheses of CaTiO$_3$. These methods included: (a) conventional solid state

reaction between TiO$_2$ and CaCO$_3$ or CaO at a high temperature (Redfern, 1996; Chen et al., 2009), (b) mechanochemical methods (Mi et al., 2009; Brankovic et al., 2007; Palaniandy and Jamil, 2009), (c) chemical co-precipitation method (Gopalakrishna et al., 1975), (d) hydrothermal method (Wang et al., 2007; Li et al., 2009), (e) sol–gel route (Holliday and Stanishevsky, 2004; Zhang et al., 2008), and (f) polymeric precursor method (Pan et al., 2003). Among these methods, mechanical alloying (MA) is a solid-state powder process at ambient temperature and has been applied to synthesize different kinds of materials, such as crystalline, nanocrystalline, quasicrystalline and amorphous materials (Zoz, 1995; Suryanarayana, 2001).

Mechanical alloying, high-energy ball milling, has been used for many years now in producing ultra fine powders in the range of a sub-micron to a nanometer. Aside from size reduction, this process causes severe and intense mechanical action on the solid surfaces, which was known to lead to physical and chemical changes in the near surface region where the solids come into contact under mechanical forces (Venkataraman and Narayanan, 1998). These mechanically initiated chemical and physicochemical effects in solids were generally termed as the mechanochemical effect. In this work, Calcium titanate (CTO) is mostly prepared by a mechanical alloying between $CaCO_3$ and $TiO_2$ without heat treatment. The mechanical synthesis process is carried out in high intensity grinding mills such as vibro mills, planetary mills, and oscillating mills. It has been noticed that the size reduction process and the microstructural evolution of the $CaTiO_3$ during milling process were mainly influenced by the type of impulsive stress applied by the grinding media, which can either be an impact or shear type. Moreover, other parameters such as milling time, mill rotational speed and ball to powder at 3 different ratios affect the mechanical process. In fact, when the mechanical synthesis of the $CaCO_3$ and $TiO_2$ was carried out in planetary mills at higher ball to powder (70 : 1) ratio to produce $CaTiO_3$, the impact stress was dominant, and not much attention was given on the mechanochemical mechanism itself. The aim of this work, therefore, is to give additional contribution in understanding the influence of milling conditions on the mechanical synthesis of $CaTiO_3$ nanoparticles without any the deleterious phase and heat treatment.

## EXPERIMENTAL

Oxide powders of $TiO_2$ (99% < 1 μm, 99% purity) and $CaCO_3$ (99%<1 μm, 99% purity) were used as raw materials which were mechanically ground in a purified air atmosphere. The ball-to-powder weight ratio was used at different ratios (20 : 1, 30 : 1 and 70 : 1). Mechanical alloying (MA) was carried out at ambient temperature and at a rotational speed (cup speed) of 350 rpm in a planetary ball mill. The mechanical alloying process was interrupted at regular intervals with a small amount of the MAed powder taken out from the vial to study changes in the microstructures at selected milling duration. The crystal phase was determined with powder X-ray diffraction. For these experiments, a Siemens diffractometer (30 kV and 25 mA) with the $K_{α1}$, radiation of copper (λ = 1.5406 Å), was used. The structural and compositional information of the product materials was obtained with scanning election microscopy (SEM). The crystalline size (D) and lattice strain were estimated by Williamson-Hall (Williamson and Hall, 1953):

$$\beta \cos \theta = 2\varepsilon \sin \theta + 0.9 \frac{\lambda}{D}$$

Where λ is the wavelength of the X-ray, ß the full width at half-maximum (FWHM), θ the Bragg angle, and ε is the microstrain. Finally, the particle size distribution of the powders was measured by zetasizer instrument (Malvern Co, HS C1330-3000, England).

## RESULTS AND DISCUSSION

The XRD patterns of the samples consisting of $TiO_2$ and

$CaCO_3$ that had been ball milled for 0, 15, 20, 25, 40, 50, 60 and 70 h are illustrated in Figure 1. In the time of zero, only the $TiO_2$ and $CaCO_3$ peaks are observed. As shown in Figure 1, at 25 h, nothing significant takes place and only starting materials peaks are observed but in 40 h, all the peaks disappear because the material has become amorphous. In 40 and 50 h, we see the same situation. Because of the decrease of the particle size in milling the diffusion paths are shortened. Additionally, high energy is stored in the particles due to the cold work. Thus, the amorphous phase begins to grow around the crystals until all the material become amorphous. Due to the above, it seems that the mechanism of changing crystalline to amorphous in MA is diffusion controlled. There are reports showing that in some cases after amorphization, the crystalline phase has engendered again. As a result of the fact, that with the increase of milling time the kinetic energy of the systems intensifies, hence, the temperature increases which provides the needed energy for the reappearance of the stable state, i.e. crystallization (Koch, 1991). While rising the milling time, after 15 h, $TiO_2$ peaks disappear and $CaTiO_3$ peaks emerge. In this situation, the only distinguishable phase is $CaTiO_3$. It seems that like an SHS reaction that needs a critical amount of energy to start and perform, in this case also, all $TiO_2$ and $CaCO_3$ have been transformed into $CaTiO_3$ due to the energy gained from milling. The thermal analysis of the 70 h milled sample showed no $TiO_2$ or $CaCO_3$ in the final composition of the synthesized powder to participate in reaction and therefore it seems that all reactants have changed to $CaTiO_3$. Using the XRD patterns, the grain sizes were calculated. Figure 2 shows Williamson–Hall diagram of the system for 70 h and the mean size of the grains and the strain percentages are shown in Figure 2. In Figure 2, y represents bcosθ and x represents 2sinθ in Williamson–Hall equation. Hence, a as the slope represents the strain (η) and b as the y-intercept identifies 0.9λ/d from which the grain sizes (d) can be calculated. The grain size of $CaTiO_3$ were 69 nm for the milling time of 70 h. It is predicted that if the milling process continues, the grains become finer until they reach a critical value for the reason that MA process is the result of the competition of cold fusion and breaking of the components that causes the fineness and activation of the particles (Wang et al., 2001). At the critical point, the speeds of fusion and breaking balance out and the particles will no longer be fined (Ko et al., 2002).

According to SEM micrographs of the powders mechanically milleded for 70 h in air atmosphere are shown in Figure 3a to h, that MAed powders are an ultra-agglomeration powder with approximately 100 ± 20 nanometers in size. Because, highly chemically active particles, these are strongly agglomerated. Interestingly, Figure 4a to d show that the product obtained after heat-treatment at 2 different temperatures (500 and 600°C) for 1 h is are mainly uniform special structures with suitable

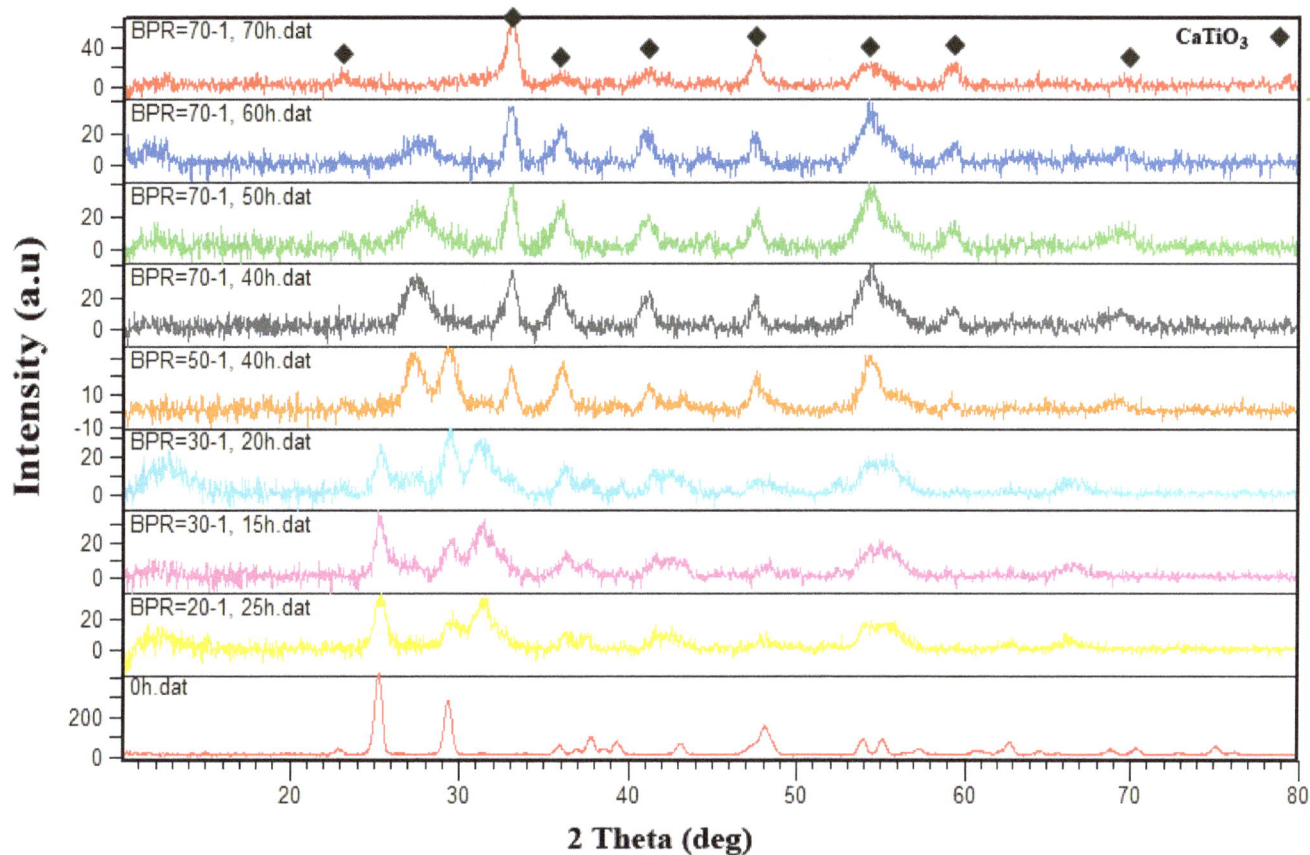

**Figure 1.** The X-ray diffraction spectra of mechanically alloyed $CaCO_3/TiO_2$ powders at different milling times.

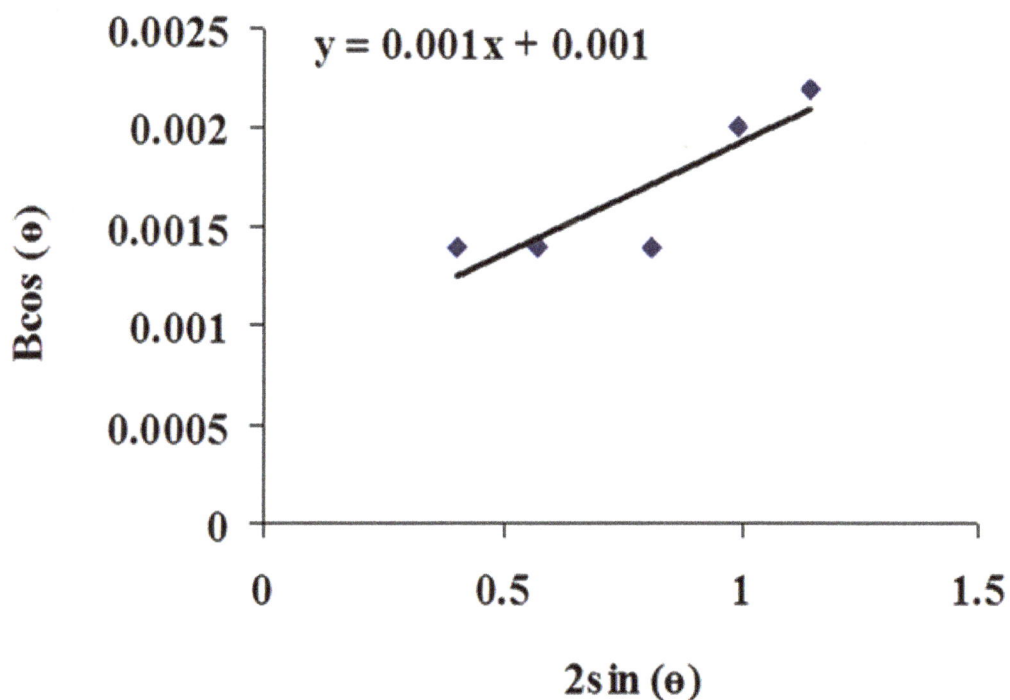

**Figure 2.** Calculation of strain and particle size in accordance to Williamson–Hall equation for CTO after 70 h of ball milling.

**Figure 3.** SEM images of milled samples in 70 h at different magnifications.

crystallinity grades with a diameter of 60 to 90 nm, which is of very extraordinary uniform morphologies. Finally, in this investigation, an effective method was developed for the formation of ultra-crystallinity with uniform morphologies. As the matter of fact, this method (MA) guarantees its production in the synthesis of CTO for different applications.

The nanoparticle size of CTO milled (70 h) product was

**Figure 4.** SEM different images after heat-treatment at different annealing temperatures with different magnifications a-b) 500°C, c-d) 600°C.

analyzed using a zetasizer method. These measurements reveal the particles to be highly wide distribution (Figure 5a). The milled CTO powders were particles with diameters 2 ranging from 55 to 100 nm and 300 to 550 nm. Figure 5b shows the zetasizer curves of the CTO powders obtained from the heat treatment for 2 h and 500°C. As can be observed in these images, the particle sizes grow up with increasing the aging time. The average particle sizes of powders aged for 2 were regular and uniform.

**Conclusion**

CTO with different morphologies was synthesized by a MA method. The purity and good quality of CTO obtained

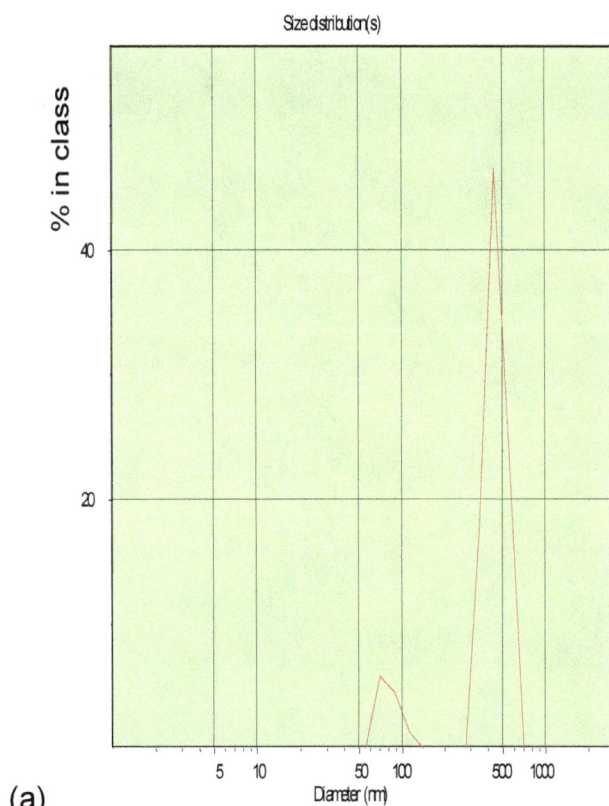

(a)

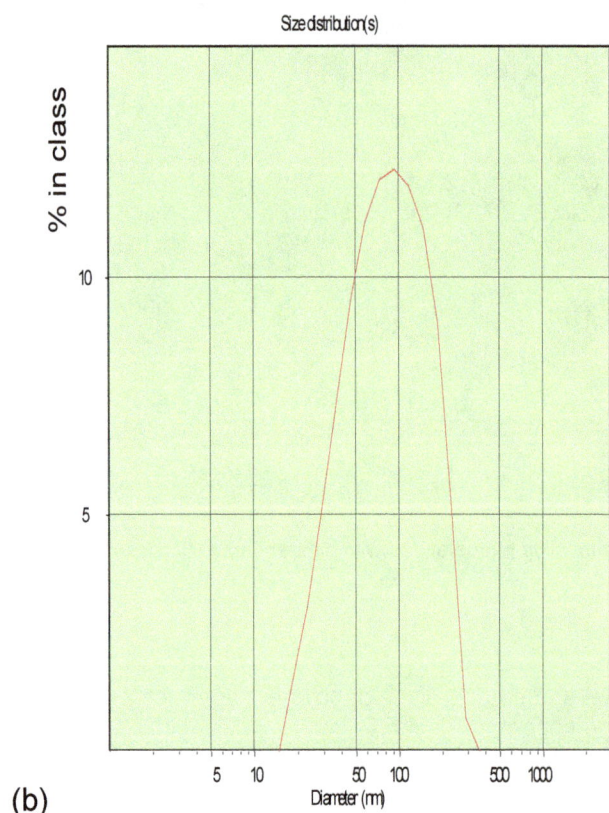

(b)

**Figure 5a and b.** Zetasizer images of the CTO powders obtained from the MA a) only 70 h of ball milling, b) 70 h of ball milling with heat treatment in 500°C.

by MA make it a promising method for the production of CTO. The synthesis of CTO was strongly dependant on the experimental parameters such as milling time and ball to powder ratio. Optimal conditions of CTO synthesis were selected as 70 : 1 ratio and 70 h of milling time. This simple approach should promise us a future large-scale synthesis of this nanostructured materials for many important applications in nanotechnology in a controlled manner.

## REFERENCES

Brankovic G, Vukotic V, Brankovic Z, Varela JA (2007). Investigation on possibility of mechanochemical synthesis of $CaTiO_3$ from different precursors. J. Eur. Ceram. Soc. 27:729-732.

Chen R, Song FL, Chen DH, Peng YH (2009). Improvement of the luminescence properties of $CaTiO_3$:Pr obtained by modified solid-state reaction. Powder Technol. 194:252-256.

Evans IR, Howard JAK, Sreckovic T, Ristic MM (2003). Variable temperature in situ X-ray diffraction study of mechanically activated synthesis of calcium titanate $CaTiO_3$. Mater. Res. Bull. 38:1203–1213.

Ganesh R, Goo E (1997). Dielectric and ordering behavior in $PbxCa_{1-x}TiO_3$. J. Am. Ceram. Soc. 80:653-662.

Gopalakrishna MHS, Subbarao M, Narayan Kutty TR (1975). Thermal decomposition of titanyl oxalates IV. Strontium and calcium titanyl oxalates. Thermochim. Acta. 13:183-191.

Holliday S, Stanishevsky A (2004). Crystallization of $CaTiO_3$ by sol–gel synthesis and rapid thermal processing. Surf. Coat. Technol. 188:741-744.

Kim WS (2000). Microwave dielectric properties and far-infrared reflectivity characteristics of the $CaTiO_3–Li_{(1/2)-3x}Sm_{(1/2)+x}TiO_3$ ceramics. J. Am. Ceram. Soc. 83:2327–2329.

Ko SH, Park BG, Hashinoto H, Abe T, Park YH (2002). Effect of MA on microstructure and synthesis path of in-situ TiC reinforced Fe–28at%Al intermetallic composites. Mater. Sci. Eng. A329–A331:78-83.

Koch CC (1991). Processing of Metals and Alloys. Mater. Sci. Technol. 15:193-245.

Li Y, Gao XP, Li GR, Pan GL, Yan TY, Zhu HY (2009). Titanate Nanofiber Reactivity: Fabrication of $MTiO_3$ (M = Ca, Sr, and Ba) Perovskite Oxides. J. Phys. Chem. C 113:4386-4394.

Mi G, Saito F, Suzuki S, Waseda Y (1998). Formation of $CaTiO_3$ by grinding from mixtures of CaO or $Ca(OH)_2$ with anatase or rutile at room temperature. Powder Technol. 97:178-182.

Palaniandy S, Jamil NH (2009). Influence of milling conditions on the mechanochemical synthesis of $CaTiO_3$ nanoparticlesJ. Alloys Compd. 476:894-902.

Pan YX, Su Q, Xu HF, Chen TH, Ge WK, Yang CL, Wu MM (2003). Synthesis and red luminescence of $Pr^{3+}$-doped $CaTiO_3$ nanophosphor from polymer precursor. J. Solid State Chem. 174:69-74.

Redfern SAT (1996). High-Temperature Structural Phase Transitions in Perovskites. ($CaTiO_3$). J. Phys.: Condens. Matter. 8:8267-8275.

Suryanarayana C (2001). Mechanical alloying and milling. Prog. Mater. Sci. 46:1-184.

Venkataraman KS, Narayanan KS (1998). Energetics of collision between grinding media in ball mills and mechanochemical effects. Powder Technol. 96:190-201.

Vukotic VM, Sreckovic T, Marinkovic ZV, Brankovic G, Cilense M, Aranelovic D (2004). Mechanochemical synthesis of $CaTiO_3$ from $CaCO_3–TiO_2$ mixture. Mater. Sci. Forum. 453/454:429-434.

Wang C, Qi B, Bai Y, Wu J, Yang J (2001). Dispersion strengthened alloy due to the precipitation of carbide during mechanical alloying. Mater. Sci. Eng. A. 308:292–294.

Wang DA, Guo ZG, Chen YM, Hao JC, Lin WM (2007). In Situ Hydrothermal Synthesis of Nanolamellate $CaTiO_3$ with Controllable Structures and Wettability Inorg. Chem. 46:7707-7711.

Williamson GK, Hall WH (1953). X-ray line broadening from filed aluminium and wolfram. Acta. Metallurgica. 1:22-31.

Zhang HW, Fu XY, Niu SY, Xin Q (2008). Synthesis and photoluminescence properties of $Eu^{3+}$-doped $AZrO_3$ (A=Ca, Sr, Ba) perovskite. J. Alloys Compd. 459:103-114.

Zoz H (1995). Attritor Technology-Latest Developments. Mater. Sci. Forum. 179:419-423.

# Experimental determination of layers films thicknesses

**S. Ourabah[1], A. Amokrane[1, 2] and M. Abdesselam[1]**

[1]Faculty of Physics, University of Sciences and Technology, Houari Boumediène, BP 31 El Alia, Bab Ezzouar, 16111 Algiers, Algeria.
[2]Preparatory National School for Engineer Studies, Rouiba, Algiers, Algeria.

The determination of particle induced x-ray emission (PIXE) cross sections and the concentration of elements in a material require the knowledge of the target sample thickness. In this aim, measurements of the thickness by three different methods have been performed. These are absorption of X-rays by a [55]Fe source, transmission of alpha particles by a [241]Am source and Rutherford backscattering of alpha particles produced by Van de Graff Accelerator with the use of the RUMP simulation code. The results give a thickness with uncertainties ranging from 1 to 8% according to the experimental technique used. The comparison between these methods gives an advantage for the X-rays absorption for its simplicity and accuracy, when backscattering spectrometry is preferred for thin target on backing or as a complementary technique for PIXE analysis.

**Key words:** Thickness, particle induced x-ray emission (PIXE), Rutherford backscattered (RBS), cross section, rump.

## INTRODUCTION

The technique of samples analysis by charged particles induced x-ray emission (PIXE), requires the knowledge of the targets thicknesses in order to determine the concentrations of the elements present in the sample and for the matrix effect correction. The same applies for the calculation of the ionization cross section:

$$d\sigma/d\Omega = dN/AdxI$$

where dN, Adx and I represent respectively the number of emitted X-rays, of target atoms and the intensity of the beam of incident particles.

In the aim of selecting a technique allowing the thickness determination of a target with the best possible precision, several methods of measurement have been undertaken and these are:

(a) Transmission of alpha particles given by a radioactive source or produced by an accelerator.
(b) Rutherford backscattered (RBS) of alpha particles produced by an accelerator.
(c) The attenuation of X-rays resulting from an iron source ([55]Fe).

Some of these methods are usually used in PIXE measurements (Johansson and Campbell, 1970; Tran et al., 2002; Ekinici and Valles, 2001). The various measurements were carried out at the Nuclear Research Centre of Algiers (CRNA) of the Commission of Atomic Energy (COMENA), in the division of the nuclear techniques. Self-supported targets whose thicknesses were measured by piezoelectric quartz during the evaporation process, commercial targets with thickness is given by the manufacturer and finally targets deposited on a substrate of silicon and of unknown thickness.

### Preparation of the targets

The preparation (Ourabah and Amokrane, 2006) of the

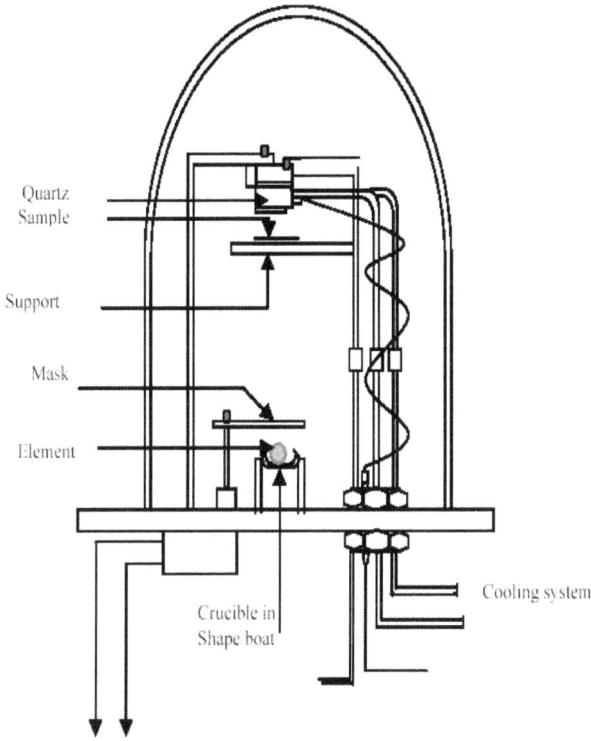

**Figure 1.** Evaporator with piezoelectric quartz.

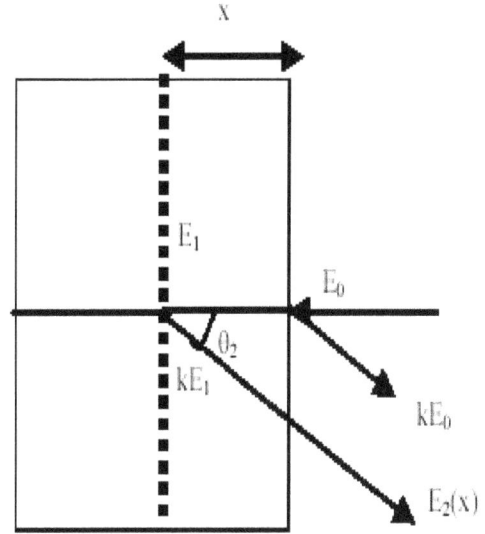

**Figure 2.** Principle of backscattering.

targets was carried out in an evaporator composed of a bell provided with a system of pumping and piezoelectric quartz for the measurement of the thicknesses (Figure 1). Two types of targets were elaborated out with and without backing (self-supported target). The self-supported targets were produced by the use of a taking off agent which dissolves easily in distilled water. This agent depends on the material deposited. It should be pointed out that the crystalline shapes of both agent and material to deposit have to be similar.

## Determination of the thickness by piezoelectric quartz

Thickness can be measured during the evaporating procedure by a piezoelectric quartz crystal (silicon dioxide crystal) put on the sample in the enclosure of the evaporator. The quartz is subjected to a mechanical pressure during evaporation, giving appearance of an electric potential on its face. The measurement of the resonance frequency of this quartz which varies as function of the thickness allows the determination of the thickness of the target.

## RUTHERFORD BACKSCATTERING

Particles backscattering principle is shown in Figure 2.

When a target of thickness x is bombarded with incident particles of energy $E_0$, their energy after diffusion, at an angle $\theta$, by the nuclei located at the surface of the target is $kE_0$, where k is the kinematic factor given by:

$$k = \left( \frac{M_1 \cos\theta + \sqrt{M_2^2 - M_1^2 \sin^2\theta}}{M_1 + M_2} \right)^2 \tag{1}$$

$M_1$ ; $M_2$ are the masses of incident particle and nucleus of the target, $\theta$ being the diffusion angle.

After crossing the sample, the energy of the particle at a depth x is:

$$E_1 = E_0 - \int_0^x \frac{dE}{dx} dx \qquad \text{for the ingoing path} \tag{2}$$

After backscattering on a nucleus of the target at the depth x, its energy will be:

$$E_2(x) = kE_1 - \int_0^{\frac{x}{|\cos\theta_2|}} \frac{dE}{dx} dx \qquad \text{for the outgoing path} \tag{3}$$

$\theta_2$ is the angle of the backscattered ion with the target's normal. The lost energy is then $\Delta E_i = k E_0 - E_2$ .
Using Equations 2 and 3, we found:

$$\Delta E = k \int_0^x \frac{dE}{dx} dx + \int_0^{\frac{x}{|\cos\theta_2|}} \frac{dE}{dx} dx \tag{4}$$

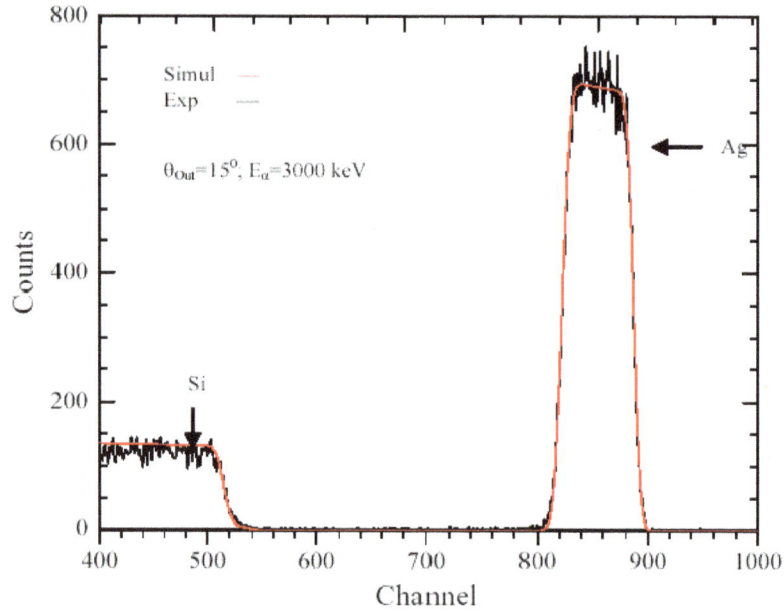

**Figure 3.** Backscattered spectrum of alpha particles on silver target of 1990°A thickness with silicon backing.

or $\quad \Delta E = k\,\Delta E_{in} + \Delta E_{out}$

Introducing the stopping power [S]= dE/dx and assuming that the energy lost dE/dx is constant and calculated at $\overline{E}_{in}$ and $\overline{E}_{out}$, the integrals give:

$$\Delta E_{in} = \left\lfloor S\!\left(\overline{E}_{in}\right)\right\rfloor\!\Delta x \qquad \text{and} \qquad \Delta E_{out} = \frac{1}{\cos\theta_2}\left[S\!\left(\overline{E}_{out}\right)\right]\!\Delta x \tag{5}$$

Many approximations (Chu et al., 1978) allow calculating [S] and finally $\Delta x$.
- Approximating the surface energy for thin target as:

$$\overline{E}_{in} = E_0 - \frac{\Delta E_{in}}{2} \qquad \text{and} \qquad \overline{E}_{out} = kE_0 - \frac{\Delta E_{out}}{2} \tag{6}$$

- On the other hand, approximating the average energy for appreciable target thickness as:

$$\overline{E}_{in} = \frac{1}{2}\left(E_0 + E_1\right) \qquad \text{and} \qquad \overline{E}_{out} = \frac{1}{2}\left(kE_1 + E_2\right) \tag{7}$$

$E_1$ being unknown, one can suppose that the energy loss can be split symmetrically between the ingoing and the outgoing paths, so that $\Delta E_{in} \approx \Delta E_{out}$ and thus the average energies will be:

$$\overline{E}_{in} = \left(E_0 + \frac{\Delta E}{4}\right) \qquad \text{and} \qquad \overline{E}_{out} = \left(E_2 + \frac{\Delta E}{4}\right) \tag{8}$$

Our measurements were done with alpha particles, produced by the Van de Graff accelerator. The backscattered particles where detected with a surface barrier detector. A typical backscattered spectrum is represented in Figure 3, showing the signal of the silicon backing and that of silver. The width at half maximum (FWHM) of the backscattered peak represents the total energy loss $\Delta E$ of he ingoing and outgoing paths. The target thickness can be obtained from:

1. The ratio of the surface of the RBS spectrum of the element over the height of signal of the backing.
2. The analysis of the RBS spectrum with the RUMP code (Doolittle, 1985).
3. The determination of the energy $E_1$ at depth x by a calculus code using different methods.

However, in this work the determination is limited to cases 1 and 2. Several samples were used:

1. Two samples Ag/Si of different thicknesses, a sample of Au/Ti/Si, and all three with two systems of detection to see if the detection angle influences the thickness determination of the target.
2. Three samples of Ag/Si of different thicknesses, two self supported targets of nickel and aluminium with only one detection system.

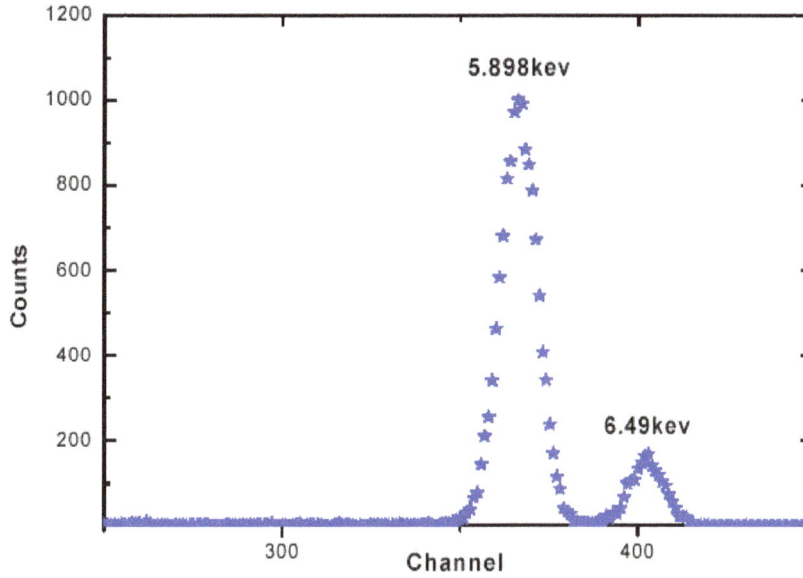

**Figure 4.** $^{55}$Fe spectrum without absorber.

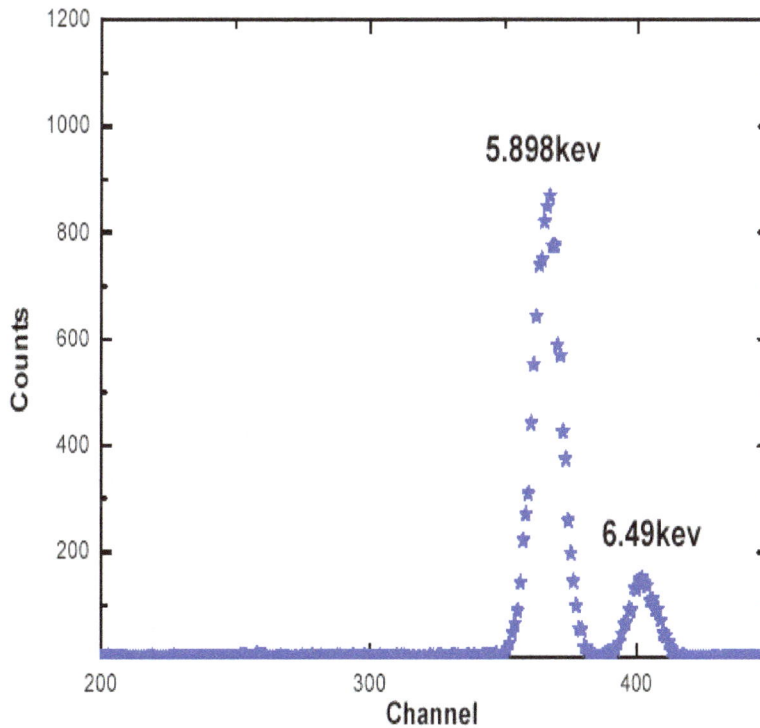

**Figure 5.** $^{55}$Fe spectrum with aluminium absorber.

## METHOD OF ATTENUATION OF X-RAYS IN MATTER

Measurements of target thickness were also made from the attenuation of the photons. It is deduced from the Lambert's law (Davisson and Evans, 1952) according to which intensity I of the transmitted photons is given by the relation $I=I_0 \exp(-\mu x)$ where $I_o$ is the initial intensity, $\mu$ the linear attenuation coefficient and x the thickness of the absorber. The thickness is then $x = (1/\mu) \ln(I_o/I)$. We performed the experiment on two films of nickel and aluminium.

The X-rays are provided by a 25mCi sealed source of Iron ($^{55}$Fe), emitting the 5,898 kev and 6,49 kev lines of manganese. The transmitted photons are collected in Si(Li) detector of 220 eV of resolution at 5,898 keV energy. The measurements were conducted three times and gave similar results. Figures 4 and 5

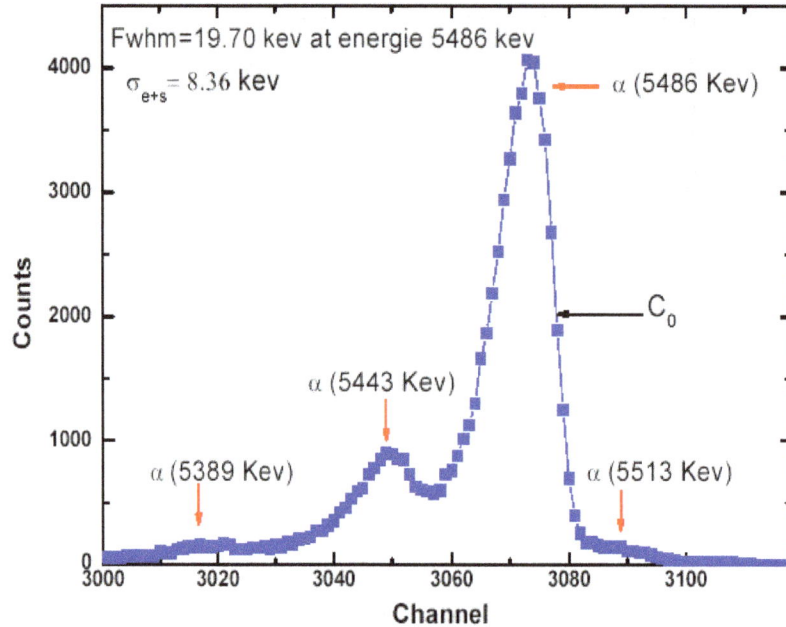

**Figure 6.** Spectrum of alpha particles issued from the[241]Am source without absorber.

show typical X-ray spectra.

**Technique by transmission of the He[++] particles from a [241]Am source**

The technique consists in measuring the energy loss $\Delta E$ of the alpha particles in self supported targets. The stopping power in the approximation of average energy $E_M$ is obtained from code SRIM 2003 (Ziegler et al., 1985), used for the determination of targets thickness. The experimental energy loss is given by:

$$\Delta E[kev] = E_0 - E_1 = a*(C_0 - C_1) \tag{9}$$

where $E_0$, $E_1$ and $C_0$, $C_1$ are energies and the corresponding channels measured without and with the target, a[ keV/channel ] is the slope of the calibration straight line. The average energy is

$$E_M = E_0 - \frac{\Delta E}{2} \tag{10}$$

By using the stopping power $[\varepsilon]$, one can write

$$\varepsilon(E_M)[keV/micron] = (\Delta E/x) = a*(C_0 - C_1)/x \tag{11}$$

$[\varepsilon]$ is the stopping power at average energy $E_M$ .We thus have

$$x[\mu m] = (\Delta E / \varepsilon(E_M)) \tag{12}$$

**Measurements**

The experimental set up is composed by an enclosure, the source, a pumping system and a chain of detection constituted by a 50 mm[2] surface barrier detector with 12 keV of resolution at the 5486 keV

energy, a preamplifier and an amplifier. Alpha particles are provided by 1μCi [241]Am source of 5486 keV energy. Measurements were carried out on targets of nickel and Aluminium manufactured and other aluminium and silver targets that we have realized in the Laboratory for Targets of CRNA. Figures 6 and 7 show spectra of alpha particles resulting from the [241]Am source without absorber and after crossing a nickel target of 1.27 μm thickness.

**Transmission of the alpha particles provided by an accelerator**

The same principle used for transmission for alpha particles resulting from the radioactive source is applied. The surface barrier detector is placed at a detection angle of 30° in order to avoid its deterioration. The experiment is carried out in combination with the RBS. The spectrum obtained is represented in Figure 8.

**RESULTS**

The resulting thicknesses determined by the two methods; the ratio of the surface of the RBS spectrum of the element over the height of signal of the backing and the analysis of the RBS spectrum with the RUMP code (Doolittle, 1985), are reported in Table 1. We should note that the measurements, carried out with the method of the ratio of the height of the spectrum of the backing over the surface of the target, are not in agreement with the results obtained using the Rump code for samples 1 and 2; the signal of the silicon backing being not well defined because of its bad quality. However, results obtained for the sample 3 are of the same order of magnitude as those obtained with the Rump code. During the simulation by the Rump code for sample 1 and in the range of quoted energies (Table 1), we noticed a light

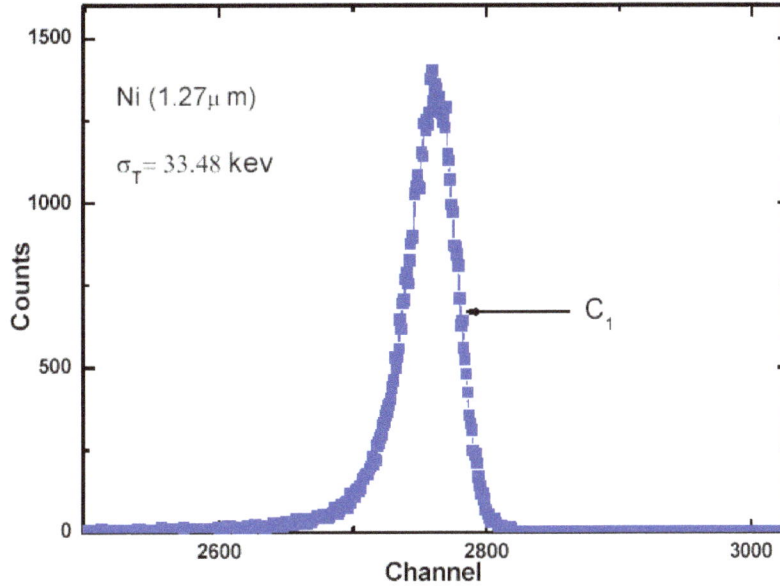

**Figure 7.** Spectrum of alpha particles after transmission through the absorber.

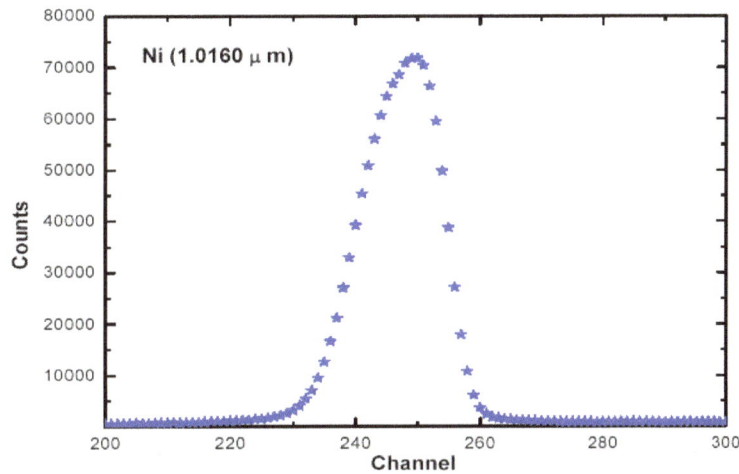

**Figure 8.** Transmitted alpha spectrum through Nickel target.

**Table 1.** Thicknesses determined by the method of the ratio of the height of the signal of the backing over that of the element and by simulation with Rump code.

| Sample | Target | Energy range of alpha particles | Detection angle $\theta$ | (surface of element target) /(Height of backing signal) (Å) | Rump(Å) |
|--------|--------|-------------------------------|--------------------------|-----------------------------------------------|---------|
| 1 | Ag/Si | [700kev-1100kev] | 150° | 691.5 | 1030 |
| | | [1200kev-1600kev] | | 663.9 | 988 |
| | | [700kev-1600kev] | 165° | 722.5 | 1055 |
| 2 | Ag/Si | [1600kev-3000kev] | 150° | 1587 | 1930 |
| | | [1600kev-3000kev] | 165° | 1591 | 1990 |
| 3 | Au/Ti/Si | [1800kev-3400kev] | 150° | 1265 | 1230 |
| | | | 165° | 1202 | 1210 |

**Table 2.** Thicknesses obtained with the piezo electric quartz and by the simulation with Rump code. (Q) measured by the piezo electric quartz, (C) given by the manufacturer.

| Sample | Target | Energy (keV) | Detection angle θ | Rump (Å) | Given thickness Quartz (Å) | Uncertainties (%) |
|--------|--------|--------------|-------------------|----------|----------------------------|-------------------|
| 4 | Ag/Si | 2000 | 160° | 207 | 214 (Q) | 4.02 |
| 5 | Ag/Si | 2000 | 160° | 433 | 450 (Q) | 3.73 |
| 6 | Ag/Si | 2000 | 160° | 822 | 859 (Q) | 4.31 |
| 7 | Al | 2200 | 165° | 6836 | 7500 (C) | 3.45 |
| 8 | Ni | 2200 | 165° | 6150 | 6350 (C) | 3.00 |

**Table 3.** Values for the attenuation coefficients μ.

| Elements | Z | μ(m⁻¹) |
|----------|-----|----------|
| Aluminium | 13 | 32721.22 |
| Nickel | 28 | 101586.61 |
| silver | 47 | 504693.87 |

**Table 4.** Measured thicknesses by the attenuation technique compared with those given by manufacturer (C).

| Element | Z | Given thickness (μm) | Measured thickness (μm) | Relative uncertainties (%) |
|---------|-----|----------------------|-------------------------|----------------------------|
| Nickel | 28 | 0.635 (C) | 0.627±0.019 | 3 |
| | | 0.762 (C) | 0.757±0.024 | 3 |
| | | 1.905 (C) | 1.879±0.025 | 1.3 |
| | | 3.750 (C) | 3.678±0.055 | 1.5 |
| Aluminium | 13 | 2.000 (C) | 1.921±0.038 | 2 |
| | | 4.000 (C) | 3.780±0.109 | 3 |
| Silver | 47 | 5.000 (C) | 4.818±0.094 | 2 |

variation of the thickness, which can be explained by the inclination and the non uniformity of the target. We can also see that the detection angle (θ=150° or θ =165°) does not influence the thickness.

In Table 2, we report the results obtained with the Rump code for self supported targets of nickel, aluminium and for Ag on Si backing. The uncertainties given for the simulation of the spectrum are estimated from the uncertainty on the channel; the resolution of the detector and on the stopping power. The comparison between the values measured with piezoelectric quartz, those obtained with RUMP code and those given by the manufacturer for Ag/Si targets and Nickel indicates a good agreement, except for the aluminium foil for which the light difference can be attributed to the value given by the manufacturer.

Measurements of target thickness were made from the attenuation of the photons using the values for the attenuation coefficients (Berger and Hubbell, 1987) given in the Table 3 for the energy 5,898 keV, we find the results in Table 4. The uncertainties were calculated using the Lambert's law, taking into account the precision on the intensity of the source before and after attenuation (<3%), and the error on the attenuation coefficient (1% for aluminium). They were also made from the technique by transmission of the HE⁺⁺ particles from a ²⁴¹AM source; the results are reported in the Table 5. The uncertainties were calculated using the equation:

$$\frac{\Delta x}{x} = \frac{\Delta \varepsilon}{\varepsilon} + \frac{\Delta(\Delta E)}{\Delta E}$$

taking into account the precisions of the stopping power (2%) and the energy lost (<0.5%).

**Comparison between the thickness measured by transmission of alpha particles given by radioactive source and produced by accelerated particles**

The results of the measurements are reported in Table 6. We can notice that the values obtained by transmission of the particles alpha produced by the ²⁴¹Am radioactive source (Table 5) or coming from the accelerator (Table 6)

**Table 5.** Measured thicknesses by transmission of alpha particles: (•) prepared in this work (C) manufactured (P) measured by piezoelectric quartz.

| Element | Z | ΔE (keV) | Given thickness (μm) | Thickness measured by transmission (μm) | Uncertainties (%) |
|---------|---|----------|----------------------|-----------------------------------------|-------------------|
|         |   | 264.799 | 0.635 (c) | 0.68±0.03 | 4.4 |
|         |   | 321.290 | 0.762 (c) | 0.82±0.04 | 4.8 |
| Ni | 28 | 434.271 | 1.016 (c) | 1.10±0.05 | 4.5 |
|         |   | 524.303 | 1.270 (c) | 1.32±0.06 | 4.5 |
|         |   | 835.001 | 1.905 (c) | 2.07±0.09 | 4.3 |
| Ag | 47 | 54.725 | •0.155 (P) | 0.16±0.008 | 5.0 |
|         |   | 81.205 | •0.500 (P) | 0.52±0.03 | 5.7 |
| Al | 13 | 112.981 | 0.750 (C) | 0.73±0.04 | 5.4 |
|         |   | 631.988 | 4.000 (C) | 3.95±0.2 | 5.0 |

**Table 6.** Measured thicknesses by transmission of 2.2 MeV alpha particles coming from an accelerator (C) thickness given by the manufacturer.

| ΔE (kev) | ΔE/E0 (%) | Given thickness (μm) | Measured thickness (μm) | Relative uncertainties (%) |
|----------|-----------|----------------------|-------------------------|----------------------------|
| 434.696 | 19.759 | 0.635 (c) | 0.65±0.03 | 4.6 |
| 547.865 | 24.903 | 0.762 (c) | 0.81±0.05 | 6.1 |
| 744.422 | 33.837 | 1.016 (c) | 1.08±0.08 | 7.4 |
| 833.766 | 37.899 | 1.270 (c) | 1.21±0.06 | 5.0 |
| 1298.354 | 59.016 | 1.905 (c) | 1.81±0.17 | 9.3 |

are in agreement with the values given by the manufacturer or those measured by the piezoelectric quartz.

We can see on the Table 5 that for the thicknesses lower than 1.2 μm, the results by transmission of the alpha particles given by the radioactive source and those produced by the accelerator (Table 6) are similar. For the thicknesses above 1.2 μm, the difference between the two measurements can be explained by the use of the approximation of average energy in the calculation of the stopping power for the 2200 keV energy of the alpha particles given by the accelerator, the total energy loss is of 833 keV, the approximation of average energy for the calculation of the stopping power is not good whereas for the alpha particles provided by the source, the energy loss being small, the approximation is more suitable.

## Conclusion

In the aim of selecting the technique allowing the thickness determination of a target with the best possible precision, several methods of measurement have been investigated. According to the relative uncertainties made

in the determination of the targets thicknesses, the followings can be concluded:
1. The method by attenuation of X-rays is preferable to the other methods for its precision and its simplicity, mainly for large thicknesses, since we found that the uncertainties on the thickness are lower than 3%. This technique is currently used in industry for the measurement of the thicknesses of different materials.
2. In the case of a target deposited on a backing, where the method by attenuation cannot be employed, RBS technique remains the method suitable compared to the two techniques by the attenuation of X-ray and the transmission of the alpha. Only one must take into account that the uncertainties made on the thickness of the target varies between 4 to 6%. This technique is used simultaneously with the PIXE analysis technique (which requires the knowledge of the thicknesses to obtain the absolute concentrations).

In conclusion, the attenuation of X-rays remains the best technique for the determination of large thicknesses targets with a better precision for the self supported ones whereas for very thin targets deposited on a backing, RBS technique remains a good method, since the difficulty of the self supported thin targets lies in their brittleness to handle them manually and in the fact that certain metal elements of the periodic table cannot be put always in the form of self supported targets.

## ACKNOWLEDGEMENT

Special thanks to E.K. Si Ahmed, MIT's Phd, Professor in our Faculty for his kindness in reading the paper.

## REFERENCES

Berger MJ, Hubbell JH (1987). XCOM: Photon Cross Sections. NBSIR pp. 87-3597.

Chu WK, Mayer JM, Nicolet MA (1978). Backscattering Spectrometry. Academic Press. New York.

Davisson CM, Evans BD (1952). Gamma-ray absorption coefficients. Rev. Mod. Phys. 24(2).

Doolittle LR (1985). Code of simulation RUMP. Nucl.Inst.and Meth. B. 9:291.

Ekinici N, Valles Jr JM (2001). Determination of thin film thickness by X-ray transmission. 16th International Conference on X-ray Optics and Microanalysis (ICXOM). Vienna.

Johansson SE, Campbell JL (1970). (Provide work title). John Wiley and Sons, P. 87.

Ourabah S, Amokrane A (2006). Conference Nationale de la Physique et de ses Applications, Béchar Algérie.

Tran CQ, Chantler CT, Barnea Z, de Jonge MD, Dhal BB (2002). Accurate determination of the thickness of thin specimens and applications in X-rays Attenuation measurements . European Conference on Energy Dispersive X ray Spectrometry. Berlin, Germany. P.12.

Ziegler JF, Biersack JP, Littmark U (1985). The stopping and ranges of ions in matter.1. New York. www.srim2003.org.

# 3

# Heavy metal removal by ambient-temperature argon plasma modified polyethylene terephthalate (PET) fibers with surface acrylic acid grafting

Juu-En Chang[1], Yi-Kuo Chang[2]*, Min-Her Leu[3], Ying-Liang Chen[4] and Jing-Hong Huang[1]

[1]Department of Environmental Engineering, National Cheng Kung University, Taiwan.
[2]Department of Safety Health and Environmental Engineering, Central Taiwan University of Science and Technology, 666 Po-Tze Road, Peitun District, Taichung city 406, Taiwan.
[3]Department of Environmental Engineering, Kun Shan University of Technology, Taiwan.
[4]Sustainable Environment Research Center, National Cheng Kung University, Taiwan.

This study utilized the capability of ambient-temperature plasma in modifying the surface properties of materials to activate the nonwoven polyethylene terephthalate (PET) fiber surfaces. The effects of different plasma treatment parameters (such as plasma power, treatment time) and grafting parameters (such as grafting temperature, acrylic acid monomer concentration, grafting time) on the activation and grafting of the PET fibers were studied. The feasibility of applying ambient-temperature plasma combined with grafting technology for the preparation of ion exchangers in the wastewater treatment was evaluated. The results showed that the optimal modification effect of PET fibers by Argon plasma and the highest hydrophilicity was achieved at a plasma power of 800 W and treatment time of 10 s. During the grafting procedure, the optimal grafting yield of 4.45% was observed at a grafting temperature of 90°C, an acrylic acid monomer concentration of 6 M, and a reaction time of 5 h. The above surface modification and acrylic acid grafting conditions were used to prepare the ion exchangers for the investigation of subsequent adsorption behaviors. Suitable adsorption performance was achieved at a solution pH of 6, with the adsorption of Copper ion of 0.073 mmole Cu/g-polymer, Lead ion of 0.037 mmole Pb/g-polymer, and Nickel ion of 0.012 mmole Ni/g-polymer. After calculation, the reaction heat was found to be 13.74 kJ/mol, indicating that the adsorption was an endothermic reaction.

Key words: Ambient-temperature plasma, polyethylene terephthalate (PET) fibers, acrylic acid grafting, heavy metal removal.

## INTRODUCTION

The electroplating industries have advanced and prospered in recent years. As a result, the release of wastewater containing heavy metals produced in the manufacturing process has increased gradually year after year. According to the statistical data from the Bureau Industrial Development of the Ministry of Economic Affairs, Taiwan (2007), the discharge of electroplating wastewater was 415,700 m$^3$ per day (CMD). To handle such an enormous quantity of wastewater containing heavy metals, the development of economical and environmentally friendly materials to remove the heavy metals in wastewater has become an important subject worth further investigations.

*Corresponding author. E-mail: ykchang@ctust.edu.tw.

The traditional methods for the treatment of wastewater containing heavy metals include chemical precipitation, ion exchange, adsorption, reverse osmosis, and membrane filtration (Bailey et al., 1999). Among these methods, the addition of base reagent with coagulation and precipitation is the most commonly used method. By the addition of an alkaline solution, the metal ion is reacted with hydroxide to form the fairly insoluble metal oxide compound, followed by coagulation and precipitation for metal removal. However, due to the dense urban environment with limited land resources in Taiwan, the traditional precipitation method involving the addition of chemical reagents has become less and less appropriate for dealing with the enormous quantity of wastewater. Moreover, the disposal and treatment of sludge generated by such method have another challenge to overcome. Without proper treatment, the heavy metals may be released (McDonald et al., 2006) and be a secondary environmental hazard. From the perspectives described above, the objective of the present study was focused on the invention of a material capable of adsorbing heavy metals based on the economical, environmental and efficiency-related considerations.

During the last two decades, the literature report many beneficial plastic waste applications such as recycled as fiber reinforced mortar and concrete aggregates (Bayasi and Zeng, 1993; Pereira de Oliveria and Castro-Gomes, 2011; Foti, 2011). The fiber-reinforced concrete performs well in the mechanical properties and makes the possibility of recycling waste plastic materials. Otherwise, an attractive development of plastic material application is surface modification. The most common preparation methods for reactive fiber have induced chemical conversion and grafting of various monomers on the fiber by grafting copolymerization. The adsorption of heavy metals ions from aqueous solution to adsorbents is usually affected by surface functional groups of the adsorbents. The wet surface modification of polyethylene terephthalate (PET), polypropylene (PP) fiber could be done by initiator activation [such as Benzoyl peroxide (BPO) and Ceric ammonium nitrate (CAN)] and then grafted with carboxyl materials. The process is complicated and the initiator is readily active. An alternative to wet chemical processing is plasma activation. Ambient-temperature plasma is easy to operate and possesses the clean way technique with low energy consumption. Therefore, its development is needed from either an economical or an environmental viewpoint, and it has become one of the most enthusiastically studied areas. It can be applied to a wide range of fields such as the improvement of the hydrophilic/hydrophobic property of the surface of material and the adsorption of heavy metals (Lin and Hsieh, 1997; Gupta et al., 2002). Ambient-temperature plasma has many unique features: 1) The plasma surface

modification only works on the surface layer (with a thickness of a few micrometers). Consequently, the properties of the fiber itself will not be destroyed. 2) The plasma device occupies a small area with not much energy required. 3) The plasma modification is a dry method producing very small quantities of hazardous wastes having little harm to the environment. 4) The modification purpose can be achieved within a short period of time with extremely high efficiency. 5) Under different plasma conditions, the organic or inorganic materials of large areas can be modified effectively (Lee et al., 1996; Svorcik et al., 2006). By utilizing the unique properties aforementioned, the economical and environment-friendly objectives for the medication of material are expected to be reached.

In present study, the PET fiber was used and the PET fiber surface was modified by ambient-temperature plasma activation followed by acrylic acid monomer grafting. The formed ion exchanger was capable for heavy metals adsorbing. Moreover, the optimal modification effect of PET fibers by Argon plasma and the highest hydrophilicity, was achieved from various plasma operating conditions. The grafting yield and COOH quantitative method was used to evaluate the grafting effects under different grafting parameters. The adsorption models were investigated to simulate the application for the heavy metal removal in wastewater plants.

## MATERIALS AND METHODS

### Activation of PET fiber applying ambient-temperature plasma treatment

The nonwoven PET was 30 $g/m^2$ with 0.15 mm thickness. The PET was used as the substrate under various ambient-temperature plasma treatment parameters such as plasma power (400, 800, and 1200 W) and plasma treatment time (10, 30, 60, and 120 s). The plasma chamber size is 425 × 350 × 150 mm. The operating characteristics were set at pressure of 200 mtorr (millitorr) in Argon gas with mass flow control (MFC) of 43.8 scc/m (standard cubic centimeter per minute). The plasma excitation was performed by a microwave generator (~2.45 GHz) with 12 × 12 antenna array.

The water loss rate experiment was applied to determine the activation effects of substrates for selecting the optimal plasma treatment parameters. In the water loss rate experiment (MOEA, 2010), the fiber specimen (4 × 4 cm) was put in the moisture analyzer and then 0.2 ml distilled water was added on it. Under the conditions of temperature 20 ± 2°C, relative humidity (RH) 65 ± 2%, the water loss rate was measured by a variety of time for water to be lost on the fiber's surface.

An increasing water loss rate indicated a shorter time for water to be lost on the fiber's surface, and a greater degree of water spread required for a large heating area. This phenomenon illustrated the effect on hydrophilicity increase by plasma modification. Therefore, the comparison of water loss rates using this experiment could reveal the effects on hydrophilicity increase after the treatments under various parameters and the optimal plasma modification conditions were thus investigated.

## Modified PET fiber (ion exchanger) preparation applying grafting treatment

The PET fiber after plasma treatment was subjected to grafting treatment under different acrylic acid monomer concentrations (2, 4, and 6 M), grafting times (1, 3, and 5 h), and grafting temperatures (70, 80, and 90°C). The grafting experiments were performed in glass vessels, and the distilled water and the acrylic acid monomer were then added. The plasma-treated PET fiber was immersed in the monomer solution. The grafting reaction was carried out by placing the glass vessel in a water bath which was set at relevant temperature and time. After grafting reaction, the grafted samples were taken out from the monomer solution and washed. The grafting percentage, GP (also known as the degree of grafting) was determined as follows:

$$GP\ (\%) = \frac{W_1 - W_0}{W_0} \times 100 \qquad (1)$$

Where, $W_0$ and $W_1$ are the weight of PET fiber samples before and after grafting, respectively.

The titration method was used to analyze the COOH functional group for the determination of grafting effects in order to identify the optimal grafting parameters. This method was based on the reaction between the carboxyl group and NaOH. The weighted PET fiber (grafted) was put in 100 ml beaker, and 25 ml 0.01 N standardized NaOH solution was then added and stirred for 24 h to neutralize the carboxyl group. The excess NaOH was then back titrated by a standardized solution of HCl reagent. Experimental error due to dissolved $CO_2$ was minimized by performing the titration experiment under a nitrogen atmosphere.

In theory, there is a linear relationship between the amount of COOH and the grafting yield. Assuming one unit of acrylic acid grafted is associated with one unit of COOH, the relationship is as follows:

$$COOH\ amount\ (mmole/g) = \frac{(W_1 - W_0) \times 1000}{72.06 \times W_0} = \frac{GP \times 10}{72.06} \qquad (2)$$

Where, 72.06 is the molecular weight of the acrylic acid monomer. Furthermore, the Fourier-transform infrared-attenuated total reflectance (FTIR-ATR) analysis was taken with Perkin Elmer Spectrum GX model spectrometer and was applied to functional group analysis. The characteristic peak of C=O of COOH in acrylic acid fall at 1720 $cm^{-1}$ (Kormunda and Pavlik, 2010), and the characteristic peak of $COO^-$ of the ion exchanger after alkalization is present at 1650 to 1540 $cm^{-1}$ (Colthup et al., 1990). The presence of these characteristic peaks in FTIR spectra could verify the grafting treatment successfully.

## Adsorption experiments

The ion exchangers prepared according to the two points described above were used in the experiments of heavy metal adsorption. Dried grafted PET fiber samples were added into 100 ml beaker containing 50 ml of each metal ion solution (50 mg/L) and adjusted to desired temperature and pH. The mixture solution was stirred and then filtrated. The ion concentration of the filtrates was analyzed with inductive coupled plasma-optical emission spectrometer (ICP-OES) (Perkin Elmer, Optima 2000). The adsorption amount was calculated as follows:

$$Q = \frac{(C_1 - C_0) \times V}{W} \qquad (3)$$

Where Q is the adsorption amount (mg/g), W the weight of grafted PET fiber (g), V the volume of solution (L), and $C_1$ and $C_0$ are the concentrations of each ion (mg/L) before and after adsorption, respectively. The adsorption behaviors were observed under various adsorption environments. Moreover, liquid and solid phase analyses were conducted to study the adsorption results and properties.

## RESULTS AND DISCUSSION

### Investigation of the optimal parameters for the plasma modification of PET fiber surface - water loss rate

Under various plasma treatment time and power, the results of the water loss rates are illustrated in Figures 1 and 2. Figure 1 shows that the hydrophilicity increased significantly from 3.2%/min (untreated PET fiber) to 10.5%/min owing to the plasma activation reaction with the optimal treatment time of 10 s. It is considered that, due to the proper treatment time, the fiber's surface reacted with the plasma gas to generate the hydrophilic active sites. Consequently, the resulting hydrophilicity increase was performed. However, longer exposure to plasma may also cause a loss of hydrophilicity, which contributes to the improvement of interfacial adhesion. Thus, the hydrophilic sites were destroyed and consequently the hydrophilicity was reduced compared to that treated for 10 s (Dogue et al., 1995). From Figure 2, the greater hydrophilicity was obtained at the power of 800 W. When the treatment power was raised from 400 to 800 W, the hydrophilic effect improved due to the increase of power provided. On the other hand, a further increase of the power to 1200 W enhanced the collision probability correspondingly due to the higher electron density in the chamber, leading to an etching effect of the active sites on the fiber's surface. Therefore, the hydrophilicity and grafting effect were slightly reduced. Such similar result could be found in the study by Li et al. (2006) who used radio frequency (RF) plasma to modify polyphenylene sulfide (PPS) at the power of 40 to 70 W, with the best grafting result observed at 50 W. To optimize the plasma treatment power and time, application of a plasma power of 800 W for 10 s may be considered as the suitable one in terms of surface modification of PET fibers.

### Investigation of the conditions for the acrylic acid grafting of plasma modified PET- grafting yield and COOH quantitative determination

The acrylic acid with a COOH functional group was selected as the grafting monomer for the grafting reaction to form the ion exchangers with the ability of adsorbing heavy metals. The effects of the factors, including grafting temperature, acrylic acid monomer concentration and

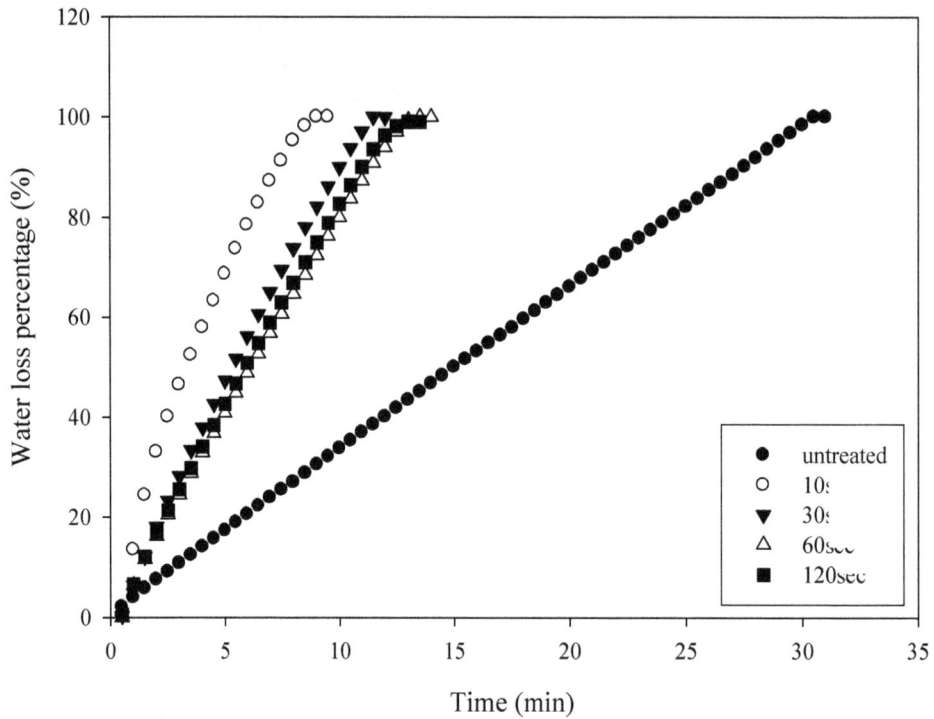

**Figure 1.** Water loss rates of original (untreated) PET fiber and the fibers treated with argon plasma of 10, 30, 60, and 120 s. The argon plasma power is 800 W and the gas pressure is 200 mtorr.

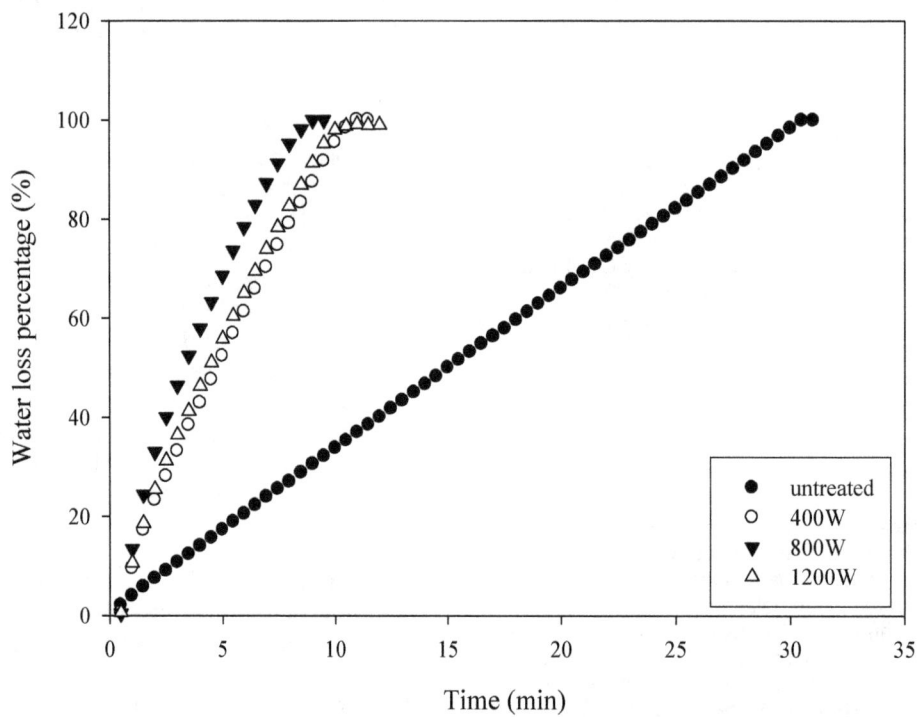

**Figure 2.** Water loss rates of original (untreated) PET fiber and the fibers treated with argon plasma of 400, 800, and 1200 W. The plasma treatment time is 10 s and the gas pressure is 200 mtorr.

grafting time on the grafting processes were investigated. Figure 3 reveals that the grafting yields altered at different grafting temperatures, with the optimal grafting yield of 4.45% obtained at 90°C. It was believed that, under elevated temperature, the peroxide groups or the initiators on the fiber's surface were easily decomposed thermally to form free radicals leading to a higher grafting probability with a higher grafting yield. Another cause for the increased grafting yield at elevated temperature was that the mobility of the acrylic acid monomer in the solution was enhanced owing to the increase of temperature. As a result, the probability of the monomer to come into contact with the free radicals was raised and giving a higher grafting yields (Xu et al., 2002). It was also found in Figure 3 that the grafting effect at the plasma power of 800 W was superior to that at 400 and 1200 W, thus representing the use of water loss rate for determining the plasma modification effect.

In the experiment of acrylic acid monomer concentrations, various monomer concentrations of 2, 4, and 6 M were used, and the results of the grafting yield were 1.3, 2.8, and 4.0%, respectively. Thus, the concentration of 6 M acrylic acid monomer was used in the experiments. As observed in Figures 4, the grafting yield increased with increasing grafting time. It was thought that the monomers reacted more easily with the free radicals on the fiber's surface as the acrylic acid monomer concentration increased, resulting in a higher grafting probability with an improved grafting yield. This trend was comparable to the result of the study by Guo et al. (1999) who discovered that the grafting yield of acrylic acid increased with increasing time within the grafting time of 150 min.

In addition to the weight method to calculate the grafting yield of acrylic acid monomer, we also used the titration method to analyze the amount of the COOH functional group on the surface. Because COOH can react with metal ions, it was used to remove the metal ions in aqueous solution. Analysis of the amount of COOH on the surface could further verify the relationship of COOH and the grafting yield. Therefore, we used the titration method to measure the amount of COOH on the surface of the PET fibers under various grafting conditions, then compared by the theoretical amount of COOH calculated from the grafting yield (weight method) to observe the difference in the amounts of COOH from the theoretical calculation and the titration method. The amounts of COOH from the analysis of the titration method and those from the calculated theoretical value of the weight method under various grafting parameters are summarized in Table 1.

The graph where the x-axis is the grafting yield and the y-axis is the COOH amount measured by the titration method is illustrated in Figure 5. It was found that there was a close correlation between the grafting yield and the COOH amount, indicating that either the grafting yield or

the COOH amount was a rational option to determine the grafting result.

Figure 6 illustrates the results using FTIR-ATR to analyze the functional groups on the surface of PET fibers before and after the plasma treatment. From Figure 6a, b, and c, it was found that the characteristic peak of $C=O$ ($1720$ $cm^{-1}$) appeared in all spectra. It was understood that the acrylic acid contained COOH functional group, yet the PET fibers already contained $C=O$. As a result, the difference in the position of the characteristic peak was less likely to be observed. Therefore, we further used the ion exchanger in the alkalization process to convert the COOH functional group into COONa by NaOH. Since the degree of ionization of COONa was higher, the characteristic functional group of $COO^-$ would be present. From Figure 6c, it reveals that the characteristic peak of $COO^-$ is present at $1566$ and $1580$ $cm^{-1}$. Colthup et al. (1990) demonstrated that the carboxyl salts have a strong asymmetric $CO_2$ stretching vibration at 1650 to 1540 $cm^{-1}$. Similar research results from Kondo et al. (2006) also showed the $COO^-$ observed at $1580$ $cm^{-1}$ by FTIR spectroscopy. The presence of these characteristic peaks confirmed the grafting of acrylic acid and characteristic functional group of $COO^-$.

## Investigation of the adsorption of heavy metal ions by the modified material

PET fiber with plasma-grafting-alkalization treatment was used in present study. It was prepared prior to the subsequent heavy metal adsorption experiment. Figure 7 shows the effects of different solution pH values on the adsorption of heavy metals. From the trend, we discovered that the adsorption of $Cu^{2+}$ increased significantly with increasing pH values. It was believed that the $H^+$ ions of higher concentration at a low pH competed with the positively charged heavy metal ions for the adsorption sites (Coskun et al., 2000; Karakisla, 2003) of $COO^-$ on the surface of the ion exchanger, resulting in the observed trend in the figure. From the view of competitiveness, we found that the adsorption effects were $Cu^{2+} > Pb^{2+} > Ni^{2+}$, demonstrating that the $Cu^{2+}$ ion has more favorable competitiveness in the aqueous solution. This result is similar to that of the study by Gérente et al. (2000) using $Cu^{2+}$, $Ni^{2+}$, $Pb^{2+}$, and consistent with that of the study by Çavuş et al. (2006) who pointed out that $Cu^{2+}$ is less likely to compete with $H^+$ ions than $Pb^{2+}$ and $Cd^{2+}$, leading to a higher adsorption effect. The highest adsorption of $Cu^{2+}$ could go as high as 0.073 mmole Cu/g-polymer, $Pb^{2+}$ 0.037 mmole Pb/g-polymer, and $Ni^{2+}$ 0.012 mmole Ni/g-polymer.

The effects of various temperatures on the adsorption of $Cu^{2+}$ were further investigated and the adsorption heat was calculated from the Van't Hoff equation

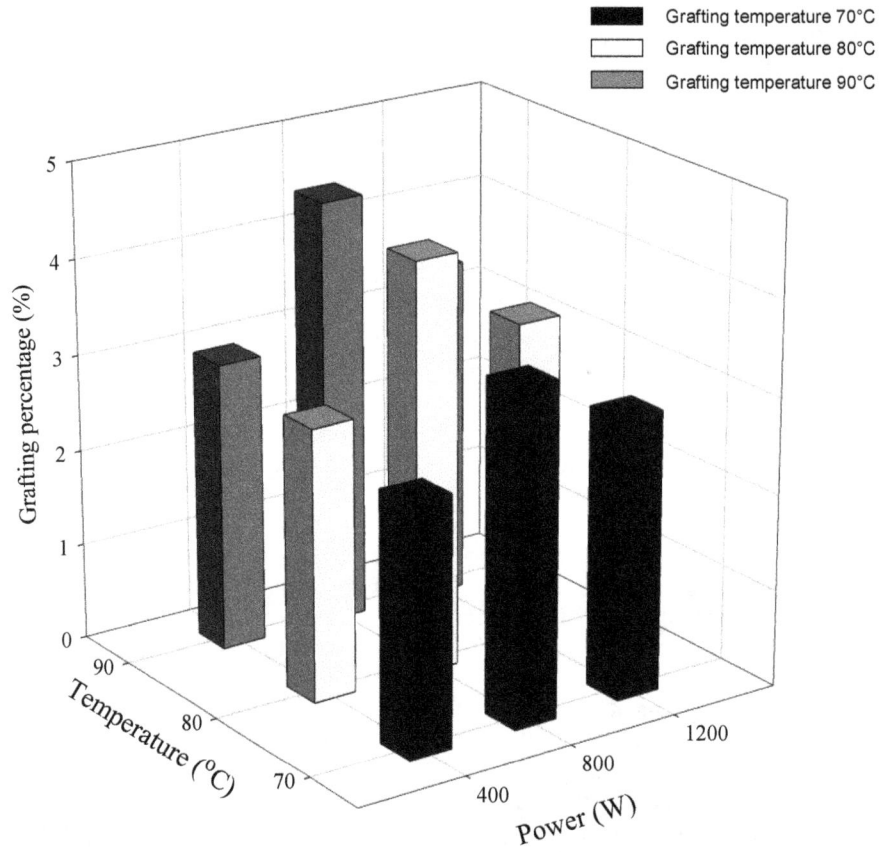

**Figure 3.** The grafting yields of PET after Argon plasma treatment under various parameters (plasma treatment time: 10 s; gas pressure: 200 mtorr).

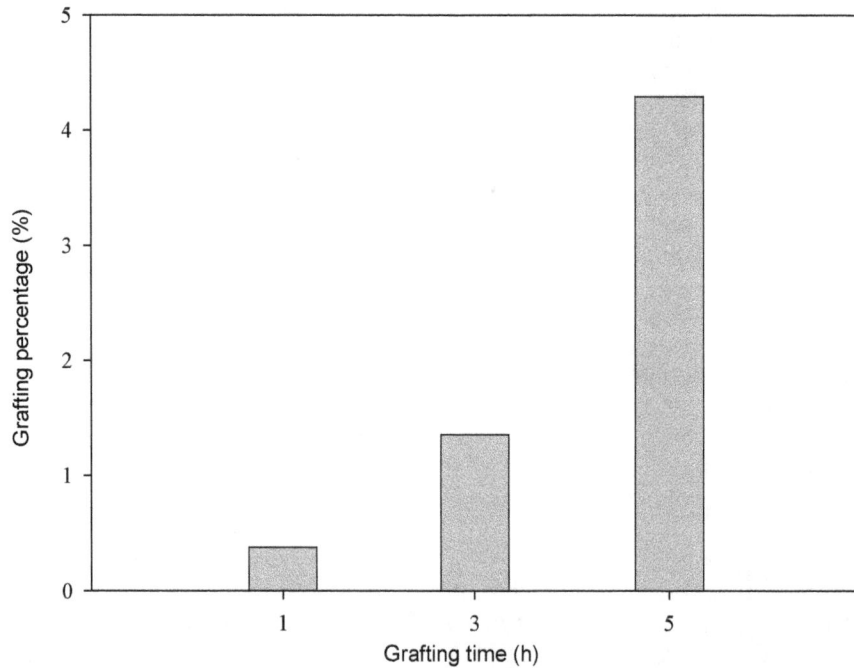

**Figure 4.** The grafting yields at different grafting time.

**Table 1.** Acrylic acid grafting yields and COOH amounts on the PET fiber surface.

| Grafting parameter | Grafting yields | COOH amount (mmole/g) | |
| --- | --- | --- | --- |
| | | Titration method | Weight method |
| 800 W, 200 mtorr, 70°C, 10 s | 2.40 | 0.213 | 0.333 |
| 800 W, 200 mtorr, 90°C, 10 s | 4.45 | 0.405 | 0.618 |
| 1200 W, 200 mtorr, 70°C, 10 s | 2.91 | 0.239 | 0.404 |
| 1200 W, 100 mtorr, 80°C, 10 s | 3.22 | 0.296 | 0.447 |
| 1200 W, 200 mtorr, 80°C, 10 s | 4.82 | 0.446 | 0.669 |

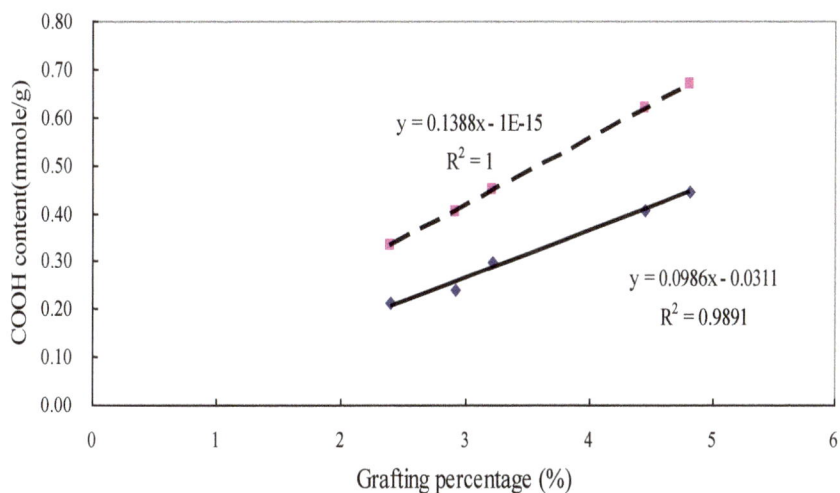

**Figure 5.** Correlation of COOH amount and grafting yield under various COOH measurement methods (solid line: titration method; broken line: weight method).

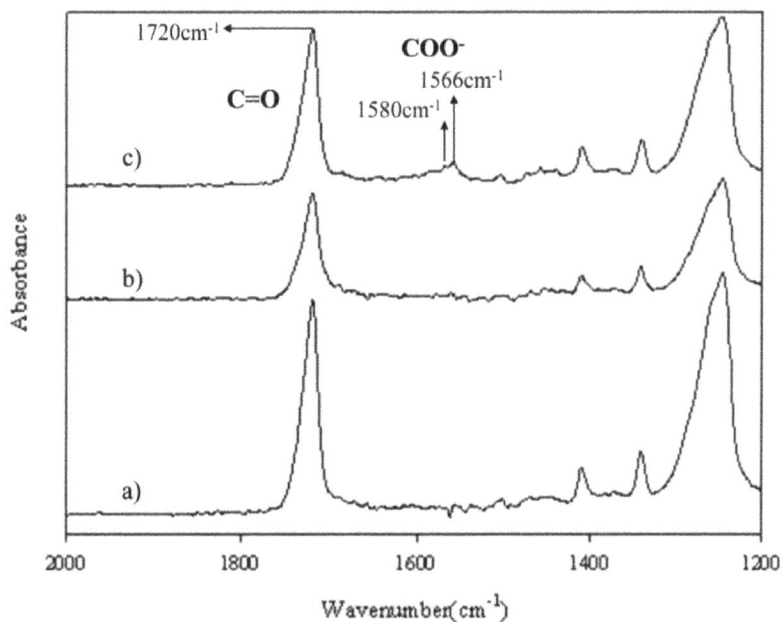

**Figure 6.** FTIR absorption spectra of PET fibers under various treatments (a, Grafting treatment only; b, plasma treatment and grafting treatment; c, plasma treatment, grafting treatment and followed by alkalization treatment).

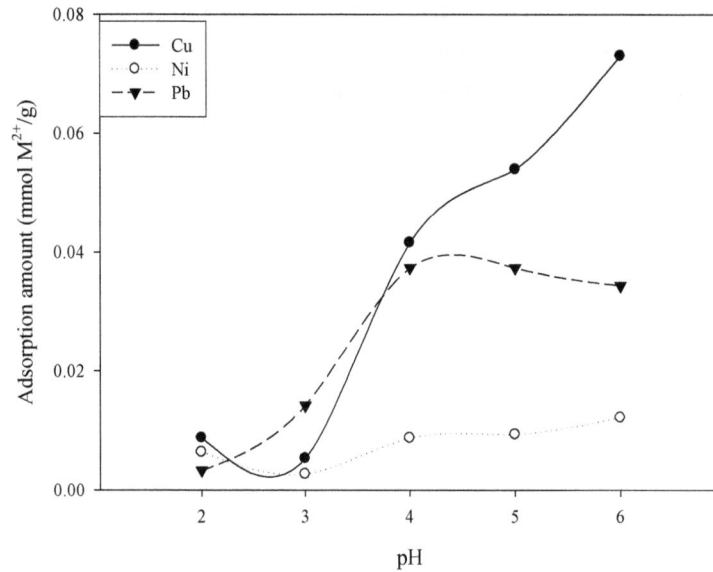

**Figure 7.** Effects of solution pH on the adsorption of heavy metal ions.

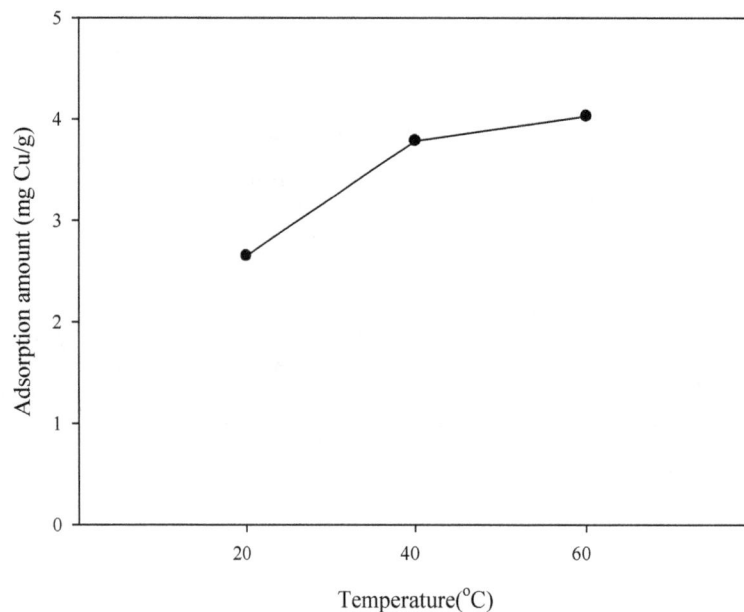

**Figure 8.** Effects of different reaction temperatures on the adsorption of $Cu^{2+}$.

(Atkins and Paula, 2006) illustrated below:

$$\log\frac{Q}{C_e} = -\frac{\Delta H}{2.303RT} + \frac{\Delta S}{2.303R} \qquad (4)$$

Where $C_e$, Concentration of solution at equilibrium (mg/L); Q, concentration of adsorbed solid (mg/L); $\Delta H$, adsorption heat (J/mol); R, 8.314 J/mol K; T, temperature

(K); $\Delta S$, entropy (4.184 J/mol K).

From the equation listed above, a graph was plotted using 1/T as the x-axis and log (Q/Ce) as the y-axis to calculate the adsorption heat $\Delta H$ from the slope, which was used to determine the reaction to be endothermic and its adsorption characteristics. Figure 8 illustrates the effects of different temperatures on the adsorption of $Cu^{2+}$. It was observed that the adsorption of $Cu^{2+}$

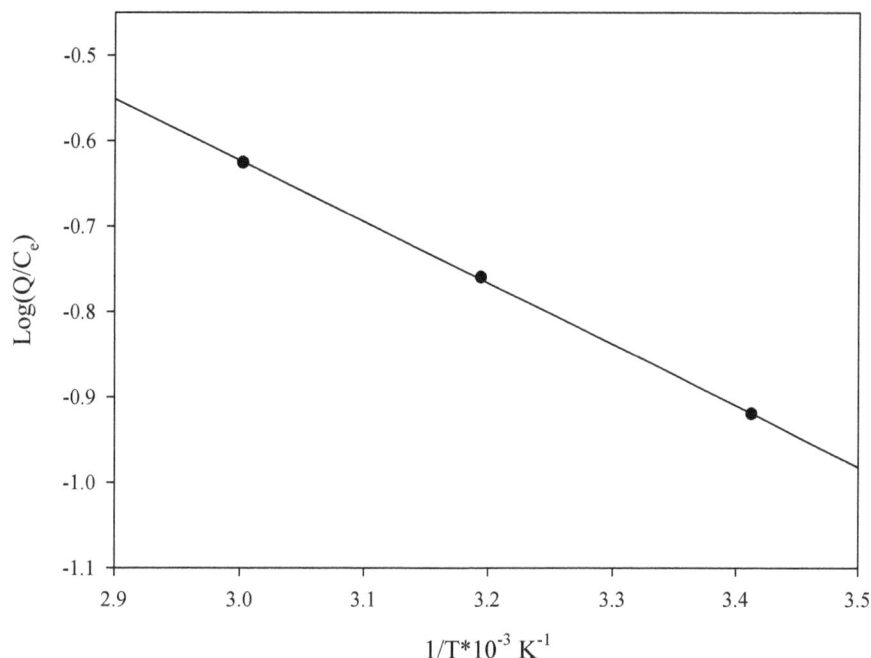

**Figure 9.** Adsorption curve of $Cu^{2+}$ from Van't Hoff equation.

improved at higher temperatures. Similar observation from Hegazy et al. (2001) demonstrated that this might be attributed to the increase in kinetic energy of metal ions with the temperature and the increase in the flexibility of the grafted chains as well. The result was factored into the Van't Hoff equation to generate Figure 9. From the graph, the slope and the reaction heat $\Delta H$ were calculated to be -717.5 and 13.74 kJ/mol, revealing that the reaction was endothermic ($\Delta H > 0$) and the reaction heat was less than 20 kJ/mol. Therefore, the reaction is believed to be a physical adsorption process.

## Conclusions

The present study used PET fibers as the substrate and ambient-temperature plasma to modify the surface properties in order to increase the hydrophilicity of the PET fibers. Afterwards, an ion exchanger capable of adsorbing heavy metals was prepared by acrylic acid grafting. Moreover, the ion exchange properties of the COOH functional group were explored to investigate the adsorption behavior of heavy metal ions. The potential of the application of the plasma surface modification combined with an ion exchanger developed by grafting technology was evaluated for wastewater treatment in an actual plant. The results are summarized as follows:

1) After the surface modification and activation of the PET fibers by ambient-temperature plasma, hydrophilicity could be improved to be favorable for acrylic acid grafting. The optimal modification conditions were a power of 800 W and a modification time of 10 s.

2) The optimal conditions under various environments for the grafting procedure of PET fibers after the plasma treatment were a grafting temperature of 90°C, an acrylic acid monomer concentration of 6 M, and a grafting time of 5 h. The grafting yield could reach 4.45% under such conditions.

3) The pH value of the adsorption environment posed a significant impact on the adsorption effect of the ion exchanger. In the condition pH of 6, Copper, Lead, and Nickel showed more favorable adsorption effects, with the adsorptions of 0.073 mmole Cu/g-polymer, 0.037 mmole Pb/g-polymer, and 0.012 mmole Ni/g-polymer.

4) The use of the Van't Hoff equation to investigate the adsorption model under different adsorption temperatures gave the reaction heat $\Delta H$ of 13.74 kJ/mol, indicating that the reaction was endothermic and a physical adsorption reaction.

5) The use of ambient-temperature plasma to modify PET fibers could improve the grafting yield and successfully prepare the ion exchanger to adsorb heavy metals. This technology is unique due to its convenience, quickness, low energy consumption, and environmental friendliness. The adsorption behavior was understood to be favorable for adsorption based on its adsorption model. The process is believed to be a physical adsorption reaction useful for the subsequent desorption and recycling of heavy metals. The developed technology displays a great potential for wastewater treatment in an actual plant and thus requires further advanced study.

## ACKNOWLEDGEMENT

The authors would like to thank the National Science Council, R.O.C. (NSC 97-2221-E-166-008) for the financial support of this research.

## REFERENCES

Atkins P, Paula JD (2006). Physical Chemistry (8th edition). W.H. Freeman and Company Printing, United States of America. P. 212.

Bailey SE, Olin TJ, Bricka RM, Adrian DD (1999). Review of potentially low-cost sorbents for heavy metals. Water Res. 33(11):2469-2479.

Bayasi Z, Zeng J (1993). Properties of polypropylene fiber reinforced concrete. ACI Mater. J. 90(M61):605-610.

Bureau of Industrial Development, the Ministry of Economic Affairs (MOEA), Taiwan (2007). www.moeaidb.gov.tw.

Bureau of Standards, Metrology and Inspection, The Ministry of Economic Affairs (MOEA), Taiwan (2010). CNS 12915-Method of test for fabrics. pp. 1-56.

Çavuş S, Gurdag G, Yasar M, Guclu K, Gurkaynak MA (2006). The competitive heavy metal removal by hydroxyethyl cellulose-g-poly(acrylic acid) copolymer and its sodium salt: The effect of copper content on the adsorption capacity. Polym. Bull. 57(4):445-456.

Colthup NB, Daly LH, Wiberley SE (1990). Introduction to Infrared and Raman spectroscopy (3rd edition). Academic Press Inc. Publisher, United States of America. pp. 317-318.

Coskun R, Yigitoglu M, Sacak M (2000). Adsorption behavior of copper(II) ion from aqueous solution on methacrylic acid-grafted poly(ethylene terephthalate) fibers. J. Appl. Polym. Sci. 75:766-772.

Dogue ILJ, Foerch R, Mermilliod N (1995). Plasma-induced hydrogel grafting of vinyl monomers on polypropylene. J. Adhes. Sci. Technol. 9(12):1531-1545.

Foti D (2011). Preliminary analysis of concrete reinforced with waste bottles PET fibers. Constr. Build. Mater. 25:1906-1915.

Gérente C, Couespel du Mesnil P, Andres Y, Thibault JF, Le Cloirec P (2000). Removal of metal ions from aqueous solution on low cost natural polysaccharides. Sorption mechanism approach. React. Funct. Polym. 46(2):135-144.

Guo Y, Zhan J, Shi M (1999). Surface graft copolymerization of acrylic acid onto corona-treated poly(ethylene terephthalate) fabric. J. Appl. Polym. Sci. 73(7):1161-1164.

Gupta B, Plummer C, Bisson I, Frey P, Hilborn J (2002). Plasma-induced graft polymerization of acrylic acid onto poly(ethylene terephthalate) films: Characterization and human smooth muscle cell growth on grafted films. Biomaterials 23(3):863-871.

Hegazy EA, Kamal H, Khalifa NA, Mahmoud GA (2001). Separation and extraction of some heavy and toxic metal ions from their wastes by grafted membranes. J. Appl. Polym. Sci. 81:849-860.

Karakisla M (2003). The adsorption of Cu(II) ion from aqueous solution upon acrylic acid grafted poly(ethylene terephthalate) fibers. J. Appl. Polym. Sci. 87(8):1216-1220.

Kondo Y, Miyazaki K, Takeuchi N, Sakurai K, Kaneko J (2006). Hydrophilization of PET wire mesh in paper manufacture by electron beam irradiation induced graft polymerization (in Japanese). Sen'i Gakkaishi 62(5):95-99.

Kormunda M, Pavlik J (2010). Characterization of oxygen and argon ion flux interaction with PET surfaces by in-situ XPS and ex-situ FTIR. Polym. Degrad. Stabil. 95:1783-1788.

Lee SD, Hsiue GH, Kao CY (1996). Preparation and characterization of a homobifunctional silicone rubber membrane grafted with acrylic acid via plasma-induced graft copolymerization. J. Polym. Sci. Part A: Polym. Chem. 34(1):141-148.

Li YN, Sun Y, Deng XH, Yang Q, Bai ZY, Xu ZB (2006). Graft polymerization of acrylic acid onto polyphenylene sulfide nonwoven initiated by low temperature plasma. J. Appl. Polym. Sci. 102(6):5884-5889.

Lin W, Hsieh YL (1997). Ionic Absorption of Polypropylene Functionalized by Surface Grafting and Reactions. J. Polym. Sci. Part A: Polym. Chem. 35(4):631-642.

McDonald DM, Webb JA, Taylor J (2006). Chemical Stability of Acid Rock Drainage Treatment Sludge and Implications for Sludge Management. Environ. Sci. Technol. 40(6):1984-1990.

Pereira de Oliveira LA, Castro-Gomez JP (2011). Physical and Mechanical behaviour of recycled PET fibre reinforced mortar. Constr. Build. Mater. 25:1712-1717.

Svorcik V, Kolarova K, Slepicka P, Mackova A, Novotna M, Hnatowicz V (2006). Modification of surface properties of high and low density polyethylene by Ar plasma discharge. Polym. Degrad. Stabil. 91(6):1219-1225.

Xu Z, Wang J, Shen L, Men D, Xu Y (2002). Microporous polypropylene hollow fiber membrane. Part I. Surface modification by the graft polymerization of acrylic acid. J. Membr. Sci. 196(2):221-229.

# 4

# Application of Cyanex® extractant in Cobalt/Nickel separation process by solvent extraction

Olushola S. Ayanda[1*], Folahan A. Adekola[2], Alafara A. Baba[2], Bhekumusa J. Ximba[1] and Olalekan S. Fatoki[1]

[1]Department of Chemistry, Faculty of Applied Sciences, Cape Peninsula University of Technology, P. O. Box 1906, Cape Town, South Africa.
[2]Department of Chemistry, University of Ilorin, P. M. B. 1515, Ilorin, Kwara State, Nigeria.

This study presents a detailed review on the advanced hydrometallurgical treatment involving the extraction and separation of Cobalt (Co) and Nickel (Ni) by Cyanex® extractants. The structures, properties and applications of various Cyanex® extractants such as Cyanex®272, Cyanex®301, Cyanex®921, Cyanex®923 and Cyanex®421X were discussed and compared. Cyanex®272 thus proved to be the most appropriate solvent extractant for the separation of Co and Nickel from sulphate and chloride media due to its stability to common oxidant, better physicochemical properties and its ability to avoid gypsum crystallization in stripping-electrowinning circuit. Finally, various solvent extraction (SX) techniques for the extraction and separation of Co and Ni using Cyanex® extractants were discussed as well as newer processes of extraction and separation.

Key words: Hydrometallurgy, Cobalt, Nickel, solvent extraction, Cyanex® extractants, Cyanex®272, Cyanex®301, Cyanex®921, Cyanex®923, Cyanex®421X.

## INTRODUCTION

The separation of Cobalt (Co) from Nickel (Ni) in aqueous solution has always been a problem for hydrometallurgists. Their adjacent positions in the transition metal series in the periodic table result in aqueous chemical behaviour that is too similar for the development of easy separation routes (Adekola et al., 2010). Co and Ni were separated traditionally by processes based on the selective oxidation and/or precipitation of Co from either sulphate or chloride solution and such processes are still in use today. While alkyl amines are the extractant of choice for the separation of Co from Ni from chloride liquors (Flett, 2004), for the weakly acidic sulphate liquors the alkyl phosphorus acids have found significant commercial application at various location around the world (Hofirek and Nofal, 1995). Because of the high Ni to Co ratio encountered in liquors produced in sulphate-based high Ni matte leach processes or those produced in the acid pressure leaching of Ni laterites, very high separation

factor (>1000) are required. Cyanex®272 has become the reagent of choice for such purposes (Rickelton and Nucciarone, 1997).

The separation of Co and Ni by precipitation processes has been and is still carried out commercially by a number of processes. Thus, sulphide precipitation can be used to completely remove Co from Ni process leached liquor. It can also be used to precipitate Ni from Co rich liquors. Co and Ni can also be separated by oxidative precipitation; strong oxidants such as Chlorine, Ammonium persulphate, caro's acid or ozone is needed in practice (Nishimura and Umetsu, 1992; Wyborn and McDonagh, 1996). The use of air under pressure has also been reported. This formed the basis for cobaltic amine process for Co-Ni separation (Nyman et al., 1992).

Ni powder is also precipitated selectively by reduction of aqueous solution containing Ni and Co ammines in concentrated ammonium sulphate solution at around 240°C with hydrogen gas at a total pressure of up to 3103 kPa. When the concentration of Ni in the solution is lowered to around that of Co, the reaction is stopped and the solution discharged from the autoclave leaving Ni powder inside (Burkin, 1987).

---

*Corresponding author. E-mail: osayanda@gmail.com.

Ion exchange separation of Co and Ni is most readily accomplished from chloride solution where advantage can be taken of the tendency for Co to form complex chloride anions, that is, $[CoCl_3]^-$, $[CoCl_4]^{2-}$, which Ni does not. These complexes are quite weak, however, relatively high concentrations of chloride ion are needed to produce the $[CoCl_4]^{2-}$ species. Although no great degree of selectivity between Ni(II) and Co(II) is achievable by ordinary cation exchange resins, chelating ion exchangers such as XFS4195, XFS4196 and XFS43084 can offer separation opportunities (Grinstead and Tsang, 1983).

Separation of Ni and Co is possible by direct hydrogen reduction of Ni and Co loaded di-ethylhexyl phosphoric acid (DEHPA) solution. Ni can be selectively reduced in the presence of Co from a metal-loaded DEHPA phase in an autoclave at 140°C and an initial pressure of 120 atm (Monhemius, 1994). It is also possible to recover Ni selectively from aqueous solutions by direct hydrogen reduction in the presence of Co.

## SOLVENT EXTRACTION (SX)

Solvent extraction (SX) is one of the most useful techniques that are used for the selective removal and recovery of metal ions from aqueous solution and it is largely applied in the purification process in chemical and metallurgical industries (Thurman and Mills, 1992; Dean, 1998). SX makes use of an organic compound capable of extracting the metal ion of interest, or a complex of it, from the aqueous phase into an immiscible organic solution. Conventional SX techniques however needs large amount of organic solvents and often creates environmental problems (Rudberg et al., 1992). The advantages of SX are simplicity and rapidity, the solvents are easily recoverable, and solvents are stable, transparent to ultraviolet (UV), not emulsifying during extraction and as selective as possible. In these processes, metal ion containing solution is contacted with a selective solvent. After extraction, stripping follows the process. SX is very difficult for quantitative separation of metal ions because of low driving force, and then a large amount of solvent is required. These make the extraction and stripping of desired species very expensive (Alfassi and Wai, 1992). More environmental friendly technology is needed nowadays, and there have been increasing attentions to extract metal ions by supercritical fluid extractants, solid phase extraction and bioleaching.

### Principles of solvent extraction

SX procedures utilize non-uniform distribution of substances between two immiscible liquid phases. Enrichment of the substance in one of the phases is dependent on many factors, such as pH, metal concentration and ionic strength in aqueous phase, salt concentration, reagent concentration in organic phase, contact time and temperature. Under suitable conditions, a substance of interest can be transferred to one of the phase, while unwanted substances are retained in the other. The transfer of the solute from one liquid phase to another involves extraction reactions, which permit the establishment of liquid-liquid equilibrium. The distribution of solute M (equilibrate between an aqueous phase and an organic solvent) can be described by an equilibrium equation (Rydberg, 1992):

$$[M]_{aq} <=> [M]_{org} \tag{1}$$

Thus, when this distribution reaches equilibrium, the distribution ratio (D) of the solute concentrations between the two phases is:

$$D = [M]_{org} / [M]_{aq} \tag{2}$$

Where $[M]_{org}$ is the concentration of a solute in the organic phase and $[M]_{aq}$ is concentration in the aqueous phase.

If the aqueous: organic (A:O) ratio is equal to 1, then the percentage of metal extracted (%E) would be:

$$\%E = 100D / 1 + D \tag{3}$$

An important extraction is characterized by a high value of D>>1, whilst a very small value of D<<1 characterizes a very feeble extraction.

### SOLVENT EXTRACTION OF COBALT AND NICKEL

Presently, most of the commercial Co/Ni SX plants operate using dialkyl phosphinic acid extractant, Cyanex®272 (Olivier, 2011). In contrast to the resin ion exchange (development) and precipitation process briefly discussed above, SX with this reagent does offer the opportunity for better Co/Ni separation with high yields and purity of the separated metals. Depending on the leach liquors composition, different reagents can be used for Co/Ni SX; the anion exchangers (extractants) from chloride solution and cation (acidic chelating) extractants from sulphate solutions. For Co-Ni separation by anion exchange the same situation exists as for resin anion exchangers with the most important ligand in the aqueous phase being chloride. The extracted anion species has been shown to be $[CoCl_4]^{2-}$. For cation exchangers, only the alkyl phosphoric, phosphonic and phosphinic acids show selectivity for Co over Ni, all the rest that is, carboxylic acids, β-ketone, 8-hydroxylquinolines (8-HQs) and hydroxyoximes, show marginal selectivity for Ni(II) over Co(II). The separation factor of Co from Ni in weakly acidic sulphate solutions

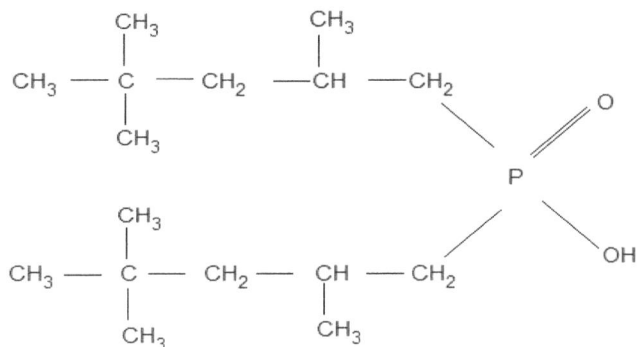

**Figure 1.** Bis(2,4,4-trimethylpentyl) phosphinic acid.

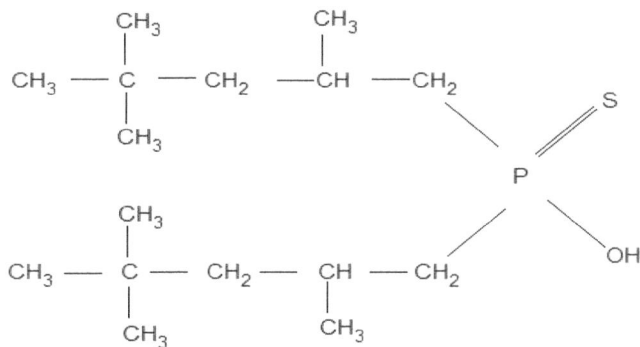

**Figure 3.** Bis(2,4,4-trimethylpentyl) monothiophosphinic acid.

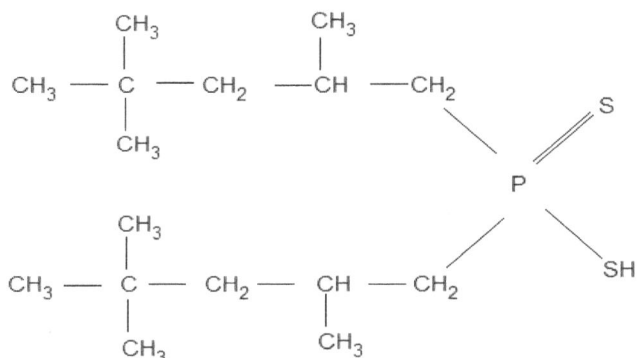

**Figure 2.** Bis(2,4,4-trimethylpentyl) dithiophosphinic acid.

increases in the following series: phosphoric < phosphonic < phosphinic acid (Luo et al., 2006).

## CYANEX® EXTRACTANTS

Cyanex® extractants are phosphorus-based products that act either as a chelating extractants or as a solvating extractants (Cyanex Industries Inc., 2007). Chelating extractants include Cyanex®272 and Cyanex®301. Cyanex®272 solvent extractant reagent was developed specifically for the separation of Co from Ni by SX. It is estimated that 40% of Co in the western hemisphere is produced using Cyanex®272 SX reagent, at plants in America, Canada, Africa, China and Australia. Cyanex®272 can also be used to separate the rare earth elements from one another. The acid concentration required for metal stripping is lower than when phosphoric or phosphonic acids are used as extractant. Cyanex®301 SX reagent is an analogue of Cyanex®272. Dithiophosphinic acid (Cyanex®301) also exhibit interesting extraction characteristics for the recovery of Co and Ni. A potential advantage of Cyanex®301 is the ability to extract both Co and Ni under very acidic conditions, thus, avoiding the need of adding alkali for pH

adjustment of the acidic leach liquors. However, the stability, better physicochemical properties and ability to avoid gypsum crystallization in stripping-electrowinning circuit makes Cyanex®272 the extractant of choice.

Solvating extractants include Cyanex®921, Cyanex®923 and Cyanex®471X. Cyanex®921 and Cyanex®923 have the potential in a wide range of applications. Specific applications are the recovery of organic solutes and/or inorganic acids from waste effluents, and metal extraction processes. Unlike its phosphine oxide analogues, Cyanex®471X is a Lewis base. It will only complex readily with metals that exhibit the characteristics of soft Lewis acids. Examples of metals falling upon those criteria are Pd(II), Pt(II), Ag(I), Cd(II), Hg(I), Hg(II) and Au(III) (Cyanex Industries Inc., 2007).

## Cyanex®272 extractant

The active component of the Cyanex®272 extractant is a bis(2,4,4-trimethylpentyl) phosphinic acid (Figure 1). Metal ions are extracted through a cation exchange mechanism. Although Cyanex®272 is selective for Co in the presence of Ni, a variety of other cations can also be extracted depending on pH of the solution (Saragi et al., 2009; Zhang et al., 2001).

## Cyanex®301 and Cyanex®302 extractants

Cyanex®302 and Cyanex®301 are the monothio- and dithio- derivative of Cyanex®® 272 with chemical formula $R_2PS(OH)$ and $R_2PS(SH)$, respectively (Gotfryd, 2005). The active component of Cyanex®301 extractant is bis(2,4,4-trimethylpentyl) dithiophosphinic acid (Figure 2), while the active component of Cyanex®302 extractant is bis(2,4,4-trimethylpentyl) monothiophosphinic acid (Figure 3). These sulphur-containing compounds are much stronger acids than their analogous oxy-acid, Cyanex®272. As such, they are capable of extracting

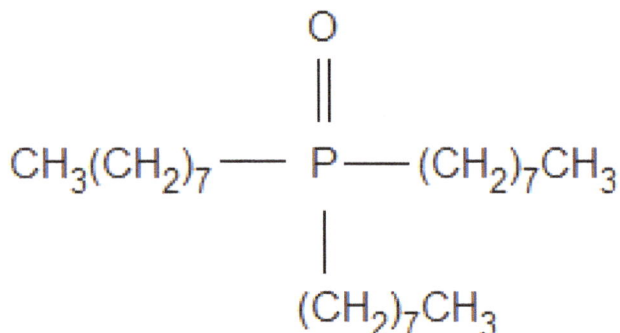

**Figure 4.** Trioctylphosphine oxide (TOPO).

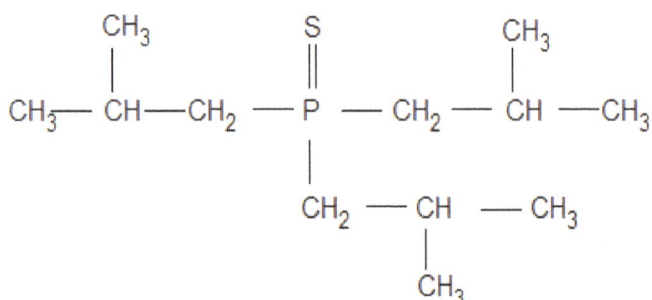

**Figure 5.** Tributyl phosphinic sulphide (TIBPS).

many metals at lower pH < 2. They do not discriminate between heavy metals at this pH range, however, a high degree of selectivity of extraction of heavy metals versus the alkaline earths is observed. Cyanex®301 was originally developed for the selective extraction of Zinc (Zn) from effluent streams also containing Calcium, such as those generated in the manufacturing of rayon by the viscose (Zhang et al., 2001).

**Cyanex®921, Cyanex®923 and Cyanex®471X extractants**

Cyanex®921 extractants (Figure 4) better known as trioctylphosphine oxide (TOPO) has been used commercially for many years to recover Uranium from wet process phosphoric acid.

Cyanex®923 is a liquid phosphinic oxide which has potential applications in the SX recovery of both organic and inorganic solutes from aqueous solution for example, carbonxylic acid and arsenic. It is a mixture of four trialkyl phosphine oxide.

$$R_3P(O) \ R_2R^IP(O) \ R^I_3P(O)$$

Where R is the normal octyl, and $R^I$ is normal hexyl Cyanex®471X is a soft Lewis base and will only complex readily with metals that exhibit the characteristics of soft Lewis acids. It is useful for the selective recovery of silver and in the separation of palladium from platinum. The active component of Cyanex®471X is tributyl phosphinic sulphide (TIBPS) (Figure 5).

Cyanex®921, Cyanex®923 and Cyanex®471X are not suitable extractants for the extraction and separation of Co and Ni (Cyanex Industries Inc., 2007). The physicochemical properties of Cyanex® extractants are presented in Table 1.

**Comparison of Cobalt and Nickel separation coefficient for various extractants**

The most important extractant is Cyanex®272. Cyanex®301 and Cyanex®302 are not very stable in contact with $Cu^{2+}$, $Cd^{2+}$ and common oxidants (Sole and Cole, 2001). Table 2 illustrates great advantages of Cyanex®272 over earlier extractants (Gotfryd, 2005).

Additionally, Cyanex®272 almost does not extract Calcium at optimal for $Co^{2+}$ extraction condition (pH 5.0 to 5.5). This allows avoiding gypsium crystallization in stripping-electrowinning circuit. Alkyl phosphonic and alkyl phosphinic acids as $Co^{2+}$ extractants have one common and important disadvantage. If saturated to level of 12 to 20 g/dm³ $Co^{2+}$, they highly increase their viscosity to over 150 cst (even to 400 cst). At such conditions, mixing and pumping can be almost impossible. Maximum concentrations of extractants, and consequently of $Co^{2+}$ extracted, should be limited to medium levels and/or elevated temperature should be applied to avoid problems of too high viscosity. The elevated temperature is moreover advantageous because the selectivity increases with increasing temperature.

**SOLVENT EXTRACTION OF COBALT/NICKEL BY CYANEX® EXTRACTANTS**

Cyanex®272 has been adopted as the reagent of choice for various laterite acid pressure leach project in Australia. Thus, the Murrin project uses SX with Cyanex®272 for Co/Ni separation from mixed sulphide pressure leach liquor. Figure 6 shows the Murrin purification flowsheet.

In the Pilot plant SX of Co and Ni for Alvin's Nkomati Project, Co was recovered from calcium-saturated solution using Cyanex®272. Ni was subsequently extracted using versatic acid. The loaded organic was stripped using spent electrolyte, producing advance electrolyte for the recovery of Ni by electrowinning. Co was then recovered with >99.5% extraction efficiency, reducing Co from 1.8 g/L to 10 mg/L. The Co/Ni ratio in the Co product solution was >1500. Ni SX was optimized to recover 99% reducing the Ni concentration from 32 to 0.3 g/L, while minimizing Calcium deportment to the Ni

**Table 1.** Physicochemical properties of Cyanex® extractants.

| Property | Cyanex®272 | Cyanex®301 | Cyanex®302 |
|---|---|---|---|
| Appearance | Colourless to light amber liquid | Green mobile liquid | Pale yellow |
| Specific gravity | 0.92 at 24°C | 0.95 at 24°C | 0.95 at 24°C |
| Viscosity | 14.2 cP at 25°C; 37 cP at 50°C | 78 cP at 24°C | 195 cP at 24°C |
| Solubility in $H_2O$ | 16 μ/ml at pH 2.6; 38 μ/ml at pH 3.7 | 7 mg/L* | 3 mg/L at 50°C |
| pKa | 6.37 | 2.61 | 5.63 |
| Boiling point | >300°C | Decomposes at 220°C | 205°C |
| Pour point | -32°C | -34°C | Approx. -20°C |
| Flash point (closed cup) | - | 165°F (74°C) | >205°F (>96°C) |
| Ignition temperature | 108°C | 74°C | 96°C |
| Specific heat | 0.48 cal/gm/°C at 52°C | - | - |
| Thermal conductivity | $2.7 \times 10\text{-}4$ cal/cm/°C | - | - |

| | Cyanex®921 | Cyanex®923 | Cyanex®471X |
|---|---|---|---|
| Appearance | Off white, waxy solid | Colourless mobile liquid | Off white crystalline solid |
| Specific gravity | 0.88 at 25°C; 0.84 at 61°C | 0.88 at 25°C | 0.91 at 22°C |
| Viscosity | 15.0 cP at 55°C | 40 cp at 25°C; 13.7 cP at 50°C | - |
| Solubility in $H_2O$ | - | >10 mg/L | 43 μg/ml at 24°C |
| Melting point | 47 - 52°C | - | 58 - 59°C |
| Boiling point | - | 310°C at 50 mm Hg | - |
| Flash point (closed up setaflash) | - | 182°C | - |
| Auto ignition temperature | - | 281°C | - |
| Vapour pressure | - | 0.09 mm Hg at 31°C | - |
| Thermal conductivity | - | 0.00302 cal/cm/s/°C at 25°C<br>0.00288cal/cm/s/°C at 120°C | - |

*Solubility will be lower in aqueous solutions containing dissolved salts; $pK_a = -\log K_a$ ($K_a$ – equilibrium constant of acidic dissociation).

**Table 2.** Comparison of Co(II) / Ni(II) separation coefficients for various extractants.

| Extractant | $\beta^{Co}_{Ni}$ | pH of optimal Co extraction | $\Delta pH_{50\%}^{Ni-Co}$ | |
|---|---|---|---|---|
| | | | 20°C | 50°C |
| DEHPA | 14 | 3.6 - 3.8 | 0.35 | 0.70 |
| PC-88A | 280 | 5.0 | 1.21 | 1.48 |
| Cyanex®272 | 7000 | 5.3 - 5.5 | 1.58 | 1.94 |

$\beta^{Co}_{Ni}$ = Coefficient of Co(II) / Ni(II) separation = D(Co) / D(Ni); D(M) = coefficient of metal ion (M) distribution; $\Delta pH_{50\%}^{Ni-Co} = pH_{50\%}^{Ni} - pH_{50\%}^{Co}$; $pH_{50\%}^{M}$, so called "pH of half extraction of ion M"; that means for D(M) =1.

electrowinning circuit (Sole et al., 2002). The overall process flowsheet is as shown in Figure 7.

The Bulong project used SX directly on the leach liquor after purification. Thus, any Iron, Aluminum and Chromium present in the leach liquor were removed hydrolytically in the two step precipitation to yield liquor at pH 4.2 to 4.5. Co together with Manganese (Mn) and Zn present in the liquor was extracted with Cyanex®272. The Ni in the raffinate was then extracted and separated from Magnesium with carboxylic acid, Versatic 10 (Scole

and Cole, 2001). The result of continuous miniplants showed that extraction with Cyanex®272 can achieve 97.5% Co recovery and >99% removal of Mn and Zn with good separation of Co and Ni with Co:Ni ratios in the strip of >1000:1. Figure 8 shows the Bulong Ni/Co purification flowsheet.

Cawse on the other hand uses mixed hydroxide precipitation followed by ammonia releach, which allowed for use of the Ni SX step. A simplified overall Cawse process flowsheet is as shown in Figure 9.

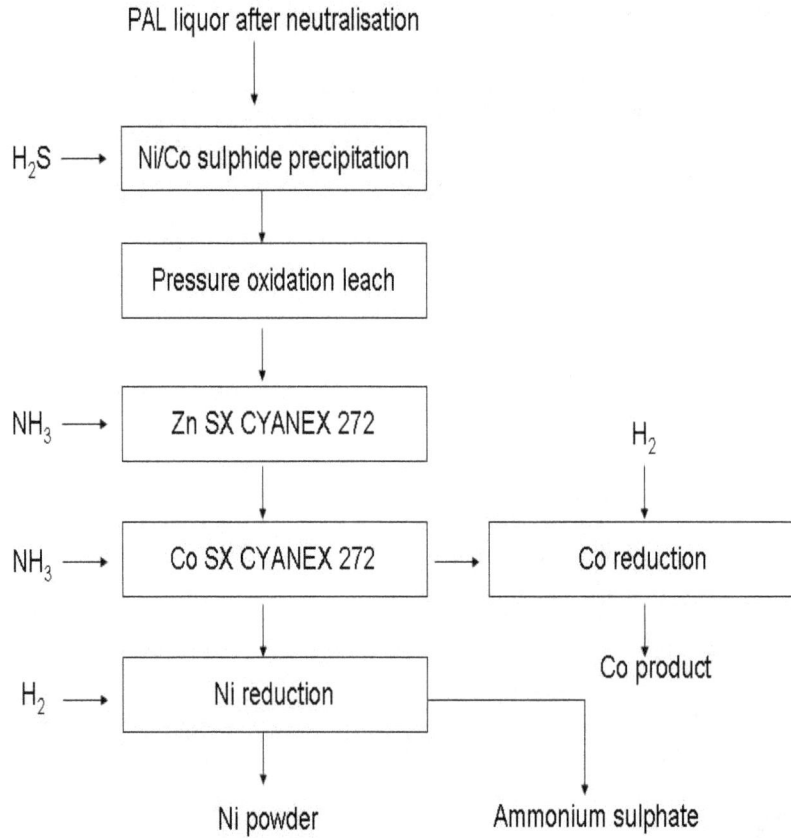

**Figure 6.** Murrin Murrin purification flowsheet (Motteram et al., 1996).

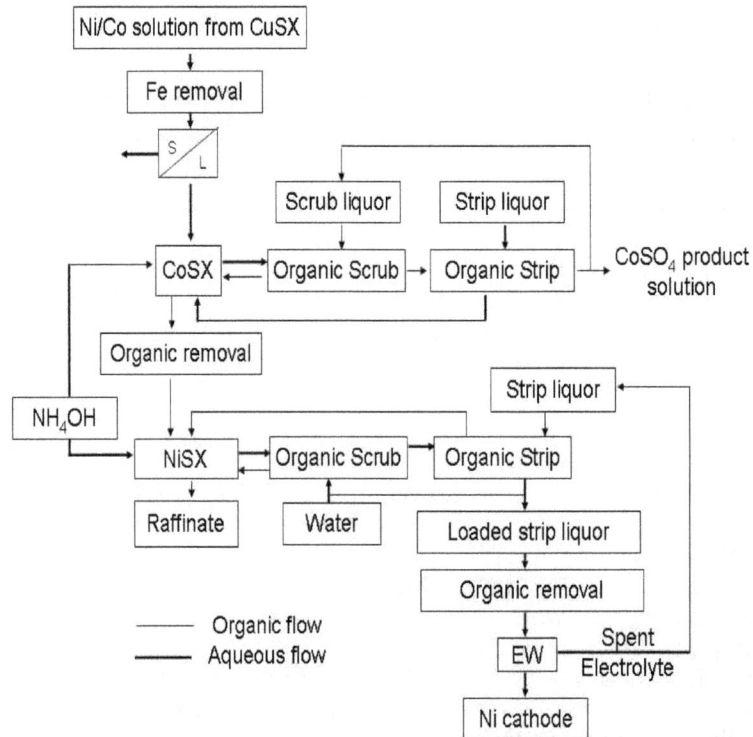

**Figure 7.** Process flowsheet for treatment of Nkomati Copper SX.

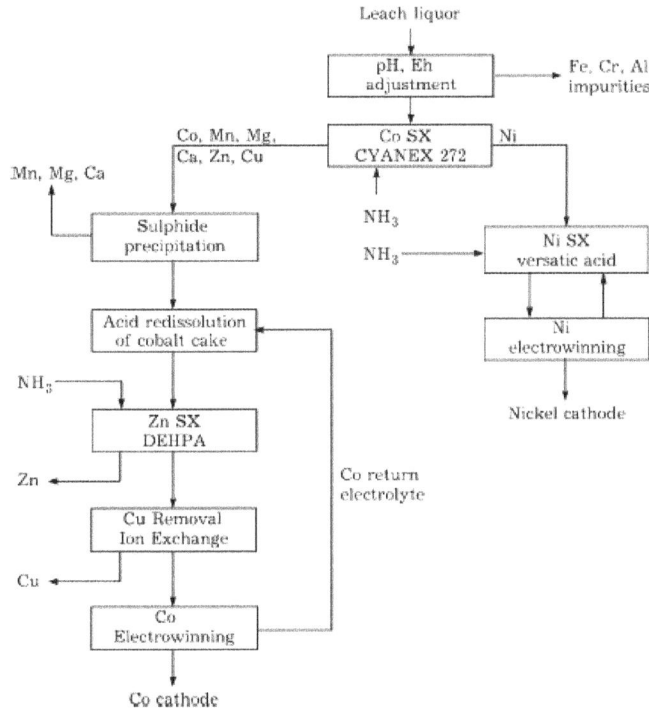

**Figure 8.** Bulong Ni/Co purification flowsheet (Flett, 2004).

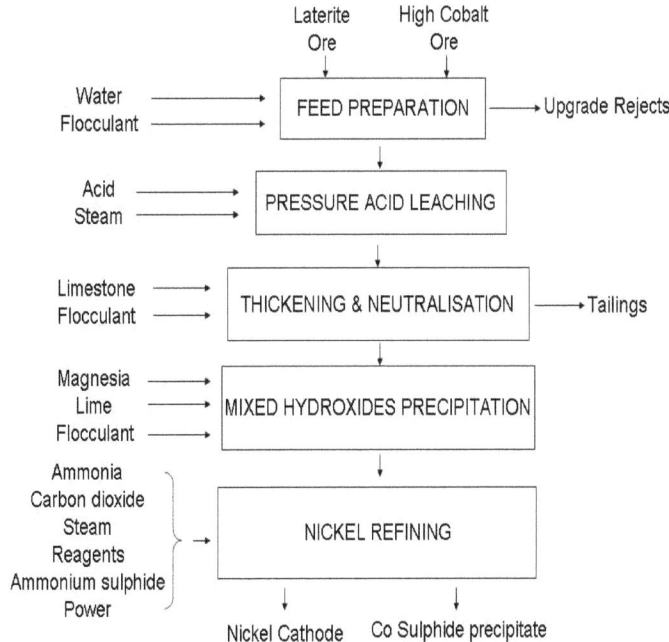

**Figure 9.** Cawse purification flowsheet (Taylor and Jansen, 2000).

Lenhard (2008) investigated the extraction and separation of Co(II) and Ni(II) from sulphate solutions with different initial volume fractions of Cyanex®302, Cyanex®272 and their mixture, in kerosene as diluent. Under the investigated range of conditions, Lenhard reported that Cyanex®302 outperformed Cyanex®272 in Co-Ni separation. In the extraction of Co and Ni with different mixtures of Cyanex®302 and Cyanex®272, no evidence for any synergistic effects was found.

Ahmed et al. (1992) in their preliminary investigations

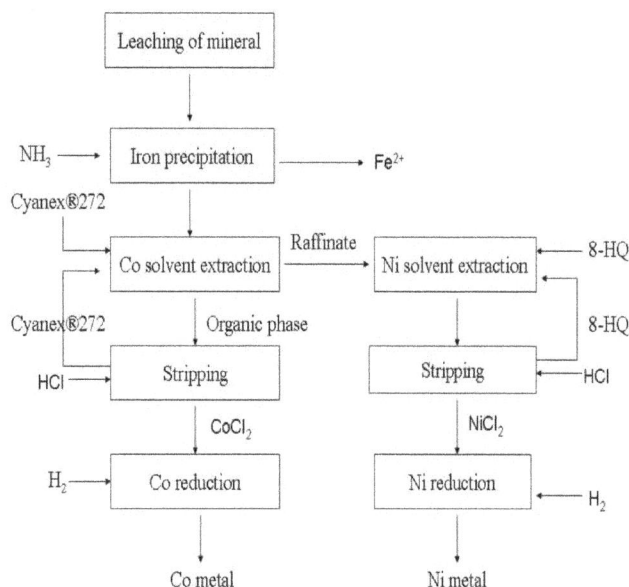

**Figure 10.** Proposed hydrometallurgical flowsheet for the extraction and separation of Co and Ni by Cyanex®272 from lateritic soil (Ayanda, 2009).

reported that only Cd(II) was extracted with Cyanex®923, while Co(II) and Ni(II) were not extracted. Different parameters affecting the extraction of Cd(II) with Cyanex®923 such as hydrochloric acid, hydrogen ion, extractant and metal concentrations and temperature were also investigated. They reported that Co(II) was found to be extracted with Cyanex®272 at pH 5.8 thereby leaving Ni(II) in the solution.

Gandhi et al. (1993) proposed the extraction of Co(II) at pH 8.0 with $5 \times 10^{-3}$ M Cyanex®272 in chloroform. Co(II) was stripped with 0.5 M nitric acid and separated from Vanadium, Chromium, Ni, Mn, Iron and Zn.

Tait (1993) investigated the use of Cyanex®301, Cyanex®302 and Cyanex®272 for the extraction of Co(II) and Ni(II) from a sulphate medium. He reported that all the reagents extracted Co selectively with Cyanex®302 exhibiting better separation characteristics than Cyanex®272, which in turn showed a higher selectivity than Cyanex®301. The separation ($pH_{0.5}^{Ni}$- $pH_{0.5}^{Co}$) found for Cyanex®302 was 2.6 pH units, compared to 1.7 pH units for Cyanex®272 and 1.1 pH units for Cyanex®301.

Ribeiro et al. (2004) studied the extraction of Co and Ni using Cyanex®302. They studied a high Ni/Co initial concentration ratio of 63.3:1 and found that the feed phase pH has a profound effect on the emulsion liquid membrane (ELM) process due to its straight relation with the extraction chemistry.

Kyung-Ho and Debasish (2006) reported the use of Cyanex®272 for the extraction of Co from a solution containing Co and Ni in a sulphate medium followed by Copper (Cu) extraction. Impurities such as Cu and Iron were removed from the leach liquor by precipitation

method before Co extraction. They reported that an increase in the concentration of Cyanex®272 increased the extraction percentage of Co due to the increase of equilibrium pH. Co extraction efficiency of > 99.9% was achieved with 0.20 M Cyanex®272 in two counter-current stages at an aqueous: organic (A:O) phase ratio of 1.5:1. Complete stripping of Co from the loaded organic containing 2.73 g/L Co was carried out at pH 1.4 by a synthetic Co spent electrolyte in two stages at an A:O ratio of 1:2. The enrichment of Co during extraction and stripping operations was reported to be carried out at about 3.5 times.

The extraction of Co from Cu-Zn-free solution was carried out using Cyanex®272 in kerosene followed by the extraction of Ni from the Co-free solution with Na Cyanex®272 in kerosene by Parhi et al. (2008). They reported that the extraction of Co and Nil increased with increasing equilibrium pH and extractant concentration. They also reported that the highest separation factor for Co and Ni was obtained with 0.1 M Cyanex®272 at pH 5.46.

Adekola et al. (2010) reported the hydrometallurgical treatment involving the SX and recovery of Co and Ni from hydrochloric acid leached solution of laterite using Cyanex®272 and 8-HQ, both diluted in kerosene. Experimental results showed that Cyanex®272 was selective for Co in 4 M HCl, while 8-HQ was found to be selective for Ni. The average percentage of Co extracted by Cyanex®272 was 94.71%, while 99.98% of Ni was extracted by 8-HQ. 1.0 M HCl was also found to be effective for the stripping of Co and Ni from both Cyanex®272 and 8-HQ, respectively. The hydrometallurgical flowsheet for the extraction and separation of Co and Ni from lateritic soil as proposed is shown in Figure 10.

## CONCLUSION

Commercial operations for separation of Co from Ni have been successfully carried out using precipitation, ion exchange resin, pressure hydrogen reduction and SX. SX does offer the opportunity of complete separation with high yields and purity of the separated metals. Under comparable conditions of the solvent extractants, the Co-Ni separation increases in the order: phosphoric < phosphonic < phosphinic acids. Among the various Cyanex® extractants, Cyanex®921, Cyanex®923 and Cyanex®471X are not suitable for the extraction and separation of Co and Ni. Cyanex®301 was developed for the selective extraction of Zn from effluent streams containing Calcium. Tait (1993) and Lenhard (2008) found Cyanex®302 to be better Co and Ni extractant than Cyanex®272 in their experimental results, Ahmed et al. (1992) also reported that Cyanex®923 is not suitable for the extraction of Co(II) and Ni(II), while Cyanex®471X will only complex readily with metals that exhibit the characteristics of soft Lewis acids. The mono and thio analogues of Cyanex®272 enable the extraction to be carried out at a much lower pH as a result of the replacement of oxygen by sulphur. However, Cyanex®272 is considered to be the most preferable Cyanex® extractant for the extraction and separation of Co and Ni from the stripping conditions point of view (Sole and Hiskey, 1992), because of its stability to $Cu^{2+}$, $Cd^{2+}$ and common oxidant (Gotfryd, 2005). Cyanex®272 is also primarily designed for the separation of Co and Ni from both sulphate and chloride media. Finally, Cyanex® extractants can be used to extract and separate vast majority of metal cations, this can be achieved by varying the experimental conditions.

## REFERENCES

Adekola FA, Baba AA, Ayanda OS (2010). Solvent extraction of Cobalt and Nickel from Nigerian lateritic soil. J. Chem. Soc. Nig. 35:123-128.

Ahmed IM, El Dessouky SI, El-Nadi YA, Saad EA, Daoud JA (1992). Recovery of Cd(II), Co(II) and Ni(II) from chloride medium by solvent extraction using Cyanex 923 and Cyanex 272. http://www.iaea.org/inis/collection/NCLCollectionStore/_Public/39/120/39120292. pdf pp. 1-11.

Alfassi ZB, Wai CM (1992). Preconcentration techniques for trace elements. CRC Press: Bica Raton, FI pp. 3-99.

Ayanda OS (2009). Leaching and Solvent Extraction of Cobalt and Nickel from Nigerian Lateritic Soil by Cyanex272 and 8-hydroxylquinoline. M.Sc. Thesis. Chemistry Department, University of Ilorin: Ilorin, Nigeria. pp. 1 - 125.

Burkin AR (1987). Extractive metallurgy of nickel. Critical Reports in Applied Chemistry 17, John Wiley and Sons pp. 51-75.

Cyanex Industries Inc. (2007). www.cytec.com/specialty-chemicals/cyanex.htm. [Accessed: 07/01/2011].

Dean JR (1998). Extraction methods for environmental analysis. John Wiley and sons, Chichester. pp. 1-225.

Flett DS (2004). Cobalt-Nickel Separation in Hydrometallurgy: A Review. Chem. Sust. Dev. 12:81-91.

Gandhi MN, Deorkar, NV, Khopkar SM (1993). Solvent extraction separation of Cobalt(II) from Nickel and other metals with Cyanex 272. Talanta 40:1535-1539.

Gotfryd L (2005). Solvent extraction of Nickel (II) sulphate contaminants. Physicochem. Probl. Miner. 39:117-128.

Grinstead RR, Tsang AL (1983). International Solvent Extraction Conference Denver, Colorado. Am. Inst. Chem. Eng. New York. pp. 1-230.

Hofirek Z, Nofal PJ (1995). Pressure leach capacity expansion using oxygen-enriched air at RBMR (Pty) Ltd. Hydrometallurgy 39:91-116.

Kyung-Ho P, Debasish M (2006). Process for Cobalt separation and recovery in the presence of nickel from sulphate solutions by Cyanex 272. Met. Mater. Int. 12:441-446.

Lenhard Z (2008). Extraction and separation of Cobalt and Nickel with extractants Cyanex 302, Cyanex 272 and their mixture. Kem. Ind. 57:417-423.

Luo L, Wei J, Wu G, Toyohisa F, Atsushi S (2006). Extraction studies of Colbalt (II) and Nickel (II) from chloride solution using PC88A. Trans. Nonferrous Met. Soc. China. 16:687-692.

Monhemius AJ (1994). Recent advances in the use of solvent extraction in hydrometallurgy. Bull. Chem. Technol. Macedonia 13(2):7-12.

Motteram G, Ryan M, Berezowsky R, Raudsepp R (1996). Murrin Murrin Nickel and Cobalt project: Project development overview. Nickel and Cobalt Pressure Leaching and Hydrometallurgy Forum, Alta Metallurgical Service, Perth, Western Australia.

Nishimura T, Umetsu Y (1992). Separation of Cobalt and Nickel by ozone oxidation. Hydrometallurgy 30:483-497.

Nyman B, Aaltomen A, Hultholm SE, Karpala K (1992). Application of new hydrometallurgical developments in Outokumpu HIKO process. Hydrometallurgy 29:461-478.

Olivier MC (2011). Developing a solvent extraction process for the separation of Cobalt and iron from Nickel sulfate solution. M.Sc. Engineering Thesis, Stellenbosch University, South Africa.

Parhi PK, Panigrahi S, Sarangi K, Nathsarma KC (2008). Separation of cobalt and nickel from ammoniacal sulphate solution using Cyanex 272. Sep. Purif. Technol. 59:310-317.

Ribeiro CP, Costa AOS, Lopes IPB, Campos FF, Ferreira AA, Salum AJ (2004). Cobalt extraction and Cobalt-Nickel separation from a simulated industrial leaching liquor by liquid surfactant membranes using Cyanex 302 as carrier. Membr. Sci. 241:45-54.

Rickelton WA, Nucciarone D (1997). In Cooper WC, Mihaylov I (Eds.), Nickel-Cobalt'97, Hydrometallurgy and refining of Nickel and cobalt. Metallurgical Soc. CIM 1(1):275.

Rudberg J, Musikas C, Choppin CM (1992). Principles and practices of solvent extraction, Marcel Dekker: New York pp. 357-412.

Saragi K, Reddy BR, Das RP (1999). Extraction studies of Cobalt (II) and Nickel (II) from chloride solutions using Na-Cyanex 272. Separation of Co(II)/Ni(II) by the sodium salts of D2EHPA, PC88A and Cyanex 272 and their mixtures. Hydrometallurgy 52:253-265.

Sole KC, Cole P (2001). Purification of Nickel by solvent extraction. In Ion Exchange and Solvent Extraction, Marcus and SenGupta, New York: Marcel Dekker, 15:143-197.

Sole KC, Cole PM, Preston JS, Robinson DJ (2002). Extraction and separation of Nickel and Cobalt by Electrostatic Pseudo Liquid Membrane (ESPLIM). International Solvent Extraction Conference proceedings, Cape Town, South Africa, pp. 730-735.

Sole KC, Hiskey JB (1992). Solvent extraction characteristics of thiosubstituted organophosphinic acid extractants. Hydrometallurgy 30:345-365.

Tait BK (1993). Cobalt-Nickel separation: The extraction of Cobalt(II) and Nickel(II) by Cyanex 301, Cyanex 302 and Cyanex 272. Hydrometallurgy 32:365-372.

Taylor A, Jansen ML (2000). Future trends in PAL plant design for Ni/Co laterites. Nickel and Cobalt 2000. ALTA Metallurgical Services, Melbourne. pp. 1-13.

Thurman EM, Mills MS (1992). Solid-phase extraction, principle and practice. In Winefordner JD (ed.), Series Chemical Analysis 147, John Wiley and Sons, New York. P. 988.

Wyborn PJ, Mcdonagh CF (1996). Minerals, metals and the environment II, IMM, London. P. 421.

Zhang P, Yokoyama T, Suzuki TM, Inone K (2001). The synergistic extraction of Nickel and Cobalt with a mixture of di(2-ethylhexyl)phosphoric acid and 5-dodecylsalicylaldoxime. Hydrometallurgy 61:223-227.

# 5

# Mechanical properties of microwave sintered 8 mol% yttria stabilized zirconia

P. Ganesh Babu[1,2] and P. Manohar[2]

[1]Department of Physics, Sri Ramanujar Engineering College, Chennai-600 048, India.
[2]Department of Ceramic Technology, A. C .Tech Campus, Anna University, Chennai-25, India.

This paper reports the synthesis and characteristics of nano sized 8 mol% yttria stabilized zirconia (8YSZ). The nano sized 8YSZ powder (15 nm) has been prepared by the sol-gel technique. The phase formation was confirmed by the X-Ray Diffraction (XRD) analysis and the particle size verified by transmission electron microscope (TEM). The prepared 8YSZ green samples were sintered in a microwave furnace at 1500°C for three different holding times, such as 5, 10 and 15 min. The density of the sintered samples was measured with the Archimedes method. This method revealed 98% of theoretical density of the 8YSZ sample. The surface morphology of the sintered samples was examined by the scanning electron microscope (SEM). The mechanical properties, including nanohardness, elastic modulus and coefficient of friction of the high density sample have been evaluated.

**Key words:** Sol-gel, nano 8YSZ, microwave sintering, nano hardness, elastic modulus, coefficient of friction.

## INTRODUCTION

8 mol% yttria stabilized zirconia (8YSZ) has found many engineering applications, because of its high strength, fracture toughness, and high thermal stability (Lange et al., 1986). It is used as an electrolyte material in oxygen pumps, and in oxide fuel cells, owing to its oxide ion conductivity at high temperatures and relatively low cost (Minh, 1993; Steele, 2001; Fukul et al., 2004). Although 8YSZ possesses high ionic conductivity, its low mechanical properties limit its applications. In order to increase the mechanical properties of 8YSZ, some important factors should be considered, that is, particle size, grain size, and different sintering methods. Nanopowder offers the possibility of manufacturing dense ceramics at a low sintering temperature, and sintering techniques can have considerable influence on the grain growth of the samples (Mazaheri et al., 2009). So the synthesis of nano 8YSZ powder is essential in research fields. Until recently, a wide variety of chemical methods

have been used to prepare 8YSZ powder, such as co-precipitation, solution combustion, hydro thermal treatment, hydrolysis of alhoxides and sol-gel synthesis (Kumar and Manohar, 2007). Usually, 8YSZ is synthesized by a mixed-oxide reaction method. However, the powders prepared by this method usually exhibit high degree of agglomeration, and an inhomogeneous particle size, thus requiring a subsequent high-temperature treatment. The sol-gel synthesis is a chemical solution process. In comparison with other techniques, the sol–gel process has shown significant advantages, including excellent chemical stoichiometric, compositional homogeneity and lower crystallization/sintering (processing) temperature, due to the mixing of the liquid precursors on the molecular level (Liu et al., 2007; Surowiak et al., 2001; Fan and Kim, 2002). Nowadays, we need minimum processing time, high density, uniform sized grains and good mechanical properties. Many researchers have suggested different types of sintering methods, to achieve the maximum density of the sintered 8YSZ sample and improve its mechanical properties. By using the hot press, and spark plasma sintering methods, Dhal et al. (2007) achieved the 98% density of the 8YSZ sample and measured its mechanical properties. Mazaheri et al. (2008) used microwave sintering, and achieved 96% density for 8YSZ sample. Microwave sintering has been performed by many researchers for different 8YSZ applications (Ciacchi et al., 1996; Upadhyaya et al., 2001; Thridandapani et al., 2009; Janney et al., 1992). Microwave sintering is a more advanced sintering technique than conventional sintering. In the conventional sintering method, an external heating element is used to generate the heat, and it is transferred to the sample via, conduction, convection, and radiation, which produces a high temperature gradient and internal stresses (Coasta et al., 2003). But, in microwave sintering, the heat is generated internally, within the test sample, by the rapid oscillations of the dipole at microwave frequency (Phani and Santucci, 2006). Since electromagnetic waves are used to generate the heat in microwave sintering, a large amount of heat can be transferred to the interior of the ceramic sample. The salient features of microwave sintering are it's volumetric and uniform heating, and short processing time. In general, nano indentation has been performed for thin films and crystals. Soyez et al. (2000) found the mechanical properties of polycrystalline YSZ, using the Nano indentation technique. Further, Voevodin et al. (2001) fabricated YSZ/Au nano composite films and evaluated their mechanical behaviour by the nano indentation technique. Gaillard et al. (2009) applied the nano indentation technique on the yttria stabilized zirconia crystal and analysed the phase transformation of the single and polycrystal. However, Mukhopadhyay et al. (2009) applied the nano indention technique and calculated the nano hardness of pressure less sintered alumina samples. Many investigators applied the microwave sintering on the commercially available 8YSZ powder pellets and studied the mechanical properties by using the micro indentation technique, but the nano indentation studies of microwave sintered 8 mol% yttria stabilized zirconia bulk sample have not been reported by the previous researchers. 3YSZ -TZP ceramics was one of the promising materials for tribological applications, because of its high wear resistance, bending strength, fracture toughness and mechanical strength and low friction coefficient than cubic zirconia (Ran et al., 2007). Many investigators found the tribological characteristics of 3 mol% of yttria stabilized zirconia (3YSZ) sintered samples with and without metal oxide additives of CuO, $MnO_2$, MgO, $B_2O_3$ and reduced the coefficient of friction (Bas et al., 2004). It is known that 8YSZ materials are mostly used as solid oxide fuel cell (SOFC) applications, during the operation solid oxide fuel tend to bend at higher temperatures so it is necessary to improve their

mechanical strength, and tribological characteristics, and hence it is necessary to give importance to the coefficient of friction of 8YSZ.

In this present investigation, the nano 8YSZ powder was synthesised by the sol-gel wet chemical route, followed by microwave sintering. The nano mechanical properties, including nano hardness, elastic modulus and one of the important tribological properties coefficient of friction were evaluated. And the results are compared with the mechanical properties reported by the micro indentation technique of previous research.

## MATERIALS AND METHODS

In the sol-gel process, zirconium oxychloride ($ZrOCl_2 8H_2O$), yttrium nitrate hexa hydrate ($Y(NO_3)_3.6H_2O$), and oxalic acid ($C_2H_2O_4$) were used as the starting materials (Kumar and Manohar, 2007). These respective salts were dissolved in 200 ml of millipore water in the stoichiometric ratio of 1 M. The initial solutions of the precursors were mixed under constant stirring by a magnetic stirrer, until a transparent viscous gel was formed. The obtained gel was dried in a hot air oven at 45°C for five days. The dried gel was calcined at 600°C for 3 h and well ground in a planetary mill for 5 h in an ethanol medium at a rotational speed of 300 rpm. The planetary jar mill and grinding balls were made of tungsten carbide material. The milled powder was dried in air at 60°C for 24 h and the powder was characterized by the X-ray diffractometer (XRD, Seifert 3000P) with Cu - Kα radiation (λ = 1.5406 Å). The XRD patterns were recorded in the 2θ scanning range of 10° to 80°. The phase formation of the nano 8YSZ was analysed by the XRD and the particle morphology was analysed by the transmission electron microscope (TEM) [JEOL-1200 EX II operated at 120 kV (max)]. The milled powder was mixed with the polyvinyl alcohol (PVA) binder and pressed in to pellets (10 mm in diameter) at a pressure of 40 MPa using a uni axial press. The 8YSZ green samples were sintered in a microwave furnace (V. B. C. C India) at 1500°C at a 1.1 kW 2.45 GHz frequency, at a heating rate of 100°C / min for three different holding times that is, 5, 10, and 15 min. In the microwave sintering furnace, a special susceptor is designed to generate heat against the microwave at a heating rate of 100°C/min. A non-contact optical sensor (RAYTEK, USA) was used to measure the temperature in the range of 600 to 1600°C; the time temperature profile is programmed by the Eurotherm temperature indicator cum programmer. The density of the samples was measured by the Archimedes method. The microstructure of the sintered and polished samples was analysed by the scanning electron microscope (SEM) (HITACHI Model S-3400 JAPAN). The average grain size of the sintered samples was measured by the linear intercept length method (Mendelson, 1969).

### Nanohardness and coefficient of friction

The nano indentation test was applied on a fully dense microwave sintered 8ysz sample using a nano-indenter (CSM open platform Switzerland) with Berkovich indenter calibrated with a standard silica specimen, by running a standard continuous stiffness measurement (CSM). The CSM continuously measures the stiffness, and allows the hardness and elastic modulus to be determined as a continuous function of the penetration depth. The nano hardness was found to be a strongly sensitive function of load and depth (Mukhopadhyay et al., 2009). In this regard the unique advantage of the nano indentation technique is that, it can measure the mechanical properties at the micro structural length scale. The

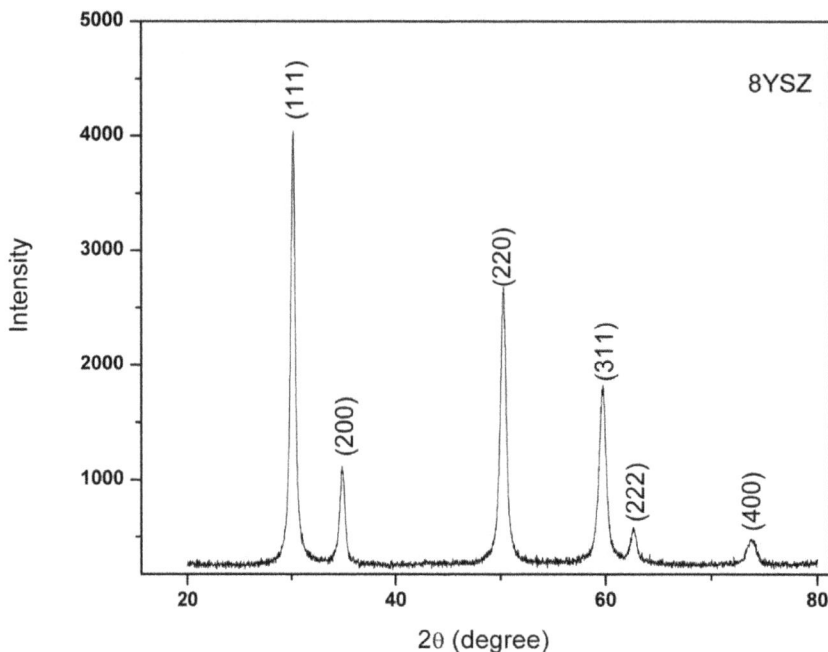

**Figure 1.** XRD pattern of 8YSZ Nano powder calcined at 600 ℃.

load is defined as the total force on the indenter and the depth is measured from the displacement of the indenter's starting position. The load and the displacement into the surface are continuously measured during loading and unloading. The load applied was 200 mN and the loading and unloading rate was 400 mN/min. The hardness and elastic modulus were calculated from the load-displacement curve using the Oliver and Pharr (1992) method. The important tribological property, coefficient of friction, was evaluated by a micro tribo meter for the fully dense 8YSZ sample. The coefficient of friction was found to be 0.2 against $Si_3N_4$ under the following conditions; the load applied was 2N and the reciprocal scratching was 2 mm.

## RESULTS AND DISCUSSION

### X-ray diffraction studies

The XRD pattern of the 8YSZ nano powder calcined at 600 ℃ for 3 h is as shown in Figure 1. The corresponding diffraction peaks coincide with the JCPDS file number: 82-1246 (Kumar and Manohar, 2007). The average crystalline size of the 8YSZ sample has been calculated, using the Debye–Scherrer formula given in Equation (1)

$$D = 0.98\lambda / \beta \cos\theta \qquad ---- \qquad (1)$$

where D is the average particle size in nm, $\beta$ is the full width at half the maximum (FWHM) of the X – ray reflection expressed in radians, and $\theta$ is the position of the diffraction peaks in the diffractogram. The average

crystalline size of the nano 8YSZ was 20 nm (Li et al., 2007).

## Morphological studies of 8YSZ using TEM and SEM

The particle morphology of the 8YSZ nano powder was analysed by the TEM, as shown in Figure 2. The TEM study confirms that the average particle size was 15 nm. The sol-gel derived nano 8YSZ particle size was very small, compared to the same particles prepared by the Co-precipitation technique (Keshmiri and Kesler, 2006). From the microwave sintering results, three types of sinterd samples of different densities were obtained that is, A (Density 92%)' B (Density 95%), C (Density 98%) corresponding to 5, 10 and 15 min holding time respectively. The surface morphology of the microwave-sintered 8YSZ samples A, B, C has been depicted in Figure 3(a), (b) and (c).

The density of the microwave sintered samples were increased due to the increase in the sintering temperature (or) holding time. Because of the holding time during the microwave sintering process, the sintered sample (C) shows higher density (98%) with an average grain size of (<900 nm) than the other two samples 'A' and "B". The SEM image of the sample 'A' shows the maximum pores and agglomerates. Sample' B' appeared with minimum pores, from this it can be concluded that the minimum holding time during sintering is reponsible for the poor densificaton and pores appearing in the microstructure of the samples A and B. Mazaheri et al. (2008) prepared 8YSZ nano powder with average

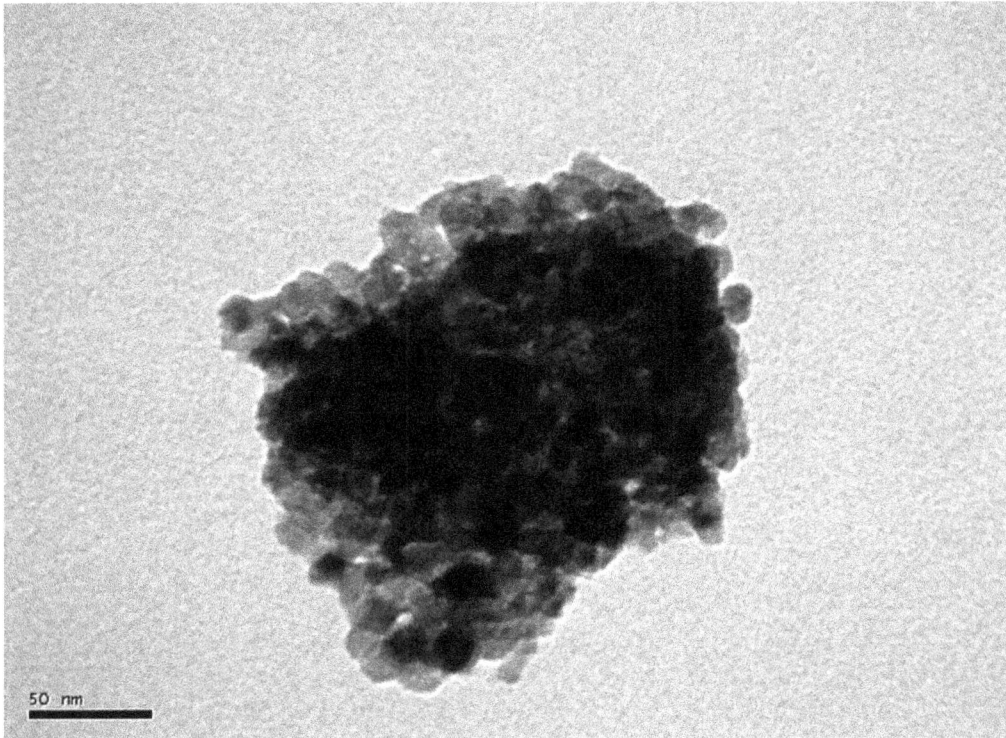

**Figure 2.** TEM image of 8YSZ NANO powder.

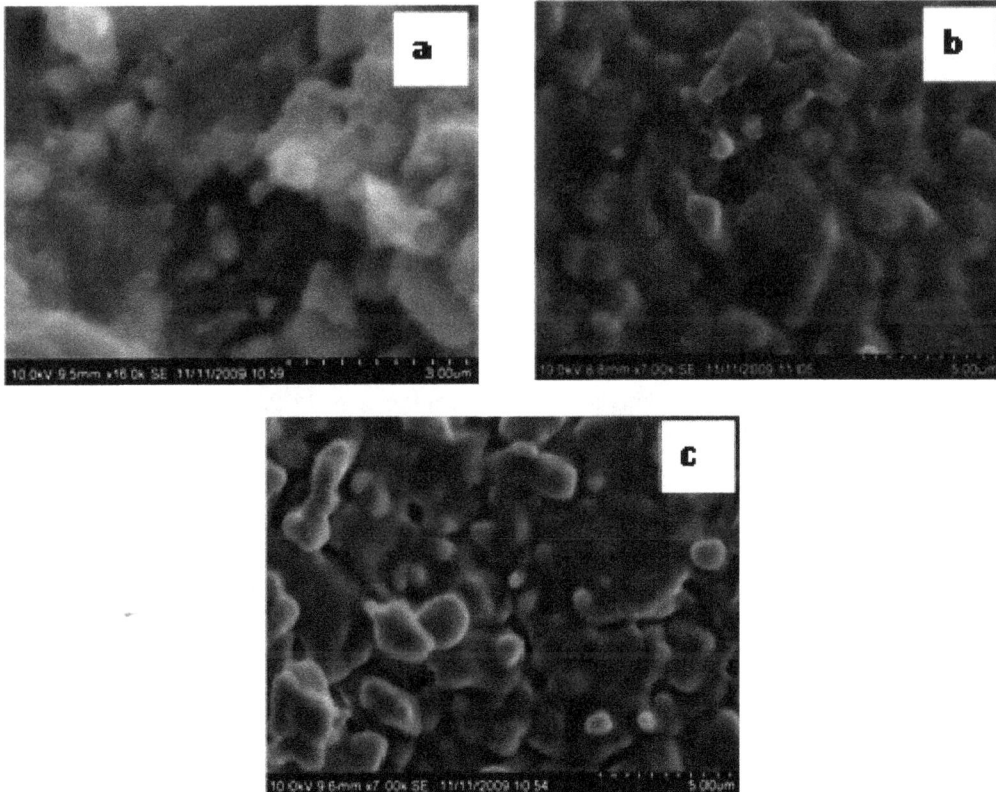

**Figure 3.** (a),(b) and (c) SEM image of microwave sintered samples at1500°C.

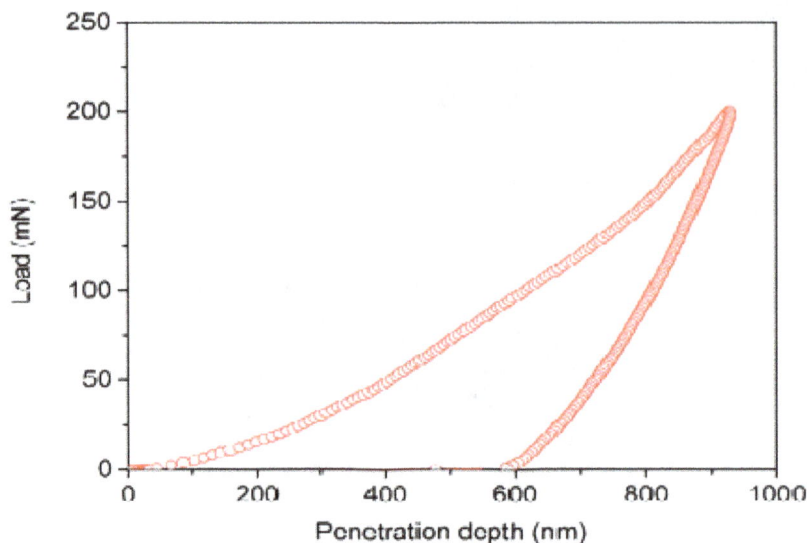

**Figure 4.** Load versus penetration depth curve of 8YSZ sample.

Particle size of 30 nm by smouldering combustion technique and applied microwave sintering on 8YSZ green samples and achieved 98% density with uniform grain size of 2.35 μm. Rajeswari et al. (2010) used commercially available 8YSZ powder (TZ-8Y Tosh.Tokyo,Japan) with an average particle size of 205 nm then performed microwave sintering method, got high density sample (98%) with an average grain size of 2.77 μm. Here the microwave sintering study results revealed the obtained grain size was (<900 nm) small in size comparared to the previous researcher's reports (Mazaheri et al. 2008; Rajeswari et al., 2010). The main reson for this minimum size grain (<900 nm) were present in the microstructure of the sample was due to the nano sized particle (15 nm) which confirmed the particle size influences on the grain size, and the microstructure of the sample. Figure 4 presents typical load-depth curves obtained by the nano indentation technique for the microwave sintered 8YSZ sample. This curve is fairly smooth and kink-free which implies that sample C' was fully sintered with minimum pores (Lu et al., 2006). The measured nano hardness, elastic modulus and coefficient of friction of the three microwave sintered samples A, B, C were shown in Table 1. Among the three samples highly densified sample, C has better enhanced mechanical properties and lower coefficient of friction than the other two sintered samples A, B. The hardness and elastic modulus were found to be 13.2Gpa and 210 Gpa, respectively for the sample 'C' these measured nano mechanical properties hardness and elastic modulus of microwave sintered 8YSZ sample agreed well with mechanical properties found by the micro indentation technique of the previous research reports. Dhal et al. (2007), Gogotsi et al. (1995) and Donzel and Roberts

(2000) proved the mechanical properties of 8YSZ sample measured by microhardness not depend on the various grain size of the samples prepared by different sintering techniques. Mazaheri et al. (2009), performed two-step sintering method to sinter 8YSZ sample and measured the, hardness value13.51 G Pa by applying the micro indentation technique. Mazaheri et al. (2008) employed the microwave sintering technique to sinter the fully stabilized cubic zirconia samples and applying the micro indentation technique, reported two types of hardness values 12.97 Gpa (for a lower heating rate of 5°C /min) and13.72 G Pa (for a higher heating rate of 50°C/min). In the present study, 100°C/min heating rate was used in the microwave furnace during the sintering of the specimen, and the obtained nano hardness and elastic modulus were 13.2 and 210 GPa.

Wellman et al. (2004), applied the nano indentation test on the bulk zirconia sample without yttria addition and reported that the hardness and elastic modulus values are 11.32 and 168 GPa. Menvie et al. (2009) conducted the Berkovitch nano-indentation tests on pristine and irradiated YSZ polycrystals, and measured the hardness values as 12 to 15 GPa. The value of the Young's modulus at the maximum indentation depth was 200 GPa. The present investigation concludes that the nano hardness results were close to the micro hardness results of previous study reports.

The nano indentations were performed mostly at various positions of sintered ceramic sample viz well inside the grain and far away from the grain boundary, close to the grain boundary and on the grain boundary interface. In this investigation the nano-indentation test was conducted inside the grain surface of 8YSZ sample (Mukhopadhyay et al., 2009). Which leads to the

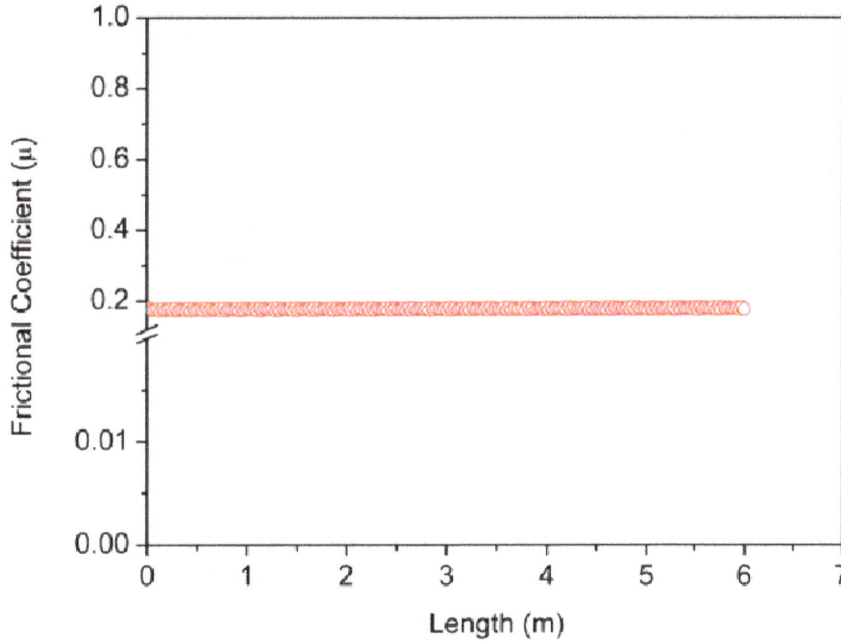

**Figure 5.** Length versus coefficient of friction diagram.

**Table 1.** Mechanical properties of Microwave sintered 8 mol% YSZ samples for different densities

| S/N | Density of the sample (%) | Hardness (G.Pa) | Elastic modulus (G.Pa) | Coefficient of friction |
|-----|---------------------------|-----------------|------------------------|-------------------------|
| 1 | 92 | 10. 12 | 185 | 0.4 |
| 2 | 95 | 11.5 | 198 | 0.35 |
| 3 | 98 | 13.2 | 210 | 0.2 |

maximum hardness, and young modulus and these results were close to the mechanical properties found by the micro indentation technique.

The nano hardness and elastic modulus results obtained by the nano indentation technique were comparable with the microhardness values reported by previous investigators. The reason for the similarity between the nano indentation and micro indentation mechanical properties of the 8YSZ sintered sample was the minimum grain size and higher micro structural homogeneity in the sample.

The coefficient of friction is an important factor in understanding the tribological behaviour. The frictional behaviour is described by the coefficient of friction (COF) which is defined as the ratio of friction force and normal force. Pasaribu et al. (2003) reported that the coefficient of friction of 1% CuO doped with alumina against the silicon nitride was 0.6. Min-Soo et al. (2008) conducted a wear test on a, $ZrO_2$ disk with three structural ceramic ball materials including $ZrO_2$, $Al_2O_3$, and SiC and reported the coefficient of friction was relatively lower in $ZrO_2$ disk and SiC ball viz 0.4 to 0.5. Shen et al. (2009) reported

that the coefficient of friction of the undoped Yttria stabilized tetragonal Zirconia poly crystals sample remains steady. They measured the coefficient of friction of pure 3YSZ-TZP, was 0.6 to 0.7, and that of dense 8 mol% CuO doped 3Y-TZP ceramics was 0.2 to 0.3 when sliding against an alumina ball under unlubricated conditions.

Winnubst et al. (2004) reported that the coefficient of friction of the 3YSZ- CuO doped sintered sample was ($\mu$= 0.2-0.3) against an alumina ball. Here in our investigation the friction coefficients of the microwave sintered 8 mol% YSZ sample measured against silicon nitrate at room temperature and under dry sliding condition was 0.2 (Figure 5). From the result it is seen that the curve is almost stable, in this investigation the obtained coefficient of friction was 0.2 which is quite low. The main reason for the low coefficient of friction obtained for the 8YSZ microwave sintered sample was the fine grains were present in the microstructure of the sintered sample. This study concludes that the decrease in the grain size of 8YSZ by the microwave sintering reduces the coefficient of friction.

## Conclusion

In this study, 8 YSZ nano powder was prepared by the sol-gel method. The cubic phase of the nano 8YSZ particle was confirmed by the X-ray diffraction studies. The particle size was about 15 nm. The SEM image of the sintered sample confirms the presence of sub-micron grains <900 nm. The achieved density of the sintered 8YSZ sample was 98%. The nano mechanical properties hardness, elastic modulus and coefficient of friction of the microwave sintered 8YSZ samples were studied by the Berkovich nano-indenter and micro tribometer and the obtained results were compared with the previous researcher's reports.The hardness and elastic modulus measured by the nano indentation technique agreed well with that of the mechanical properties reported by the micro indentation technique. The coefficient of friction of the microwave sintered 8YSZ sample value agreed well with that of the 3YSZ-CuO doped sample. This study confirmed the fine grains size in the microstructure which enhanced the mechanical properties of 8YSZ.

## REFERENCES

Bas KA, Monserrat Garc'ıa A, Werner E, van Zyl A, Louis Winnubst A, Elmer J. Mulder b, Dik J. Schipper b, Henk Verweij A (2004). Friction behaviour of solid oxide lubricants as second phase in Al2O3 and stabilised ZrO2 composites Wear. 256:182-189.

Ciacchi FT, Nightingale SA, Badwal SPS (1996). Microwave sintering of zirconia-yttria electrolytes and measurement of their ionic conductivity. Solid State Ionics 86-88:1167-1172.

Coasta TACFM, Morelli EMR, Kiminami RHGA (2003). Synthesis, microstructure and magnetic properties of Ni–Zn ferrites. J. Magn. Magn. Mater. 256:174.

Dhal P, Kaus I, Zhao Z, Johnson M, Nygren M, Wilk K, Grande T, Einarsrud MA (2007). Densification and properties of zirconia prepared by three different sintering techniques. Ceram Int. 33:1603-1610.

Donzel L, Roberts SG (2000). Microstructure and mechanical properties of cubic zirconia (8YSZ)/Si C nano composites. J. Eur. Ceram. Soc. 20:2457-2462.

Fan H, Kim HE (2002). Microstructure and Electrical Properties of Sol-Gel Derived Pb(Mg$_{1/3}$Nb$_{2/3}$)$_{0.7}$Ti$_{0.3}$O$_3$ Thin Films with Single Perovskite Phase. Jpn. J. Appl. Phys. 41:6768-6772.

Fukul T, Murata K, Sohara Abe H, Natio M (2004). Morphology control of Ni–YSZ cermet anode for lower temperature operation of SOFCs. J. Power Sour. 125:17-21.

Gaillard Y, Anglada M, Jiménez-Piqué E (2009). Nano indentation of yttria stabilized Zirconia: Effect of crystallographic structure on deformation. J. Mate. Res. 24(03):719-727.

Gogotsi GA, Dub SN, Lomonova EE, Ozersky BI (1995). Vickers and Knoop Indentation Behaviour of Cubi cand Partially Stabilized Zirconia Crystals. J. Eur. Ceram. Soc. 15:405-413.

Janney MA, Calhoun CL, kimery HD (1992). Microwave sintering of solid oxide fuel cell materials:I, zirconia–8mol% yttria. J. Am. Ceram. Soc. 75:341-346.

Keshmiri M, Kesler O (2006). Colloidal formation of monodisperse YSZ spheres: Kinetics of nucleation and growth. Acta. Mater. 54:4149-4157.

Kumar C, Manohar P (2007). Conductivity and dielectric properties of sol-gel derived porous zirconia. Ionics 13:333-338.

Lange F, Dunlop G, Davis B (1986). Degradation duringaging of transformation toughened ZrO2-Y2O3 materials at 250°C. J. Am. Ceram. Soc. 69(3):237-240.

Li Q, Xia T, Liu XD, Ma, XF, Meng XQJ (2007). Fast densification and electrical conductivity of yttriastabiize dzirconianano ceramics.Mater. Sci. Eng. B. 138:78-89.

Liu H, Fang FP, Jin L (2007). Electrical Heterogeneity in CaCu$_3$Ti$_4$O$_{12}$ Ceramics Fabricated by Sol–Gel Method. Solid State Commun. (142):573-576.

Lu XJ, Wang X, iao PX (2006). Nanoindentation and residual stress measurements of yttria-stablized zirconia composite coatings produced by electrophoretic deposition. Thin Solid Films 494:223-227.

Mazaheri M, Razavi HZ, Golistani-farad F, Mollazadeh S, Jafari S, Sadrnezhadd SK (2009). The effect of confirmation method and sintering technique on the densification and grain growth of nano crystalline 8 mol% yttria stabilized zirconia. J. Am Ceram. Soc. 92(5):990-995.

Mazaheri M, Zahedi AM, Hejazi MM (2008). Processing of nanocrystalline 8 mol% yttria-stabilized zirconia by conventional, microwave-assisted and two-step sintering. Mat. Sci. Eng. A 492:261-267.

Mendelson MI (1969). "Average grain size in polycrystalline ceramics. J. Am. Ceram. Soc. 52:443-446.

Menvie BV, Sattonnay G, Legros C, Huntz AM, Poissonnet S, Thomé L (2009). Mechanical properties of cubic zirconia irradiated with swift heavy ions. J. Nuclear Mater. 384:70-76.

Minh NQ (1993). Ceramic fuel cells. J. Am. Ceram. Soc. 76(3):563-588.

Min-Soo S, Young-Hun C, Seock-Sam Kimb (2008). Friction and wear behavior of structural ceramics sliding against zirconia. Wear 264:800-806.

Mukhopadhyay AK, Dey A, Chakraborty R, Joshi KD, Rav A, Biswas S, Gupta SC (2009). Nano hardness of sintered Alumina ceramics. National seminar on Recent Advances in Trational ceramics. pp. 11-12.

Oliver WC, Pharr GM (1992). An Improved technique for determining haedness and elastic modulus using load and displacement sensing indentation expriments. J. Mater. Res. 7:1564-1583.

Pasaribu HR, Sloetjes JW, Schipper DJ (2003). "Friction reduction by adding copper oxide into aluminaand zirconia ceramics." WEAR 255:699-707.

Phani AR, Santucci S (2006). Evaluation of structural and mechanical properties of aluminum oxide thin films deposited by a sol–gel process: Comparison of microwave to conventional anneal. J. Non-Crystsolids 352:4093.

Rajeswari K, Hareesh US, Subasri R, Chakravarty D, Johnson R (2010). Comparative Evaluation of Spark Plasma (SPS), Microwave (MWS), Two stage sintering (TSS) and Conventional Sintering(CRH) on the densification and Micro structural Evolution of fully Stabilized Zirconia Ceramics. Sci. Sinter. 42:259-267.

Ran SAJA, Winnubst H, Koster PJ, de Veen DHA (2007). Blank Sintering behaviour and microstructure of 3Y-TZP + 8 mol%CuO nano-powder composite. J. Eur. Ceramic Soc. 27:683-687.

Shen R, Louis W, Dave HAB, Henry PR, Jan-Willem S, Dik SJ (2009). Dry-sliding self lubricating ceramics: CuO Doped 3Y-TZP. Wear 267(9-10):1696-1701.

Soyez G, Eastman JA, Thompson LJ, Bai GR, Baldo PM, McCormick AW, DiMelfi RJ, Elmustafa AA, Tambwe MF, Stone DS (2000). Grain-size-dependent thermal conductivity of nanocrystalline yttria-stabilized zirconia films grown by metal-organic chemical vapour deposition. Appl. Phys. Lett. 77(8):1155-1157.

Steele BCH (2001). Material science and engineering: the enabling technology for the commercialisation of fuel cell systems. J. Mater. Sci. 36:1053-1068.

Surowiak Z, Kupriyanov MF, Czekaj D (2001). Properties of nanocrystalline ferroelectric PZT ceramics. J. Eur. Ceram Soc. 21:1377-1381.

Thridandapani RR, Folgar CE, Folz DC, Clar DE, Wheeler K, Peralta P (2009). Microwave sintering of 8 mol% yttria-zirconia (8YZ): An inert matrix material for nuclear fuel applications. J. Nucl. Mat. 384(2):153-157.

Upadhyaya DD, Ghosh A, Gurumurthy KR, Prasad R (2001). Microwave sintering of cubic Zirconia. Ceram Int. 27(4):415-418.

Voevodin AA, Hu JJ, Jones JG, Fitz TA, Zabinski JS (2001). Growth and structural charaterization of yttria stabilized zirconia--gold nanocomposite with improved toughness. Thin Solid Films 401:187-195.

Wellman RG, Dyer A, Nicholls JR (2004). Surface and Coatings Technology. 176(2):253-260. nano and micro indentation studies of bulk zirconia and eb pvd tbcs.

Winnubst AJA, Ran S, Wiratha KW, Blank DHA, Pasaribu HR, Sloetjes JW, Schipper DJA (2004). Wear resistant zirconia ceramic for low friction application. Key Eng. Mat. 264-268:809-812.

# 6

# Adsorption of Cr(III) and speciation of chromium in aqueous solution using Ambrosia beetle-generated acacia polycantha frass

**C. Mahamadi**

Department of Chemistry, Bindura University of Science Education, P. Bag 1020, Bindura, Zimbabwe.

**Acacia polycantha frass (APF) generated by boring activity of Ambrosia beetles was investigated using a batch technique for its potential to adsorb Cr(III) ions from aqueous solution. Studies were carried out to determine the effect of pH, adsorbent dose, and KBr concentration on the sorption behavior Cr(III) on the adsorbent. The adsorption process was found to be highly dependent on solution pH with maximum removal occurring at pH 11. Thermodynamic studies indicated that the sorption process was exothermic and spontaneous. Adsorption recoveries of up to 75 and 95% for a 5 mgL-1 solution were obtained for untreated and acid-treated APF respectively at room temperature. In all experiments the acid-treated adsorbent showed an apparently higher metal uptake than the untreated one. Reduction of Cr(VI) to Cr(III) in synthetic solution using hydroxylamine prior to adsorption preconcentration showed potential for studying chromium speciation.**

**Key words:** Adsorption, entropy of activation, enthalpy of activation, chromium, speciation, acacia polycantha frass.

## INTRODUCTION

The presence of heavy metals in the environment has become a major threat to plant, animal and human life due to their bioaccumulating tendency and toxicity and therefore must be removed from industrial effluents before discharge. Conventional physico-chemical methods for removing heavy metals from waste streams include electrochemical treatment, chemical reduction, ion-exchange, precipitation and evaporation recovery (Rengaraj et al., 2003; Ahluwalia and Goyal, 2007). Nevertheless, many of these approaches are marginally cost-effective or difficult to implement (Pipiska et al., 2007).

Biosorption, which utilizes inexpensive biological materials such as fungi, bacteria, algae, and agricultural waste products has increasingly been studied as a possible alterative technique for metal removal from aquatic environments (Volesky and Holan, 1995; Volesky et al., 2000; Schmuhl et al., 2001; Vasudevan et al., 2001, 2002, 2003; Aksu, 2002; Chojnacka and Nowryta, 2004; Loukidou et al., 2004; Chojnacka et al., 2005; Horsfall and Spiff, 2004b; Singh et al., 2011; Sarkar and Majumdar, 2011; Benamer et al., 2011; Zhang et al., 2011; Kumar et al., 2011). Despite such volumes of work, industrial application of the biosorption technology is still to be developed for use in routine metal-detoxification of aqueous solutions. The present study investigated the sorption properties of acid-treated and untreated frass generated by the boring activity of Ambrosia beetles on acacia polycantha for the sorption of Cr(III) ions from a

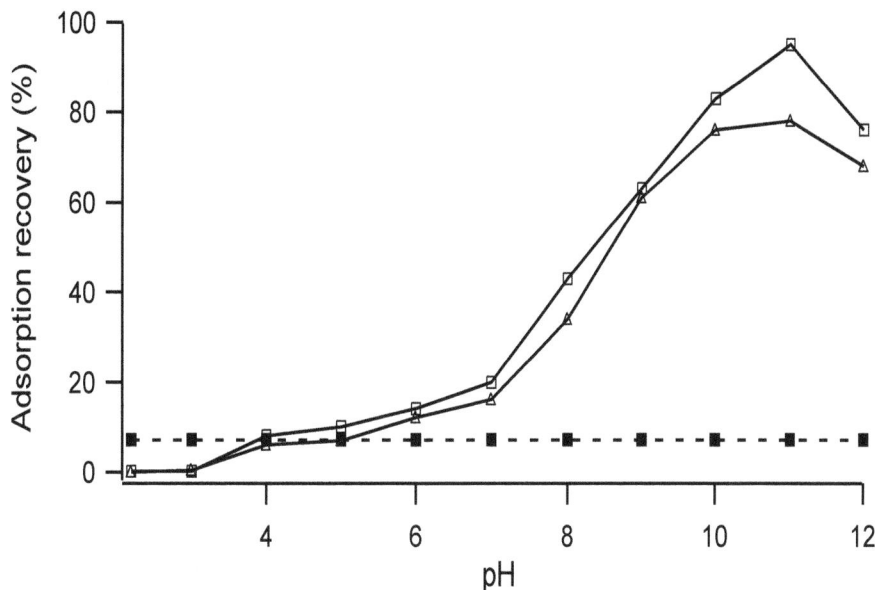

**Figure 1.** pH dependency of adsorption recovery of 5 mg L$^{-1}$ Cr(III) and Cr(VI): acid-treated APF: (☐); untreated APF: (△); Cr(VI): (■).

Cr(III)-Cr(VI) mixture for the first time and assessed the applicability of the adsorbent for chromium speciation.

## EXPERIMENTAL

Stock solutions of 500 mgL$^{-1}$ of Chromium(III) and (VI) were prepared from Chromium(III) nitrate non-ahydrate and potassium dichromate (Saarchem Muldersdrift, RSA), respectively. Working solutions were prepared daily from the stock solution by serial dilutions with double distilled de-ionised water. All glassware was soaked in 5% (v/v) nitric acid for at least 24 h and washed with water. Samples of Acacia polycantha frass (APF) were generated through the boring activity of Ambrosia beetles on acacia polycantha at room temperature. The samples were oven-dried at 65°C for 12 h, and sieved through 2.5 mm. Acid-treatment was achieved by mixing 10 g of the adsorbent with 500 ml of HNO$_3$ and agitating at 150 rpm for 4 h at room temperature. The adsorbent was then collected by centrifugation at 4500 × g for 10 min and then washed several times with distilled water to remove excess acid. The collected biomass was further agitated in acetone (500 ml) for 4 h, centrifuged, air-dried and stored in polythene bottles until use.

The concentration of metal ions in samples was determined by a Unicam 701-Emission Inductively Coupled Argon Plasma Echelle Spectrophotometer with "crossed" dispersion. The instrument was optimised using a 100-mg/L solution of Mn. For adsorption, 0.1-4 g of adsorbent was weighed into individual centrifuge tubes. To the tubes were added 10 ml of 1-5 mgL$^{-1}$ of Cr(III) and Cr(VI) pH-adjusted solutions (pH was adjusted using Na$_2$CO$_3$-NaHCO$_3$). The resulting mixture was agitated at 25°C for 30 min in a water bath shaking at 150 rpm and then centrifuged at 5000 × g for 10 min. 5 ml of the supernatant was extracted and diluted to 50 ml using deionised distilled water and analysed for Cr using ICP-AES. The adsorbent was re-suspended in metal-free buffer solution to wash out traces of metal not sorbed but present on the surface. The mixture was centrifuged 4500 × g for 10 min and the supernatant discarded. The chromium ions adsorbed were extracted twice by

suspending the biomass in 20 ml aliquots of 0.1 molL$^{-1}$ HNO$_3$ and repeating the centrifuging and metal analysis procedure.

## RESULTS AND DISCUSSION

### Effect of pH

The pH was the first critical parameter evaluated concerning its effect on the sorption of Cr(III) from aqueous solution by the acid-treated and untreated APF. Sorption experiments were performed at 5 mgL$^{-1}$ of metal solution, at room temperature in the pH range 2-12. As can be seen in Figure 1, adsorption of Cr(III) is highly depended on solution pH. At pH 11, nearly all Cr(III) is retained on the acid-treated APF (about 95%) compared to about 75% by the untreated APF. At low pH, cell wall ligands were closely associated with the hydroxonium ions [H$_3$O$^+$] and restricted the approach of metal cations as a result of the repulsive force.

### Effect of amount of adsorbent:

The results obtained for the effect of adsorbent dose on Cr(III) uptake showed that more than 87 and 97% of Cr(III) was adsorbed by untreated and acid-treated APF respectively in the 0.1-4.0 g mass range. Optimum adsorbent doses were found to be 1.0 and 1.25 g/L for adsorption onto untreated and acid-treated biomass, respectively.

**Table 1.** Results from determination of Cr(III) and total Chromium.

| Adsorbent | Added (mg/L) | | Determined (mg/L) | | |
|---|---|---|---|---|---|
| | Cr(III) | Cr(VI) | Cr(III) | Cr(VI) | Cr(total) |
| Acid-treated APF | 0.5 | 0.5 | 0.5+0.1 | 0.5+0.2 | 1.0+0.1 |
| | 1.0 | 1.0 | 1.1+0.3 | 1.0+0.1 | 1.0+0.2 |
| | 5.0 | 5.0 | 4.8+0.3 | 5.2+0.2 | 10+0.2 |
| Untreated APF | 0.5 | 0.5 | 0.4+0.1 | 0.4+0.1 | 0.8+0.1 |
| | 1.0 | 1.0 | 0.7+0.3 | 0.8+0.1 | 1.5+0.2 |
| | 5.0 | 5.0 | 3.7+0.2 | 3.2+0.1 | 6.9+0.2 |

## Speciation of Chromium

Total Chromium in solution was determined after reduction of Cr(VI) to Cr(III) with hydroxylamine. Two milliliters of 2 mol $L^{-1}$ HCl was added to 20 ml of the Cr(III)-Cr(VI) mixture followed by addition of 2 ml of 0.5 mol $L^{-1}$ hydroxylamine. The solution was left at room temperature for at least 30 min before 10 ml of 1 mol $L^{-1}$ acetate was added, pH adjusted to 4 and the volume made up to 100 ml with distilled water. The analysis of Cr(III) was performed after adsorption using acid-treated and untreated APF using the above outlined procedure. The results shown in Table 1 indicate that generally the acid treated APF was more effective in adsorption of Cr(III) and hence more suitable for the speciation study of chromium than the untreated adsorbent at metal concentrations of 0.5-5 mg $L^{-1}$.

## Effect of KBr concentration on Cr(III) adsorption at various temperatures

To study the effects of ionic strength and temperature on the adsorption of Cr(III) from aqueous solution by the adsorbents, the uptake of Cr(III) was investigated in the presence of 0, 0.08 and 0.2 mol $L^{-1}$ KBr at a temperature range 60-80°C. The initial rates method was used to determine the rate constants for adsorption process and the results are shown in Table 2. The results indicate that the apparent adsorption rate constant, $k_{ads}$, was raised in the presence of KBr, but decreases as the salt concentration is increased from 0.08 to 0.2 mol $L^{-1}$ for the acid-treated APF. However, for the untreated APF, the apparent $k_{ads}$ values increased as the KBr concentration was raised from 0.00 to 0.2 mol $L^{-1}$. It has been observed that the mechanism of metal removal from aqueous metal solution involved four steps: (i) migration of metal ions from the bulk solution to the surface of the adsorbent; (ii) diffusion through boundary layer to the biomass surface; (iii) adsorption at a binding site and (iv) intra-particle diffusion into the interior of the biomass (Horsfall Jnr and Spiff, 2004b). The boundary layer resistance will be affected by the rate of sorption and increasing

the agitation time will reduce this resistance and increase mobility of the ions.

Addition of divalent ions such as $Mg^{2+}$ and $Ca^{2+}$ may reduce metal adsorption through competitive adsorption (Benaissa and Benguella, 2004). Addition of KBr salt to the Cr(III) solution increases the overall ionic strength of the solution. Since adsorption is mainly electrostatic in nature due to presence of polar functional groups on the adsorbent cell wall, then addition of the monovalent ions is expected to affect the rate of adsorption of metal ions by adsorbents. It is proposed in this study that adsorption at a binding site may involve formation of an intermediate, with associated activation parameters. From Transition State Theory, the thermodynamic parameters relating to the activated complex can be calculated using Equations (1) and (2).

$$k_{ads} = k_B T exp(-\Delta G^0/RT) \tag{1}$$

$$ln(k_{ads}/T) = ln k_B + \Delta S^0/R - \Delta H^0/RT \tag{2}$$

where $\Delta G^0$, $\Delta S^0$, $\Delta H^0$ are the Gibbs free energy, change in entropy, change in enthalpy; $k_B$ and $R$ are the Boltzman constant and the molar gas constant respectively.

Therefore plots of $ln(k_{ads}/T)$ against $1/T$ should enable evaluation of the activation parameters relating to the adsorption activated complex. Figures 2 and 3 show plots of $ln(k_{ads}/T)$ against $1/T$ used to evaluate both $\Delta S^0$ and $\Delta H^0$ and the values obtained are shown in Table 3. A negative $\Delta G^0$ value confirms the feasibility of the process and spontaneous nature of adsorption of Cr(III) onto the biosorbent. Negative values of $\Delta H^0$ showed that the uptake of Cr(III) decreased with increasing temperature suggesting that the adsorption process was exothermic whereas the negative values of $\Delta S^0$ indicated a decrease in the degree of freedom of the adsorbed species.

## Conclusion

The adsorption of Cr(III) onto acacia polycantha frass

**Table 2.** Apparent rate constants, $k_{ads}$, for the adsorption of Cr(III) onto acid-treated and untreated APF at various temperatures as a function of added KBr electrolyte concentrations.

| Adsorbent | Temperature (°C) | $K_{ads}$ $(\times 10^{-8})$ | | |
|---|---|---|---|---|
| | | 0.00 mol L$^{-1}$ (KBr) | 0.08 mol L$^{-1}$ (KBr) | 0.2 mol L$^{-1}$ (KBr) |
| | 60 | 1.02 | 21.2 | 18.2 |
| | 70 | 2.55 | 34.5 | 29.3 |
| Acid-treated APF | 75 | 3.82 | 55.2 | 36.4 |
| | 80 | 9.90 | 63.4 | 52.6 |
| | 60 | 0.87 | 2.80 | 1.51 |
| | 70 | 1.12 | 4.51 | 8.92 |
| Untreated APF | 75 | 2.21 | 6.29 | 12.3 |
| | 80 | 2.32 | 7.91 | 12.5 |

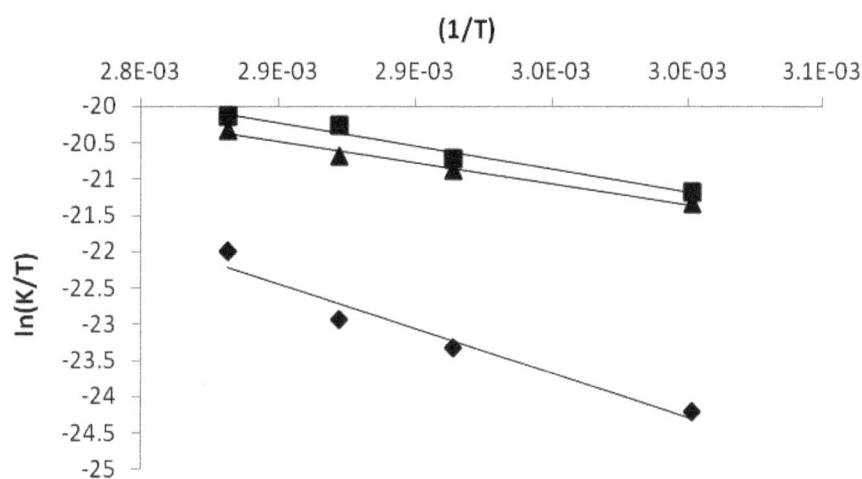

**Figure 2.** Effect of KBr on Cr(III) adsorption. 0.00 mol L$^{-1}$: (◆); 0.08 mol L$^{-1}$: (■); 0.2 mol L$^{-1}$: (▲), acid-treated APF.

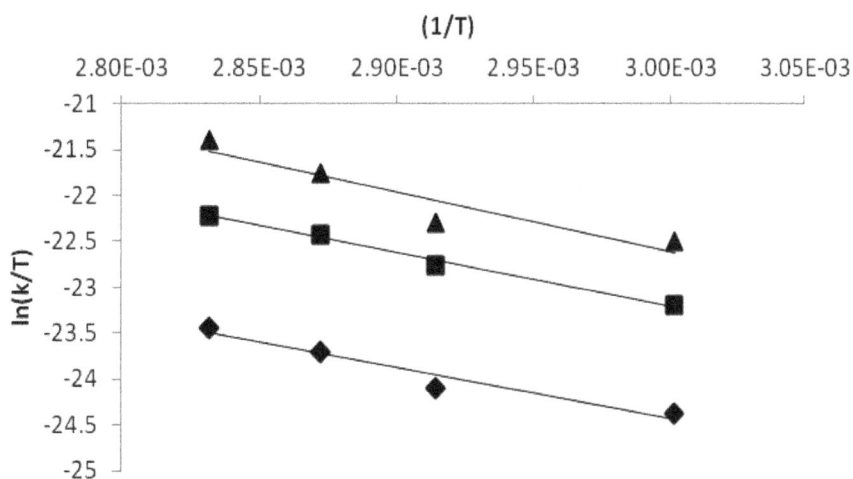

**Figure 3.** Effect of KBr on Cr(III) adsorption. 0.00 mol L$^{-1}$: (▲); 0.08 mol L$^{-1}$: (◆); 0.2 mol L$^{-1}$: (■), untreated APF.

**Table 3.** Thermodynamic parameters relating to the transition state for the effect of KBr on adsorption of Cr(III) by acid-treated and untreated APF.

| Adsorbent | [KBr] (mol L$^{-1}$) | $\Delta G^0$ (kJ mol$^{-1}$) | $\Delta H^0$ (kJ mol$^{-1}$) | $\Delta S^0$ (J mol$^{-1}$ K$^{-1}$) | $R^2$ |
|---|---|---|---|---|---|
| | 0.00 | -32.38 | -47.56 | -42.10 | 0.8624 |
| Acid-treated APF | 0.08 | -37.60 | -53.60 | -46.51 | 0.9629 |
| | 0.20 | -33.69 | -49.12 | -44.68 | 0.9841 |
| | 0.00 | -35.56 | -45.56 | -37.23 | 0.9422 |
| Untreated APF | 0.08 | -26.82 | -40.28 | -39.24 | 0.9935 |
| | 0.20 | -29.86 | -44.73 | -41.21 | 0.9821 |

generated by Ambrosia beetles was studied in a batch system with respect to the initial pH, temperature, adsorbent dose, and KBr concentration. Adsorption recovery for Cr(III) using acid-treated adsorbent was higher than for the untreated adsorbent showing that acid treatment enhanced the metal sorption capacity of the adsorbent. Thermodynamic parameters observed indicated that the sorption process was highly feasible. In the end it was possible to study the speciation of chromium after reduction of the hexavalent state to Cr(III) in hydroxylamine solution.

## ACKNOWLEDGEMENT

The BUSE research supported this research through grant number CRP-2 to C. Mahamadi.

## REFERENCES

Ahluwalia SS, Goyal D (2007). Microbial and plant derived biomass for removal of heavy metals from wastewater. Bioresour. Technol. 98(12):2243-2257.

Aksu Z (2002). Determination of equilibrium, kinetic and thermodynamic parameters of the batch biosorption of nickel(II) ions onto *Chlorella vulgaris*. Process Biochem. 38:89-99.

Benaissa H, Benguella B (2004). Effect of anions and cations on cadmium sorption Kinetics from aqueous solutions by chitin: Experimental studies and modeling. Environ. Poll. 130:157-163.

Benamer S, Mahlous M, Tahtat D, Nacer-Khodja A, Arabi M, Lounici H, Mameri N (2011). Radiation synthesis of chitosan beads grafted with acrylic acid for metal ions sorption. Rad. Phys. Chem. 80:1391-1397.

Chojnacka K, Chojnacki A, Gorecka H (2005). Biosorption of Cr$^{3+}$, Cd$^{2+}$ and Cu$^{2+}$ ions by blue-green algae *Spirulina* sp.: kinetics, equilibrium and the mechanism of the process. Chemosphere 59(1):75-84.

Chojnacka K, Noworyta A (2004). Evaluation of *Spirulina* sp. growth in photoautotrophic, heterotrophic and mixotrophic cultures. Enzyme Microb. Technol. 34(5):461-465.

Horsfall Jnr M, Spiff AI (2004). Studies on the effect of pH on the sorption of Pb$^{2+}$ and Cd$^{2+}$ ions from aqueous solutions by *Caladium bicolor* (Wild Cocoyam) biomass. Electr. J. Biotechnol. 7(3):311-320.

Kumar R, Bhatia D, Singh R, Rani S, Bishnoi NR (2011). Sorption of heavy metals from electroplating effluent using immobilized biomass *Trichoderma viride* in a continuous packed-bed column. Int. Biodeterior. Biodegrad. 65:1133-1139.

Loukidou MX, Zouboulis AI, Karapantsios TD, Matis KA (2004). Equilibrium and kinetic modeling of Chromium(VI) biosorption by *Aemonas caviae*. Colloids and Surfaces A: Physicochem. Eng. Aspects 242:93-104.

Pipiska M, Hornik M, Vrtoch L, Snircova S, Augustin J (2007). Sorption of cobalt and zinc from single and binary metal solutions by Evernia *prunastri*. Nova Biotechnol. 8:23-31.

Rengaraj S, Joo CK, Kim Y, Yi J (2003). Kinetics of removal of chromium from water and electronic process wastewater by ion exchange resins: 1200H, 1500H and IRN97H. J. Hazard. Mater. B102: 257-275.

Sarkar M, Majumdar P (2011). Application of response surface methodology for optimization of heavy metal biosorption using surfactant modified chitosan bead. Chem. Eng. J. 175:376-387.

Schmuhl R, Krieg HM, Keizer K (2001). Adsorption of Cu(II) and Cr(VI) ions by Chitosan: Kinetics and equilibrium studies. Water SA 27:1-7.

Singh L, Asalapuram RP, Ramnath L, Gunaratna KR (2011). Effective removal of Cu$^{2+}$ ions from aqueous medium using alginate as biosorbent. Ecol. Eng. 38:119-124.

Vasudevan P, Padmavathy V, Tewari N, Dhingra SC (2001). Biosorption of heavy metal ions. J. Sci. Ind. Res. 60:112-120.

Vasudevan P, Padmavathy V, Dhingra SC (2002). Biosorption of monovalent and divalent ions on baker's yeast. Bioresour. Technol. 82:285-289.

Vasudevan P, Padmavathy V, Dhingra SC (2003). Kinetics of biosorption of cadmium on Baker's yeast. Bioresour. Technol. 89:218-287.

Volesky B, Figueira MM, Ciminelli VS, Roddick FA (2000). Biosorption of metals in brown seaweed biomass. Water Res. 34:196-204.

Volesky B, Holan ZR (1995). Biosorption of heavy metals. Biotechnol. Prog. 11:235-250.

Zhang Y, Kogelnig D, Morgenbesser C, Stojanovic A, Jirsa F, Lichtscheidl-Schultze I, Krachler R, Li Y, Bernhard K, Keppler BK (2011). Preparation and characterization of immobilized [A336][MTBA] in PVA–alginate gel beads as novel solid-phase extractants for an efficient recovery of Hg (II) from aqueous solutions. J. Hazard. Mater. 196:201-209.

# 7

# Gamma-ray shielding of concretes including magnetite in different rate

## B. Oto[1] and A. Gür[2]

[1]Department of Physics, Faculty of Science, Yüzüncü Yil University, 65080 Van, Turkey.
[2]Department of Chemistry, Faculty of Science, Yüzüncü Yil University, 65080 Van, Turkey.

**The effect of magnetite ore rate on the gamma radiation shielding properties of concrete samples have been measured by using the transmission method for 59.54 and 80.99 keV gamma rays with a NaI(Tl) detector. Linear attenuation coefficients of concrete samples including magnetite in different rates were obtained by Lambert law in which measured gamma ray intensities were used. They were compared with theoretical values calculated from WinXCom computer software. It was determined that linear attenuation coefficients increased with increasing magnetite rate in concrete samples.**

**Key words:** Linear attenuation coefficients, magnetite, concrete, WinXCom.

## INTRODUCTION

There are some methods that control the intensity of radiation received from a radioactive source. One of the most significant of these methods is the radiation shielding, which is the science of protecting people and the environment from the harmful effects of radiation. The principle of the radiation shielding is to decrease the intensity of external radiation to the desired level. A good photon shielding material should have high value of photon attenuation coefficients and irradiation effects on its mechanical properties should be small. Concrete is an excellent sheilding material most widely used for radiation shielding of nuclear plants, in the walls of radiology and oncology departments in hospitals (Kaplan, 1989). In medical accelerator room's construction, high-density concrete (3.0 to 5.0 $g/cm^3$) is employed to provide shielding against radiation (Facure and Silva, 2007). Concrete density depends on aggregate weight. Heavy or high-density aggregates are used to increase the density of concrete. Magnetite ($Fe_3O_4$) having a high density (4.9-

5.2 $g/cm^3$) is an effective shielding material for γ-rays (Bashter, 1997; El-Sayed and Megahid, 2001). In recent years, several studies relevant to the measurement of linear and mass attenuation coefficients for different types of materials have been published (Akkurt et al., 2009, 2010, 2012; Demir and Keleş, 2006; Han et al., 2011; Oto et al., 2013; Seven et al., 2004; Shirmardi et al., 2013; Medhat, 2009; Mostofinejad et al., 2012).

In the present work, linear attenuation coefficients of concrete samples with magnetite ($Fe_3O_4$) at the energies of 59.54 and 80.99 gamma rays were obtained experimentally, and calculated using WinXCOM computer code. The experimental results have been compared with the calculations.

## MATERIALS AND METHODS

### Preparation of concrete samples

In this study, four concrete samples including magnetite with

**Table 1.** The result of the chemical analysis of magnetite.

| Compound | Magnetite (%) |
|----------|---------------|
| CaO | 1.15 |
| MgO | 0.50 |
| $Al_2O_3$ | 1.26 |
| $Na_2O$ | 0.195 |
| $SiO_2$ | 8.41 |
| MnO | 0.44 |
| CuO | 0.013 |
| FeO | 85.74 |
| BaO | 2.24 |

**Table 2.** Amounts of concrete composition.

| Concrete | w/c ratio | Magnetite rate (%) | Aggregate (kg/m$^3$) |
|----------|-----------|--------------------|-----------------------|
| K* | 0.5 | 0 | 1350 |
| M1 | 0.5 | 2.5 | 1350 |
| M2 | 0.5 | 5 | 1350 |
| M3 | 0.5 | 10 | 1350 |
| M4 | 0.5 | 20 | 1350 |

*Oto et al. (2013).

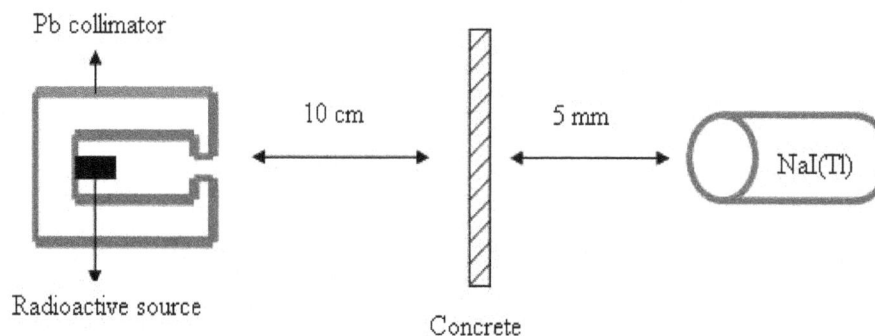

**Figure 1.** Schematic diagram of the experimental arrangement.

different rates have been produced. Cement and modular sand used in the concrete samples were provided from the Van Gölü Cement Factory in Van, Turkey and magnetite supplied from the Cataş Mine Works in Elbistan, Turkey. The result of elemental analysis of magnetite determined by X-ray fluorescent method and result is given in Table 1. The magnetite was used in fractions of 2.5, 5, 10 and 20% in weight of cement. Amounts of concrete composition are listed in Table 2. Aggregates used in this work were 0 to 4 mm and 8 to 16 mm grain sizes. In order to prepare concrete samples, magnetite sieved by a 100 mesh standard sieve. For the measurements 15×15×4 cm concrete samples were prepared.

**Experimental details**

The linear attenuation coefficients of concrete samples have been measured using NaI(Tl) detector which connected to a multichannel analyzer. The gamma photons were obtained from [241]Am (59.54 keV) and [133]Ba (80.99 keV) radioactive sources. These sources were shielded by the pin hole lead collimators to obtain a narrow beam. For each energy, all concrete samples were irradiated three times by gamma rays.

The schematic arrangement of the experimental setup used in the present work is given in Figure 1. The source-sample and sample-detector distance was adjusted as 10 cm and 5 mm, respectively. The data were collected into 4096 channels of a Desktop Inspector Digital Spectrum Analyzer, connected with the PC by Genie 2000(3.0) software. The linear attenuation cofficients ($\mu$,cm$^{-1}$) was obtained by Lambert law's:

$$I = I_0 \exp(-\mu x) \tag{1}$$

where x is material thickness (mass-per-unit area), $I_0$ is the incident

**Table 3.** Experimental (E) and theoretical (T) linear attenuation coefficients at 59.54 keV and 80.99 keV.

| Concrete | Magnetite rate (%) | Density | 59.54 keV | | 80.99 keV | |
|---|---|---|---|---|---|---|
| | | | μ (E) | μ (T) | μ (E) | μ (T) |
| K* | 0 | 1.94 | 0.640 ± 0.00110 | 0.673 | 0.420 ± 0.00100 | 0.446 |
| M1 | 2.5 | 2.11 | 0.708 ± 0.00066 | 0.736 | 0.463 ± 0.00185 | 0.485 |
| M2 | 5 | 2.12 | 0.738 ± 0.00233 | 0.742 | 0.469 ± 0.00120 | 0.489 |
| M3 | 10 | 2.16 | 0.763 ± 0.00266 | 0.768 | 0.489 ± 0.00260 | 0.503 |
| M4 | 20 | 2.22 | 0.819 ± 0.00202 | 0.812 | 0.514 ± 0.02107 | 0.528 |

*Oto et al. (2013).

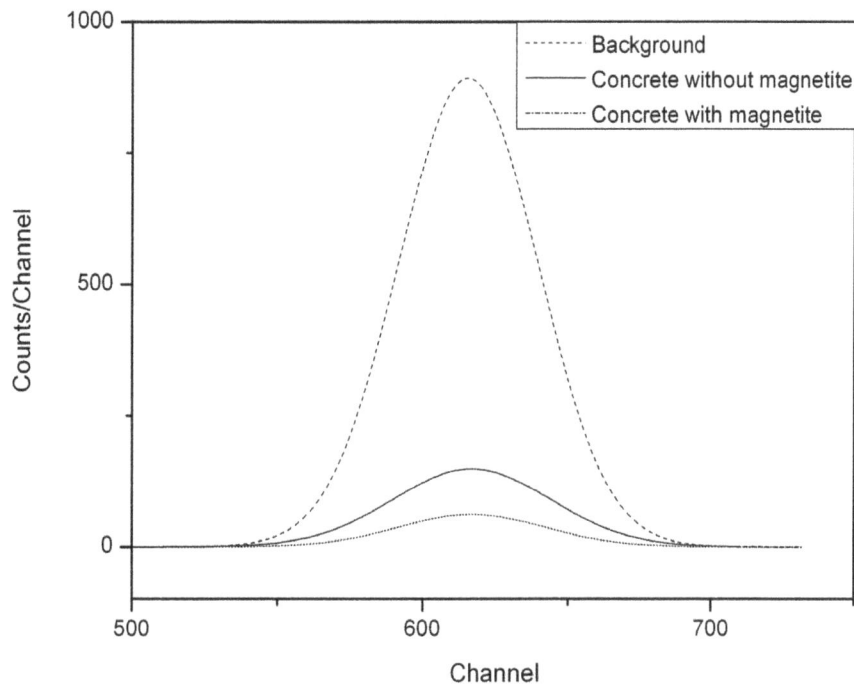

**Figure 2.** Spectrums of gamma radiation obtained background, concrete samples with and without magnetite.

gamma ray and I is the photon intensity recorded in detector. The theoretical linear attenuation coefficients which were calculated using WinXCom computer code were compared with the measured linear attenuation coefficients. WinXCom based on applying the mixture rule to calculate the partial and total mass attenuation coefficients for elements, mixtures and compounds for photon energies ranging from 1 keV to 1 GeV (Gerward et al., 2001).

## RESULTS AND DISCUSSION

In this study, the measured intensities of transmitted gamma ray (I) through the concrete samples with or without magnetite and the incident gamma ray ($I_0$) were using in Lambert law, linear attenuation coefficients were obtained experimentally and theoretical mass attenuation

coefficients were calculated from WinXCom software at the energies 59.54 and 80.99 keV. The value of linear attenuation coefficient of ordinary concrete sample without magnetite, (K, 0%), was taken from our previous study (Oto et al., 2013). The measured and calculated results are listed in Table 3. It can be seen from this table that there is a good agreement between experimental and theoretical results. A typical spectrum of 80.99 keV gamma ray transmissions through the concrete sample with and without magnetite is shown in Figure 2. It is seen from this figure, gamma rays intensity obtained from concrete samples with magnetite is lower than that of ordinary concrete samples. For all types of concrete samples, linear attenuation coefficients decrease with increasing gamma ray energy.

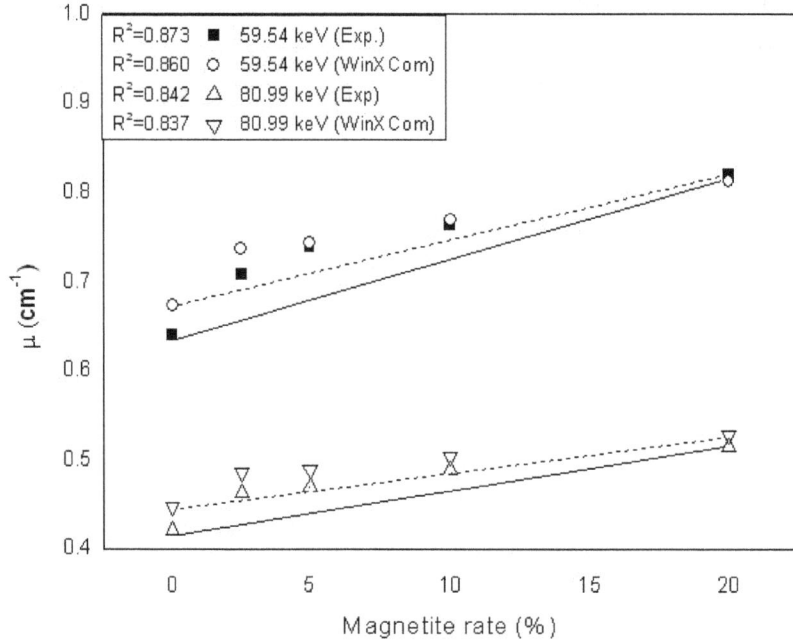

**Figure 3.** Linear attenuation coefficients of the concrete samples versus magnetite weight concentration for two photon energies.

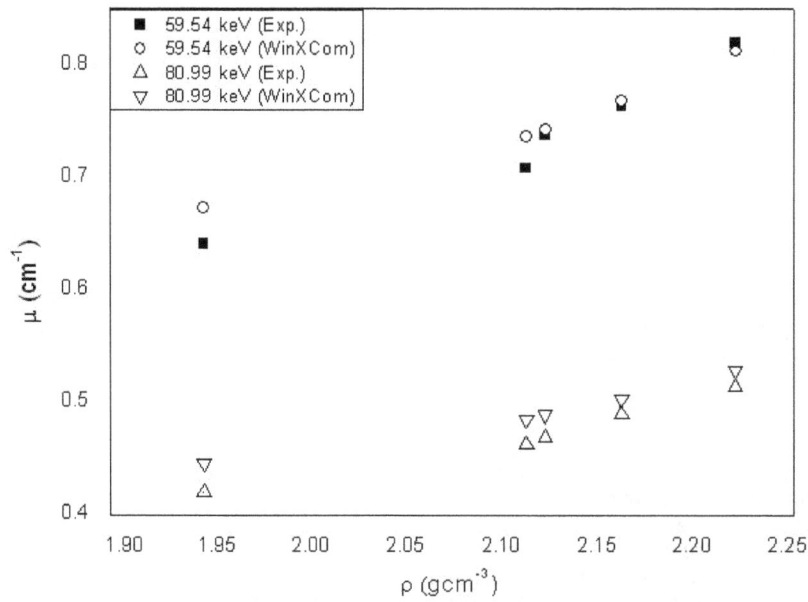

**Figure 4.** Linear attenuation coefficients of the concrete samples versus density for 59.54 and 80.99 keV energies.

The experimental and theoretical linear attenuation coefficients of concrete samples versus the sample types were plotted in Figure 3 using the least-squares methods. The correlation theory is used to confirm the linearity of the experimental and theoretical values. For 59.54 and 80.99 keV photon energies, the correlation coefficients of the experimental and theoretical values are $R^2 = 0.873$, $R^2 = 0.860$ and $R^2 = 0.842$, $R^2 = 0.837$, respectively. It can be seen from Figure 3 that the linear attenuation coefficients of concrete samples with magnetite are higher than the ordinary concrete (K) and the linear attenuation coefficients increase with increasing magnetite rate. This means that concrete samples with magnetite attenuated gamma rays more than the ordinary concrete.

Figure 4 shows the variation of linear attenuation coefficients with density of concrete samples for both experimental and theoretical values. It is clear from this figure that, with regard to gamma-ray shielding, the concentration rates of magnetite in concrete samples increase the density of the sample, because the concrete samples with magnetite have iron content. Also ($\mu$) values of concerete samples containing magnetite are greater than the ordinary concrete. For 1.5 MeV gamma energy, linear attenuation coefficients of the two concretes prepared from steel-magnetite ($\rho = 5.11$ gcm$^{-3}$) and basalt-magnetite ($\rho = 3.05$ gcm$^{-3}$) have been reported to be 0.220 and 0.139 cm$^{-1}$, respectively (Bashter et al., 1997). Akkurt et al. (2009) studied the radiation shielding of concretes containing barite in different ratios (Akkurt et al., 2010). They have found that, linear attenuation coefficients of BC0 (%0 barite), BC50 (%50 barite), BC100 (%100 barite) samples were 0.257 ± 0.012, 0.287 ± 0.014, 0.297 ± 0.014 at 662 keV gamma energy, respectively and their results showed that linear attenuation coefficient was a function of concrete density. It is clearly seen that the linear attenuation coefficients depend on the photon energy and density of the shielding material, and the concrete samples containing magnetite are remarkably effective for shielding gamma rays. It can be concluded that concrete samples with magnetite have better properties than the ordinary concrete.

## REFERENCES

Akkurt I, Akyildirim H, Mavi B, Kilinçarslan Ş, Basyigit C (2010). Photon attenuation coefficients of concrete includes barite in different rate. Ann. Nucl. Energy 37:910-914.

Akkurt I, Akyildirim H, Karipcin F, Mavi B (2012). Chemical corrosion on gamma-ray attenuation properties of barite concrete. J. Saudi Chem. Soc. 16:199-202.

Akkurt I, Kilincarslan S, Basyigit C, Mavi B, Akyildirim H (2009). Investigation of photon attenuation coefficient for pumice. Int. J. Phys. Sci. 4(10):588-591.

Bashter II (1997). Calculation of radiation attenuation coefficients for shielding concretes. Ann. Nucl. Energy 24:1389-1401.

Bashter II, El-Sayed AA, Abdel-Azim MS (1997). Magnetite ores with steel or basalt for concrete radiation shielding. Jpn. J. Appl. Phys. Part 1, 36(6A):3692-3696.

Demir D, Keleş G (2006). Radiation transmission of concrete including boron waste for 59.54 and 80.99 keV gamma rays. Nucl. Instrum. Methods Phys. Res. B. 245:501-504.

El-Sayed AA, Megahid RM (2001). Homogeneous and multilayered shields for neutrons and gamma-rays. Jpn. J. Appl. Phys. Part 1 40 (4A):2460-2464.

Facure A, Silva AX (2007). The use of high-density concretes in radiotherapy treatment room design. Appl. Radiat. Isotopes 65(9):1023-1028.

Gerward L, Guilbert N, Jensen BK, Levring H (2001). X-ray absorption in matter reengineering XCOM. Rad. Phys. Chem. 60:23-24

Han I, Kolayli H, Sahin M (2011). Determination of mass attenuation coefficients for natural minerals from different places of Turkey. Int. J. Phys. Sci. 6(20):4798-4801.

Kaplan MF (1989). Concrete radiation shielding, Longman Scientific and Technology, Longman Group UK Limited, Essex England. P. 458.

Medhat ME (2009). Gamma-ray attenuation coefficients of some building materials available in Egypt. Ann. Nucl. Energy 36:849-852.

Mostofinejad D, Reisi M, Shirani A (2012). Mix design effective parameters on γ-ray attenuation coefficient and strength of normal and heavyweight concrete. Constr. Build. Mater. 28:224-229.

Oto B, Gür A, Kaçal MR, Doğan B, Arasoğlu A (2013). Photon attenuation properties of some concretes containing bariteN and colemanite in different rates. Ann. Nucl. Energy 51:120-124.

Seven S, Karahan İH, Bakkaloğlu ÖF (2004). The measurement of total mass attenuation coefcients of CoCuNi alloys. J. Quant. Spectrosc. Radiat. Transf. 83:237-242.

Shirmardi SP, Shamsaei M, Naserpour M (2013). Comparison of photon attenuation coefficients of various barite concretes and lead by MCNP code, XCOM and experimental data. 55:288-291.

# 8

# Effects of growth rate on the physical and mechanical properties of Sn-3.7Ag-0.9Zn eutectic alloy

**U. Böyük[1] , S. Engin[2], H. Kaya[1], N. Maraşli[3], E. Çadirli[4] and M. Şahin[4]**

[1]Department of Science Education, Education Faculty, Erciyes University, Kayseri, Turkey.
[2]Department of Physics, Institute of Science and Technology, Erciyes University, Kayseri, Turkey.
[3]Department of Physics, Faculty of Science, Erciyes University, Kayseri, Turkey.
[4]Department of Electronics and Automation, Technical Vocational School of Sciences, Niğde University, Niğde, Turkey.

Sn-3.7wt.%Ag-0.9wt.%Zn alloy was directionally solidified upward under different conditions, with different growth rates (V = 3.38 - 220.12 μm/s) at a constant temperature gradient (G = 4.33 K/mm) and with different temperature gradients (G = 4.33 -12.41 K/mm) at a constant growth rate (V = 11.52 μm/s) by using a Bridgman-type directional solidification furnace. The microstructure was observed to be a rod $Ag_3Sn$ structure in the matrix of β-Sn from the directionally solidified Sn-3.7wt.%Ag-0.9wt.%Zn samples. The microhardness, tensile strength and electrical resistivity of alloy were measured from directionally solidified samples. The dependency of the microhardness, tensile strength and electrical resistivity on the solidification parameters for directionally solidified Sn-Ag-Zn eutectic alloy was investigated and the relationships between them were experimentally obtained by using regression analysis. The results obtained in the present work were compared with the previous similar experimental results.

**Key words:** Alloys, crystal growth, microstructure; mechanical properties, electrical properties.

## INTRODUCTION

Solidification and melting are transformations between the crystallographic and non-crystallographic states of a metal or alloy. These transformations are basic to such technological applications as ingot and continuous casting, and directionally solidification of composites and single crystals. An understanding of the mechanism of solidification and how it is affected by such parameters as temperature distribution, solidification condition and alloying, are important in the control of the mechanical and electrical properties of cast metals and fusion welds (Porter and Easterling, 1992). Most Sn-based lead-free solders contain only minor amounts of alloying additions. For example, in the widely studied Sn–Ag–Cu (SAC) alloy, more than 95% of the solder is Sn, as measured by weight (Anderson et al., 2001).

However, in these solders, the thick Cu6Sn5 IMC (intermetallic compound) layer and large Ag3Sn primary phase are often reported to influence the integrity and reliability of solder joints (Kang and Sarkhel, 1994; Jeong et al., 2004; Jo et al., 2008; Kang et al., 2003). Reducing Ag and Cu content, as well as adding minor alloying elements such as Zn, In, Bi, Co, Ni in Sn-based solder, was recently proposed in order to improve the reliability of the Pb-free solder joint (Choi et al., 2001; Seo et al., 2006, 2007; Kim et al., 2009; Cho et al., 2007; Kang et al., 2008). Although, Sn–Ag solder has great properties of strength, resistance to creep and thermal fatigue (Abtew and Selvaduray, 2000; Wu et al., 2004; Zeng and Tu, 2002) the small addition of Zn can improve the mechanical performance at no cost to ductility and wettability. In addition, the combination of Zn and Ag dramatically reduces corrosion potential (Knott et al., 2005).

Previous works for the Sn–Ag–Zn lead-free solder had been done mainly focused on the formation of the interfacial structure, the phase equilibria and the mechanical properties (Xu et al., 2010; Wang et al., 2009; Wei et al., 2009; Liu et al., 2008). For a system with such promising applications, however, studies considering the directional solidification of Sn-Ag-Zn alloys are rather scarce. Until now, few numbers of researches have been carried out on the characterization, physical and mechanical properties. The aim of the present work was to investigate the mechanical, electrical and thermal properties Sn-3.7wt.%Ag-0.9wt.%Zn alloy. For this purpose, the dependency of microhardness (HV), tensile strength ($\sigma_t$) and electrical resistivity (ρ) on the solidification processing parameters (G and V) for directionally solidified Sn-3.7wt.%Ag-0.9wt.%Zn alloy was investigated.

## EXPERIMENTAL PROCEDURE

In the present work, the experimental procedure consists of alloy preparation, the measurements of the microhardness, tensile strength and electrical resistivity of the directionally solidified Sn-3.7wt.%Ag-0.9wt.%Zn alloy.

### Alloy preparation

Using a vacuum melting furnace and a hot filling furnace, Sn-3.7wt.%Ag-0.9wt.%Zn eutectic alloy was prepared under vacuum atmosphere by melting tin, silver and zinc of high purity (>99.9 %). After allowing time for the melt homogenization, the molten alloy was poured into 13 cylindrical graphite crucibles (200 mm in length, 4 mm inner diameter and 6.35 mm outer diameter) held in a specially constructed casting furnace (hot filling furnace) at approximately 50 K above the melting point of the alloy. The molten alloy was directionally solidified from bottom to top to ensure that the crucible was completely full.

Solidification of Sn-Ag-Zn eutectic alloy was carried out with different growth rates (V = 3.38 - 220.12 μm/s) at a constant temperature gradient (G = 4.33 K/mm) and with different temperature gradients (G = 4.33 - 12.41 K/mm) at a constant growth rate (V = 11.52 μm/s) in the Bridgman–type growth apparatus. The temperature of water in the reservoir was kept at 283 K to an accuracy of ± 0.01 K using a Poly Science digital 9102 model heating / refrigerating circulating bath to get a well quenched solid–liquid interface. The temperature of the sample was also controlled to an accuracy of ± 0.1 K degrees with a Eurotherm 2604 type controller. The details of the apparatus and experimental procedures are given in Refs (Gündüz et al., 2004; Böyük et al., 2009). The quenched samples were removed from the graphite crucible and cut into lengths of typically 3 mm. After the metallographic process, the microstructures of the samples were revealed. Typical images of growth morphologies of directionally solidified Sn-Ag-Zn eutectic alloy are shown in Figure 1. While the

cooling rate is so slow (about 0.16 K/s) that it could be considered as an equilibrium solidification process, which corresponds to an eutectic reaction: L→ AgZn +Ag₃Sn + Sn (Wei et al., 2008). It follows that β-Sn, AgZn and Ag₃Sn phases will separate out in the slowly cooled solder and eutectic microstructure consist of a mixture of ζ-AgZn and Ag₃Sn intermetallic compounds in a matrix of β-Sn (Wei et al., 2008). But, in this work were observed only rod Ag₃Sn IMC in a matrix of β-Sn as shown in Figure 1, because the cooling rate range is fairly above 0.16 K/s.

### The measurement of microhardness

One of the purposes of this investigation was to learn the relationships between the solidification processing parameters and microhardness for the directionally solidified Sn–Ag–Zn eutectic alloy. The mechanical properties of solidified materials are generally determined by a hardness test, tensile strength test, etc. The Vickers hardness (HV) is the ratio of a load applied to the indenter to the surface area of the indentation. This is given by:

$$HV = \frac{2P \sin(\theta / 2)}{d^2} \tag{1}$$

where HV is the Vickers microhardness in kg/mm², P is the applied load (kg), d is the mean diagonal of the indentation (mm) and θ is the angle between opposite faces of the diagonal indenter (136°). Microhardness measurements in this study were made with a Future–Tech FM–700 model hardness measuring test device using 500 g load and a dwell time of 10 s giving a typical indentation depth of about 40–60 μm. The average microhardness is achieved by measuring at least 30 different points on the transverse sections. Variations of microhardness with growth rates and temperature gradients for the Sn-3.7wt.%Ag-0.9wt.%Zn eutectic alloy is plotted in Figures 2 and 3, respectively, and compared with the previous experimental results for Sn-3.5wt.%Ag-0.9wt.%Cu (Böyük and Maraşli, 2009; Çadirli and Şahin, 2012) and Sn-3.5wt.%Ag (Çadirli and Şahin, 2012; Böyük and Maraşli, 2010) eutectic alloys.

### Measurement of tensile strength

The uniaxial tensile test was performed at room temperature at a strain rate of 10⁻³ s⁻¹ with a Shimadzu Universal Testing Instrument (Type AG–10KNG) which was designed for testing the stress–strain responses of solders. In order to avoid damaging the sample surface, two seals were stuck to the sample instead of the traditional clip gauge. Strains were then measured by observing the displacement between the two seals using a video camera. A computer with a data acquisition software was used to collect the data. The data collected from the tensile test can be analyzed to determine the strength (σ) using the following formula,

$$\sigma = \frac{F}{A} \tag{2}$$

where σ is the strength in N/mm² (or MPa), F is the applied force (N) and A is the original cross sectional area (mm²) of the sample. The round rod tensile samples with a diameter of 4 mm and a gauge length of 20 mm were prepared from directionally solidified rod samples with different solidification parameters. The tensile axis was chosen parallel to the growth direction of the sample and the tests were repeated three times. Variations of tensile strength with growth rate and temperature gradient for the Sn-3.7wt.%Ag-0.9wt.%Zn eutectic alloy is plotted in Figures 4 and 5, respectively, and compared with binary Sn-3.5wt.%Ag (Çadirli and Şahin, 2012)

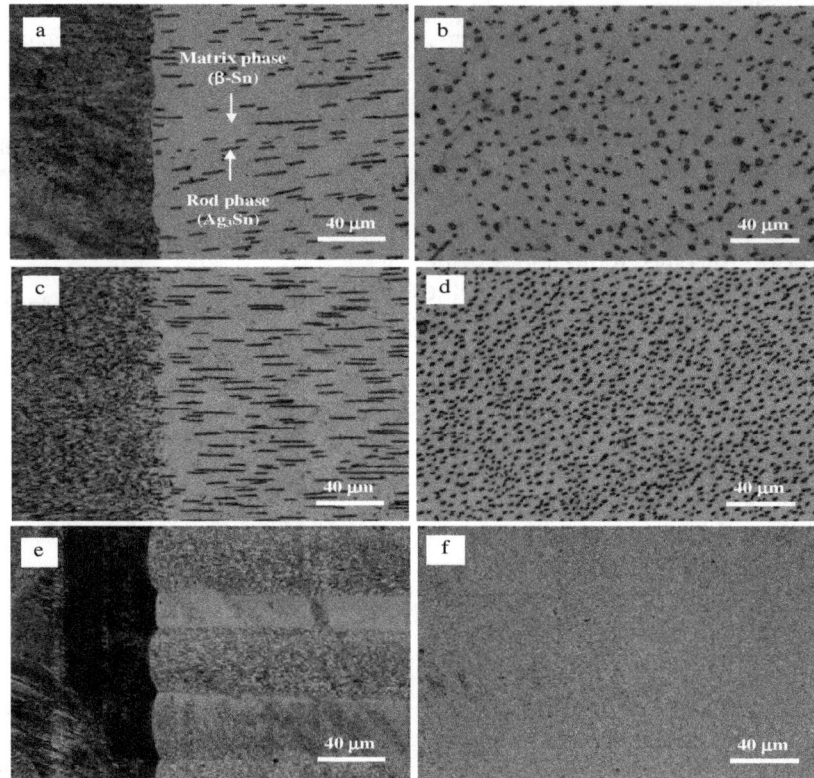

**Figure 1.** Typical optical images of the growth morphologies of directionally solidified Sn-3.7wt.%Ag-0.9wt.%Zn eutectic alloy, (a) longitudinal section (b) transverse section (V=3.38 μm/s, G=4.33 K/mm), (c) longitudinal section (d) transverse section, (V=11.52 μm/s, G=12.41 K/mm) (e) longitudinal section (f) transverse section (V=220.12 μm/s, G=4.33 K/mm).

eutectic alloys. It can be seen from Figures 4 and 5, the values of tensile strength for the Sn-3.7wt.%Ag-0.9wt.%Zn eutectic alloy increase with increasing the values of V and G.

### The measurement of electrical resistivity

Electrical resistivity is an imperative physical property. Impurities observed in metals distort the metal lattice and can affect the behavior of $\rho$ to a considerable extent. This is particularly true for metal alloys. The value of the electrical resistivity is also affected by grain size (e.g., higher $\rho$ corresponds to finer grain), plastic deformation, heat treatment, and some other factors, but to a smaller extent compared to the effect of temperature and chemical composition (Rudnev et al., 2003).

The growth rate, temperature gradient and temperature dependence of electrical resistivity for Sn-3.7wt.%Ag-0.9wt.%Zn alloys were measured by the four-point probe method. A *Keithley 2400* source meter was used to provide constant current, and the potential drop was measured by a *Keithley 2700* multimeter through an interface card, which was controlled by a computer. Platinum wires with a diameter of 0.5 mm were used as current and potential probes. The voltage drop was detected, and the electrical resistivity and conductivity were determined using a standard conversion method.

The electrical resistivities of the directionally solidified Sn-3.7wt.%Ag-0.9wt.%Zn eutectic alloys were measured by the d.c. four-point probe method at room temperature. Variations of

electrical resistivity with growth rate, temperature gradient and temperature for the Sn-3.7wt.%Ag-0.9wt.%Zn eutectic alloy is plotted in Figures 6 and 7, respectively. It can be seen from Figures 6 to 7, the values of electrical resistivity for the Sn–Ag–Zn eutectic alloy increase with increasing the values of V and G.

## RESULTS AND DISCUSSION

### The effect of solidification parameters on microhardness

It can be also seen from Figures 2 and 3 that an increase in solidification parameters leads to an increase in the HV. The dependence of the HV on V and G were determined by linear regression analysis and the relationship between them can be expressed as;

$$HV = k_1(V)^a \qquad (3)$$

$$HV = k_2(G)^b \qquad (4)$$

where k is a constant, a and b are the exponent values relating to the growth rate and temperature gradient respectively.

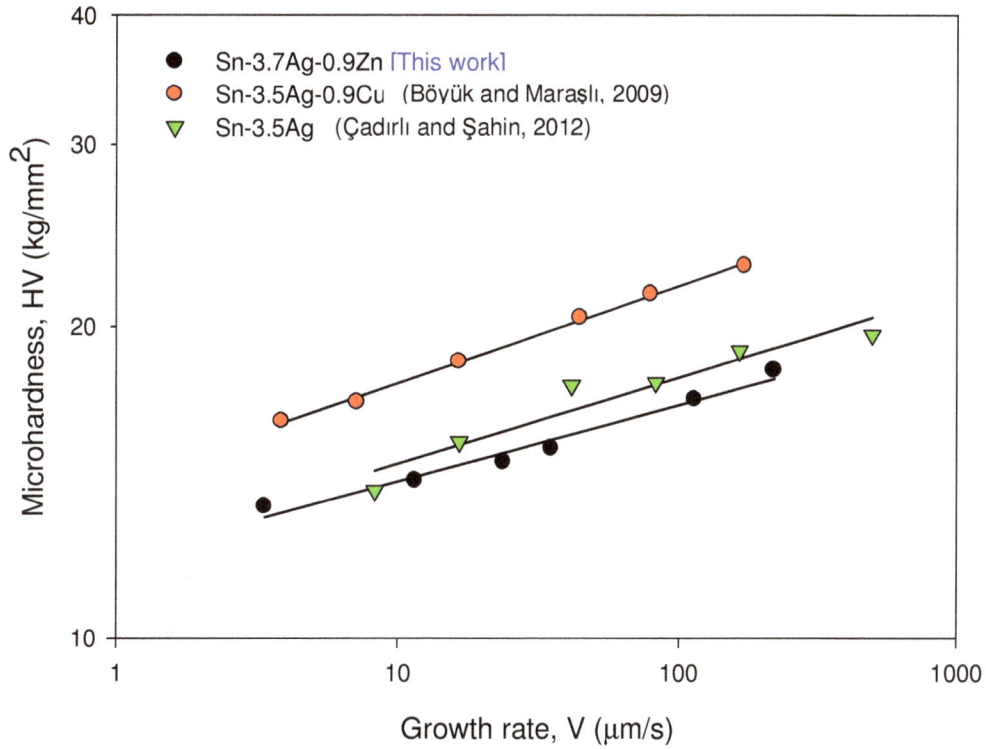

**Figure 2.** Variation of microhardness, as a function of growth rate for directionally solidified Sn-3.7wt.%Ag-0.9wt.%Zn eutectic alloy at a constant temperature gradient and compare with the Sn-3.5wt.%Ag-0.9wt.%Cu and Sn-3.5wt.%Ag eutectic alloys.

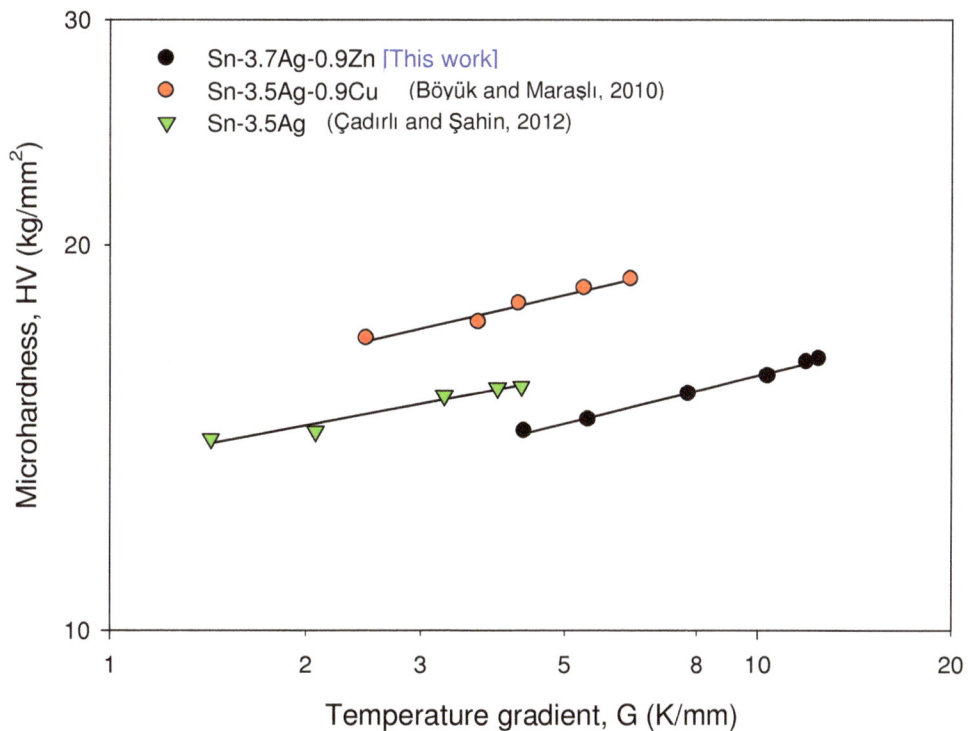

**Figure 3.** Variation of microhardness, as a function of temperature gradient for directionally solidified Sn-3.7wt.%Ag-0.9wt.%Zn eutectic alloy at a constant growth rate and compare with the Sn-3.5wt.%Ag-0.9wt.%Cu and Sn-3.5wt.%Ag eutectic alloys.

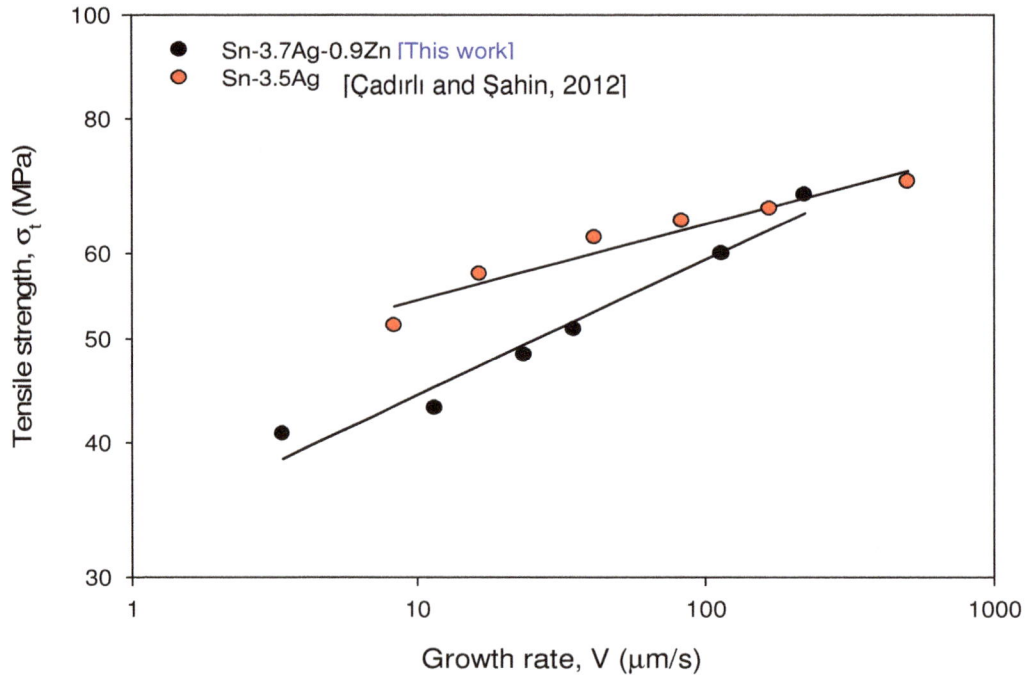

**Figure 4.** Variation of tensile strength, as a function of growth rate for directionally solidified Sn-3.7wt.%Ag-0.9wt.%Zn eutectic alloy at a constant temperature gradient and compare with the Sn-3.5wt.%Ag eutectic alloys.

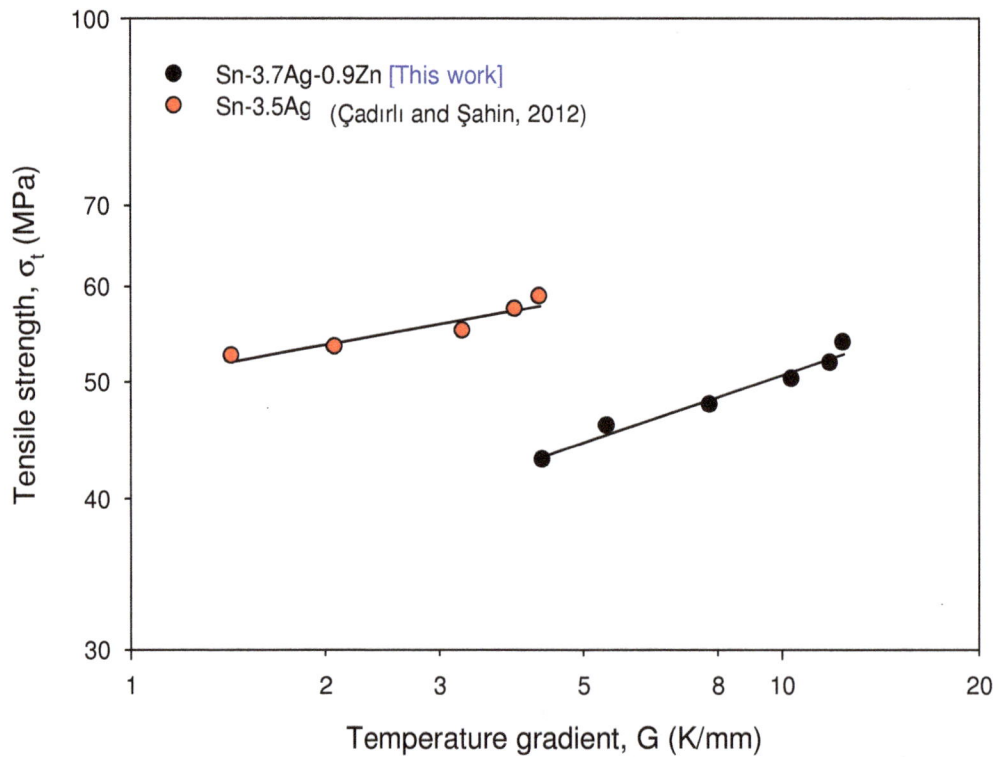

**Figure 5.** Variation of tensile strength, as a function of temperature gradient for directionally solidified Sn-3.7wt.%Ag-0.9wt.%Zn eutectic alloy at a constant growth rate and compare with the Sn-3.5wt.%Ag eutectic alloys.

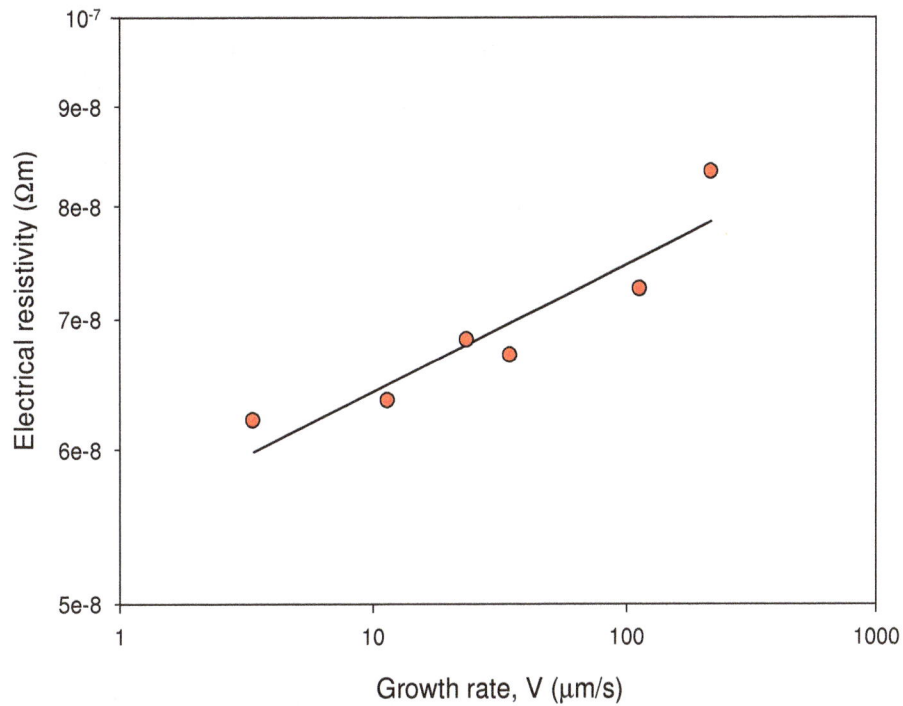

**Figure 6.** Variation of electrical resistivity, as a function of growth rate for directionally solidified Sn-3.7wt.%Ag-0.9wt.%Zn eutectic alloy at a constant temperature gradient.

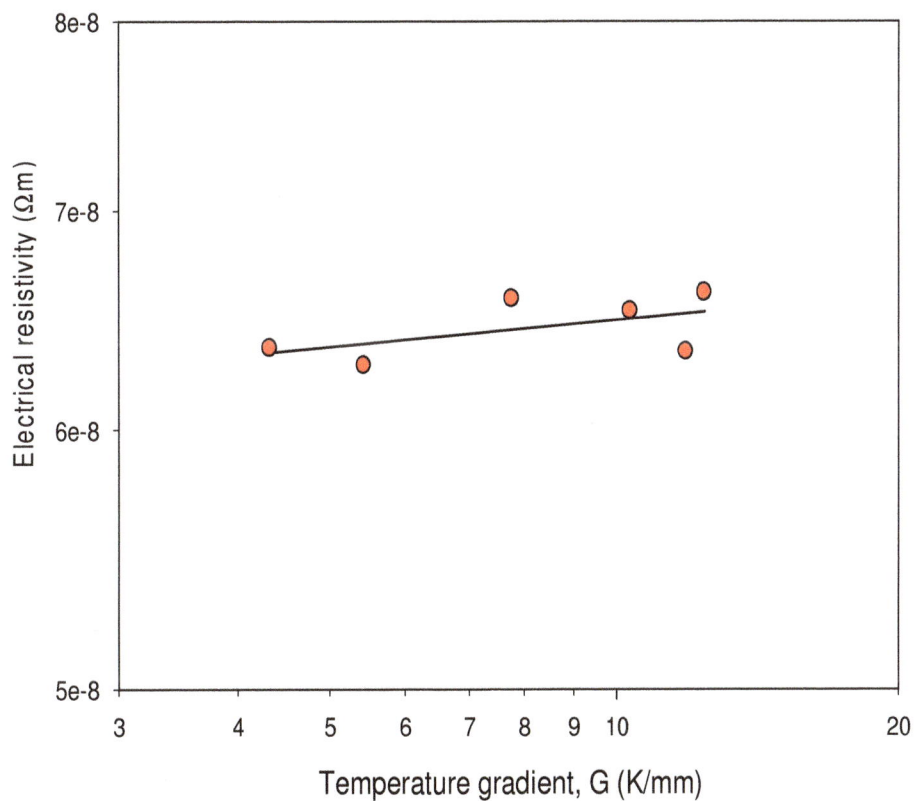

**Figure 7.** Variation of electrical resistivity, as a function of temperature gradient for directionally solidified Sn-3.7wt.%Ag-0.9wt.%Zn eutectic alloy at a constant growth rate.

Figure 2 shows the variation of HV as a function of V at a constant G. The value of HV increases with the increasing value of V. Using linear regression analysis, the relationship between the HV and V was determined as $HV = 20.65(V)^{0.09}$, and the exponent value was found to be 0.09 for Sn-Ag-Zn eutectic alloy. This exponent value (0.09) agrees with the exponent values of V (0.07–0.11) obtained by various researchers (Böyük and Maraşli, 2009; Vnuk et al., 1979, 1980; Telli and Kisakürek, 1988). for different binary and ternary eutectic alloy systems, under similar solidification conditions.

Using linear regression analysis, the relationship between HV and G was determined as $HV = 11.83(G)^{0.12}$, and the exponent value was found to be 0.12. As can be seen from Figure 3, the value of microhardness increases with the increasing the value of the temperature gradient (G) for a given constant V as well. An exponent value relating to G (0.12) generally agrees with the exponent values relating to obtained in previous experimental works (Böyük and Maraşli, 2010).

Figures 2 and 3 show the maximum and minimum values of HV for Sn-3.7wt.%Ag-0.9wt.%Zn eutectic alloy are lower than the maximum and minimum values of HV obtained by Böyük and Maraşli (2009) and Çadirli and Şahin (2012) for unidirectional solidified Sn-3.5wt.%Ag-0.9wt.%Cu and Sn-3.5wt.%Ag eutectic alloy, respectively. While the microhardness of the Sn-3.5wt.%Ag eutectic alloy decreases with the adding 0.9wt.% of Zn content, conversely increases with the adding 0.9wt.% of Cu. Besides, the microhardness of the slowly cooled Sn-Ag-Zn eutectic solder increased from 14.4 to 17 HV and the rapidly solidified one increased from 15 to 29.1 HV obtained by Wei et al. (2009).

HV values obtained by Wei et al. (2009) for Sn-Ag-Zn eutectic alloy agree with this experimental works for slowly cooled (13.4 to 18.15).

## The effect of solidification parameters on tensile strength

Figures 4 and 5 show the variation of the tensile strength values with growth rate and temperature gradient. The dependence of $\sigma_t$ on the V and G can be represented by equations as follows;

$$\sigma_t = k_3(V)^c \tag{5}$$

$$\sigma_t = k_4(G)^d \tag{6}$$

where k is a constant, c and d are the exponent values relating to the growth rate and temperature gradient, respectively.

From Figure 4, the relationship between $\sigma_t$ and V was found to be $\sigma_t = 31.62(V)^{0.14}$ by using linear regression analysis and also it can be seen that values of the tensile

strength increase with increasing growth rate. It is found that, while growth rate increasing from 3.37 to 220.12 μm/s, the tensile strength increases from 40.75 to 68.00 MPa.

Figure 5 shows the experimental results of tensile strength as a function of the temperature gradient. It can be seen that the value of the tensile strength also increases with increasing temperature gradient. It is 12.41 K/mm, the tensile strength increase from 43.06 to 53.85 MPa. Using linear regression analysis, the relationship between $\sigma_t$ and G was found to be $\sigma_t = 32.96(G)^{0.19}$.

Figures 4 and 5 show the maximum and minimum values of $\sigma_t$ for Sn-3.7wt.%Ag-0.9wt.%Zn eutectic alloy are lower than the maximum and minimum values of $\sigma_t$ obtained by Çadirli and Şahin (2012) for unidirectional solidified Sn-3.5wt.%Ag alloy. Besides, the tensile strength of the Sn–3.5%Ag eutectic alloy decreases with the adding 0.9% of Zn content.

## The effect of solidification parameters on electrical resistivity

Figures 6 and 7 show the variation of the electrical resistivity values with growth rate and temperature gradient. The dependence of ρ on the V and G can be represented by equations as follows;

$$\rho = k_5(V)^e \tag{7}$$

$$\rho = k_6(G)^f \tag{8}$$

where k is a constant, e and f are the exponent values relating to the growth rate and temperature gradient, respectively.

From Figures 6 and 7, the relationships between ρ and V, ρ and G were found to be $\rho = 5.62 \times 10^{-8}(V)^{0.07}$ and $\rho = 6.16 \times 10^{-8}(G)^{0.04}$ respectively by using linear regression analysis and also it can be seen that values of the electrical resistivity increase with increasing growth rate and temperature gradient (Figures 6 and 7).

## Conclusions

In the present work, the influence solidification parameters and temperature on the mechanical, electrical and thermal properties of Sn-3.7wt.%Ag-0.9wt.%Zn eutectic alloy was investigated. The results are summarized as follows;

(1) Values of micro hardness increase with increasing the values of V and G. The establishment of the relationships among HV, V and G can be given as; $HV = 20.65(V)^{0.09}$ and $HV = 11.83(G)^{0.12}$.

(2) The experimental expressions correlating the values of $\sigma_t$ with the values of V and G for directional solidified Sn-3.7wt.%Ag-0.9wt.%Zn eutectic alloy have shown that the values of the tensile strength increase with increasing the values of V and G. The establishment of the relationships between strength and solidification parameters can be given as; $\sigma_t = 31.62(V)^{0.14}$ and $\sigma_t = 32.96(G)^{0.19}$.

(3) The values of electrical resistivity increase with increasing the values of V and G. The establishment of the relationships among electrical resistivity and solidification parameters can be given as;

$$\rho = 5.62 \times 10^{-8}(V)^{0.07} \text{ and } \rho = 6.16 \times 10^{-8}(G)^{0.04}.$$

## ACKNOWLEDGEMENTS

This project was supported by Erciyes University Scientific Research Project Unit Contract No: FBT 07–65. The authors are grateful to Erciyes University Scientific Research Project Unit for their financial support.

## REFERENCES

Abtew M, Selvaduray G (2000). Lead-free solders in microelectronics. Mater. Sci. Eng. 27:95-141.

Anderson IE, Foley JC, Cook BA, Harringa J, Terpstra RL, Unal O (2001). Alloying effects in near-eutectic Sn-Ag-Cu solder alloys for improved microstructural stability. J. Electron. Mater. 30:1050-1059.

Böyük U, Maraşli N (2009). The microstructure parameters and microhardness of directionally solidified Sn-Ag-Cu eutectic alloy. J. Alloy. Compd. 485:264-269.

Böyük U, Maraşli N (2010). Dependency of eutectic spacings and microhardness on the temperature gradient for directionally solidified Sn-Ag-Cu lead-free solder. Mater. Chem. Phys. 119:442-448.

Böyük U, Maraşli N, Kaya H, Çadirli E, Keşlioğlu K (2009). Directional solidification of Al-Cu-Ag alloy. Appl. Phys. A–Mater. 95:923-932.

Çadirli E, Şahin M (2012). Influence of temperature gradient and growth rate on the mechanical properties of directionally solidified Sn-3.5 wt% Ag eutectic solder. J. Mater. Sci. Mater. Electron. 23:31-40.

Cho MG, Kang SK, Shih DY, Lee HM (2007). Effects of minor additions of Zn on interfacial reactions of Sn-Ag-Cu and Sn-Cu solders with various Cu substrates during thermal aging. J. Electron. Mater. 36: 1501-1509.

Choi WK, Hoi SW, Shih D-Y, Henderson DW, Gosselin T, Sarkhel A, Goldsmith C, Yoon KJ, Lee HM (2001). Effect of in addition on Sn-3.5Ag solder and joint with Cu substrate. Mater. Trans. 42:783-789.

Gündüz M, Kaya H, Çadirli E, Özmen A (2004). Interflake spacings and undercoolings in Al-Si irregular eutectic alloy. Mat. Sci. Eng. A. 369:215-229.

Jeong SW, Kim JH, Lee HM (2004). Effect of cooling rate on growth of the intermetallic compound and fracture mode of near-eutectic Sn-Ag-Cu/Cu pad: Before and after aging. J. Electron. Mater. 33:1530-1544.

Jo YH, Lee JW, Seo S-K, Lee HM, Han H, Lee DC (2008). Demonstration and characterization of Sn-3.0Ag-0.5Cu/Sn-57Bi-1Ag combination solder for 3-D multistack packaging. J. Electron. Mater. 37:110-117.

Kang SK, Cho MG, Shih DY, Seo S-K, and Lee HM (2008). Proc. 58th Electronic Components and Technology Conf. (Piscataway NJ: IEEE. 2008). P. 478.

Kang SK, Choi WK, Shih D-Y, Henderson DW, Gosselin T, Sarkhel A, Goldsmith C, Puttlitz KJ (2003). Ag3Sn plate formation in the solidification of near-ternary eutectic Sn-Ag-Cu. JOM. 55(6):61-65.

Kang SK, Sarkhel AK (1994). Lead (Pb)-Free Solders for Electronic Packaging. J. Electron. Mater. 23:701-707.

Kim DH, Cho MG, Seo S-K, Lee HM (2009). Effects of Co Addition on Bulk Properties of Sn-3.5Ag Solder and Interfacial Reactions with Ni-P UBM. J. Electron. Mater. 38:39-45.

Knott S, Flandorfer H, Mikula A (2005). Calorimetric investigations of the two ternary systems Al-Sn-Zn and Ag-Sn-Zn. Z. Metallkd. 96:38-44.

Liu YC, Wan JB, Gao ZM (2008). Intermediate decomposition of metastable Cu5Zn8 phase in the soldered Sn-Ag-Zn/Cu interface. J. Alloy. Compd. 465:205-209.

Porter DA, Easterling KE (1992). Phase transformations in metals and alloys. Second Edition. CRC Press. P. 185.

Rudnev V, Loveless D, Cook R, Black M (2003). Markel Dekker Inc. New York. P. 119.

Seo SK, Cho MG and Lee HM (2007). Thermodynamic assessment of the Ni-Bi binary system and phase equilibria of the Sn-Bi-Ni ternary system. J. Electron. Mater. 36:1536-1544.

Seo S-K, Cho MG, Choi WK, Lee HM (2006). Comparison of Sn2.8Ag20In and Sn10Bi10In solders for intermediate-step soldering. J. Electron. Mater. 35:1975-1981.

Telli Al, Kisakürek SE (1988). Effect of antimony additions on hardness and tensile properties of directionally solidified Al–Si eutectic alloy. Mat. Sci. Tech. 4:153-156.

Vnuk F, Sahoo M, Baragor D and Smith RW (1980). Mechanical-properties of the Sn-Zn eutectic alloys. J. Mater. Sci. 15:2573-2583.

Vnuk F, Sahoo M, Van De Merwe R, Smith RW (1979). The hardness of Al-Si eutectic alloys. J. Mater. Sci. 14:975-982.

Wang X, Liu YC, Wei C, Gao HX, Jiang P, Yu LM (2009). Strengthening mechanism of SiC-particulate reinforced Sn-3.7Ag-0.9Zn lead-free solder. J. Alloy. Compd. 480:662-665.

Wei C, Liu Y, Gao Z, Xu R, Yang K (2009). Effects of aging on structural evolution of the rapidly solidified Sn-Ag-Zn eutectic solder. J. Alloy. Compd. 468:154-157.

Wei C, Liu YC, Han YJ, Wan JB, Yang K (2008). Microstructures of eutectic Sn-Ag-Zn solder solidified with different cooling rates. J. Alloy. Compd. 464:301-305.

Wu CML, Yu DQ, Law CMT, Wang L (2004). Properties of lead-free solder alloys with rare earth element additions. Mater. Sci. Eng. R. 44:1-44.

Xu RL, Liu YC, Wei C, Yu LM (2010). Effects of Zn additions on the structure of the soldered Sn-3.5Ag and Cu interfaces. Solder. Surf. Mt. Tech. 22(2):13-20.

Zeng K, Tu KN (2002). Six cases of reliability study of Pb-free solder joints in electronic packaging technology. Mater. Sci. Eng. R. 38:55-105.

# 9

# Synthesis of nickel ferrite nanoparticles by co-precipitation chemical method

Aliahmad, M.[1], Noori, M.[1], Hatefi Kargan, N.[1] and Sargazi, M.[2]

[1]Department of Physics, University of Sistan and Baluchestan, Zahedan, Iran.
[2]Faculty of Science, University of Payam-e- Noor, Tehran, Iran.

In this research work, we have prepared nickel ferrite nanoparticles by using chemical route. Nanoparticle materials are characterized using X-ray diffraction (XRD), Fourier transform infrared spectroscopy (FT-IR), vibrating sample magnetometer (VSM), transmission electron microscopes (TEM) and energy dispersive X-ray (EDX) systems. We have determined magnetic properties, size, purity, stoichiometry and morphology of samples. The samples are calcinated at different temperatures, then we found that size of particles increase with heating and powder transfer from amorphous to crystalline phase of nickel ferrite. When the size of nanoparticles decreased to less than a critical grain size (10 nm), the nanomaterials transfer from ferromagnetic to super paramagnetic materials.

Key words: Nanoparticle, nickel ferrite spinel, superparamagnetic.

## INTRODUCTION

In the recent years, so much attention has been paid to the nanomagnetic materials that show very interesting magnetic properties. In this material, different properties and applications are appeared as compared to their bulk counterparts. The magnetic properties of nanomaterials are used in medical, electronic, and recording industries that depend on the size, shape, purity and magnetic stability of these materials (Maaz et al., 2009; Sellmyer and Skomski, 2006; Cullity and Graham, 2009).

In biomedical application, one can use nanomagnetic materials as drug carriers inside body where the conventional drug may not work. For this purpose, the nanosize particles should be in the superparamagnetic form with a low blocking temperature (Sellmyer et al., 2006). Ferrite nanomaterials are object of intense research because of their proper magnetic properties. It has been reported that when the size of particles reduced to small size or in range of nanomaterials, some of their fundamental properties are affected (Sellmyer and Skomski, 2006; Cullity and Graham, 2009; Billas et al.,

1994). Nickel ferrite $NiFe_2O_4$ is a cubic structure and has an inverse spinel structure. At this structure, $Ni^{2+}$ ions occupy octahedron B site and $Fe^{3+}$ ions occupy both tetrahedron A and octahedron B-sites. The spinel nanoparticles generally are prepared by using chemical route which is a proper method. Nickel ferrite is one of the most important spinel ferrites. It shows a proper ferromagnetism that originates from magnetic moment of anti-parallel spins (Martinez et al., 1998; Misra et al., 2004; Nathani et al., 2005).

In a spinel structure, there are 56 ions, 32 oxygen and 24 metal ions in a unit cell. At this structure eight molecules occupy a unit cell of spinel that they are 32 inions and 24 cations. A general formula of ferrite structure is shown as $(M_{1-x}Fe_x)[M_xFe_{2-x}] O_4$, in which M shows cations that occupy tetrahedron sites and x is degree of inversion (Abdullah et al., 2008).

In this research, work we have used co-precipitation method for making nickel ferrite nanoparticles. It is a proper technique for making small size and

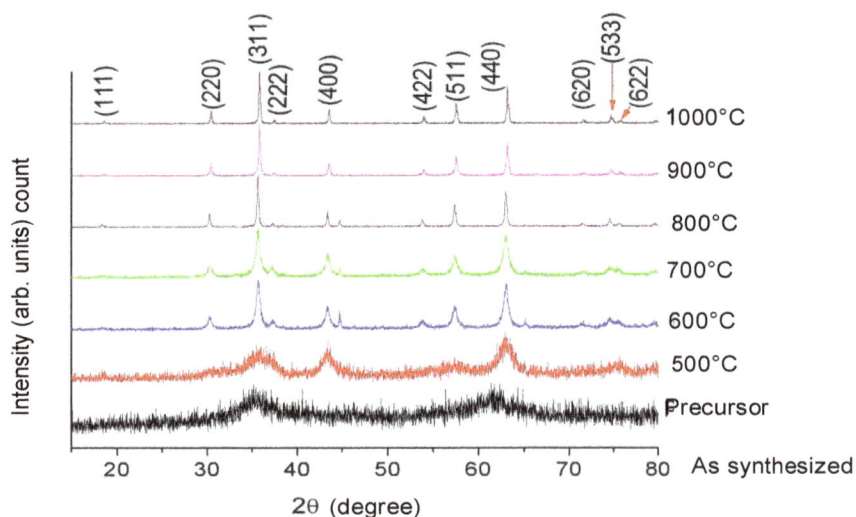

**Figure 1.** XRD Patterns of NiFe$_2$O$_4$ nanoparticles for different temperatures.

mono-dispirsity nanoparticles. Those are characterizing which are very important in application.

## SYNTHESIS PROCEEDING

Nickel ferrite nanoparticles (NiFe$_2$O$_4$) has been prepared by using co-precipitation method. Chloride salts (FeCl$_3$ and NiCl$_2$) was used as starting materials for iron and nickel sources, respectively. All chemicals were analytical grade from Merck Company. The oleic acid also is used as capping agent. Each of salts dissolved in double distilled water separately. We have used 0.2 and 0.4 M solutions from nickel and iron chloride, respectively. Then the previous solution was added to each other.

Sodium hydroxide solution (3 M) was added to mixture solution drop wise till PH received close to 13. Finally, 3 drop of oleic acid is added as a surfactant to the previous solution. Then the temperature is increased up to 80°C for 40 min. We have centrifuged and washed precipitation with double distilled water and ethanol several times. The precipitation was dried in oven at 80°C for several hours. Now we have got amorphous NiFe$_2$O$_4$ nanoparticles and also additional process is used for getting crystalline powder of nickel ferrite nanoparticles.

## RESULTS AND DISCUSSION

### X-ray diffraction (XRD) analysis

We have taken XRD patterns from as synthesized and heated samples at different temperatures. The XRD system which we have used is Xpert Philips model, made in Holand. Source of X-ray was Cu $_{k\alpha}$ with wavelength 1.54 A°. The step of scanning is 0.02° with speed of a step per second.

Figure 1 shows XRD patterns of seven samples that the calcinations temperatures are between 500 to 1000°C. It is found that width of peaks decreasing when calcinations temperature increased. This indicates that particle size increasing when temperature increasing.

The grain size of particle for sample that calcinated at 500 and 1000°C obtained 7 and 82 nm by using Sherer's formula, respectively. All crystalline size calculation have been obtained using "Xpert HighScore plus" software.

The critical grain size of NiFe$_2$O$_4$ is 10 nm for transition from ferromagnetism to superparamagnetic materials (John and Abdul, 2010). We have found that two samples are (heated at 400 and 500°C) obtained grain size less than 10 nm by calculations of Scherer's formula. One can see very nice agreement in size of particles. Size of particles for 600°C calcination temperature is obtained around 23 nm from shere's formula and Transmission electron microscopes (TEM) image.

### Fourier transform infrared spectroscopy (FT-IR) analysis

Two peaks were shown at 3448.10 and 1638.23 cm$^{-1}$ in spectrum (Figure 5) related to O-H as reported at literature (Santi et al., 2007; de Paiva et al., 2009). Presence of 3752.00 to 3650.59 stretching modes corresponding to CO$_3^{2-}$ and No$_3^-$ bonds in which have very low intensity. The stretching modes at position of 574.00 and 422 cm$^{-1}$ are showing Fe-O and Ni-O stretching modes, which indicate formation of NiFe$_2$O$_4$ nanoparticles (Figure 2).

### Vibrating sample magnetometer (VSM) analysis

For particles with large sizes multi-domain are there and becoming more bulk-like with increasing size. When particle size reduces, magnetic domains from multi transfer to a single domain. Thus, below a critical particle size domain walls will no longer form due to energy considerations and single domain particles are stable.

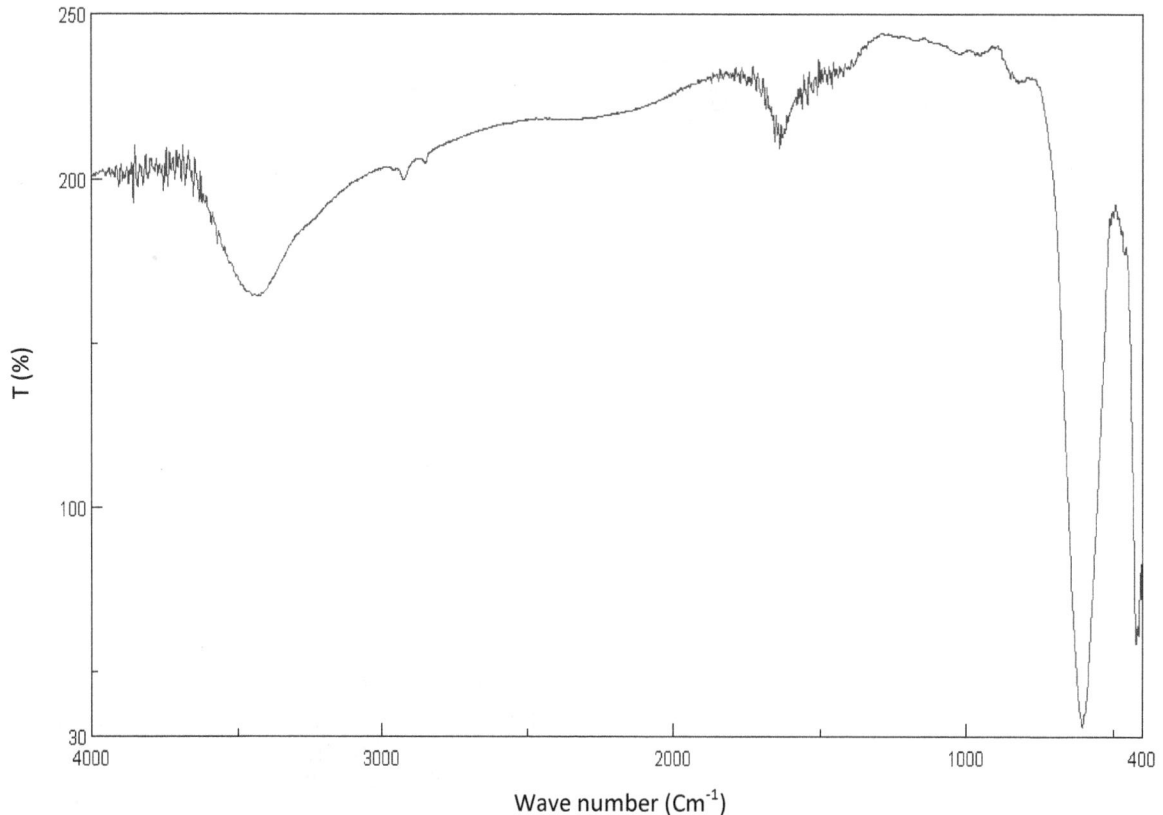

**Figure 2.** FT-IR spectrum for sample that heated at 800°C.

This critical size corresponds to the peak in the coercivity. The particles are then superparamagnetic. The superparamagnetic size strongly depends on the magnetocrystalline anisotropy of the material. In ferromagnetic and ferrimagnetic materials when size of particles decreeing, the particle transfer from multi domain to single domain and transfer to superparamagnetic (Sellmyer and Skomski, 2006; Cullity and Graham, 2009).

Some of samples are calcined at different temperatures (400, 500, 800 and 1000°C), conditions for all the samples were same except calcinations temperatures. The hysteresis loops show (Figure 3) a good magnetization. Hysteresis loops according 400 and 500°C with particle size less than 8 nm that is less than critical grain size, show superparamagnetic properties that are meaning magnetic remanence ($M_r$) and coercive force ($H_c$) are zero.

**Transmission electron microscopes (TEM) and energy dispersive X-ray (EDX) and scanning electron microscopy (SEM) analysis**

The TEM system, which we have been used for morphology and size determination was JEOL JEM-2100 FTEM model. The SEM system that has been used for

morphology of sample was CAMSCAN MV2300 model with 15 KV applied voltage.

Figure 4 shows the morphology of particles. Photograph has been taken from the samples which were calcinated at 600°C. Particle size is obtained around 23.0 nm with monodisperesed nanoparticles as one can see from the photograph. In comparison to grain size of particles from XRD results, the sizes are matching well. The voltage range which we have used was between 160 to 200 KV. The EDX pattern also is taken from the same sample (600°C). The model of this system is CAMSCANMV 2300 and 15 KV was applied. Figure 5 shows that sample is very pure and there is no impurity in the sample.

**Conclusions**

In this research work, pure nickel ferrite nanoparticles in the ranges of 7 to 82 nm is obtained, calcinations samples show that size of particles increase when calcinations temperature increase. Crystallinity of samples also increases with high temperature calcinations. Calculation of size from Sherer's formula and TEM image show a good agreement.

The VSM graphs show a good magnetization for $NiFe_2O_4$ nanoparticle and also the samples which heated

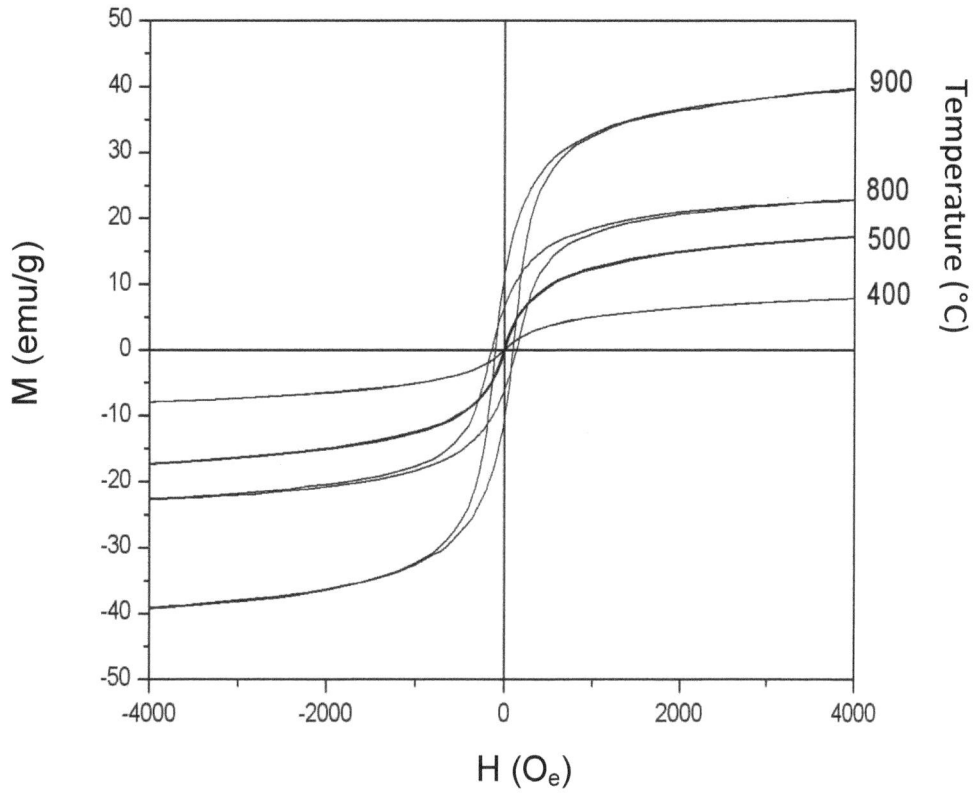

**Figure 3.** Hysteresis lops for different sizes of NiFe$_2$O$_4$ nanoparticles.

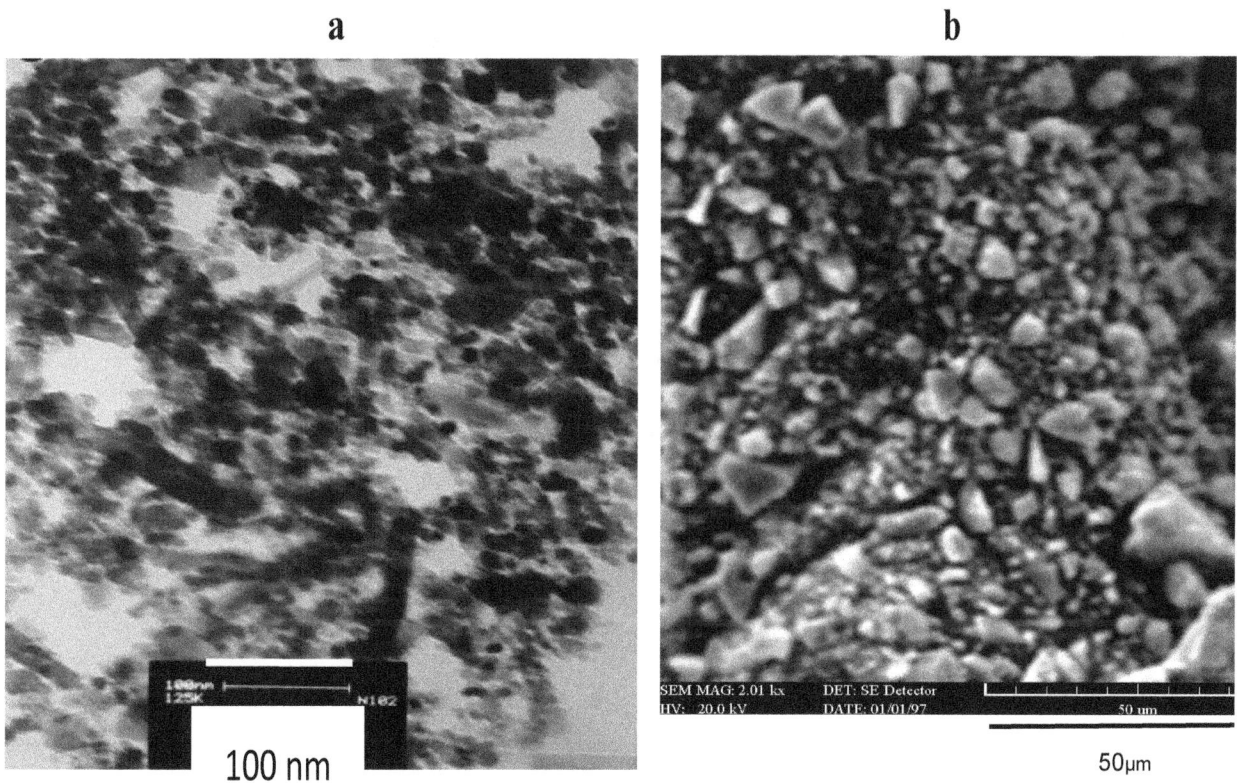

**Figure 4.** a. TEM image for sample which heated at 600°C. b. SEM image of NiFe$_2$O$_4$ heated at 600°C.

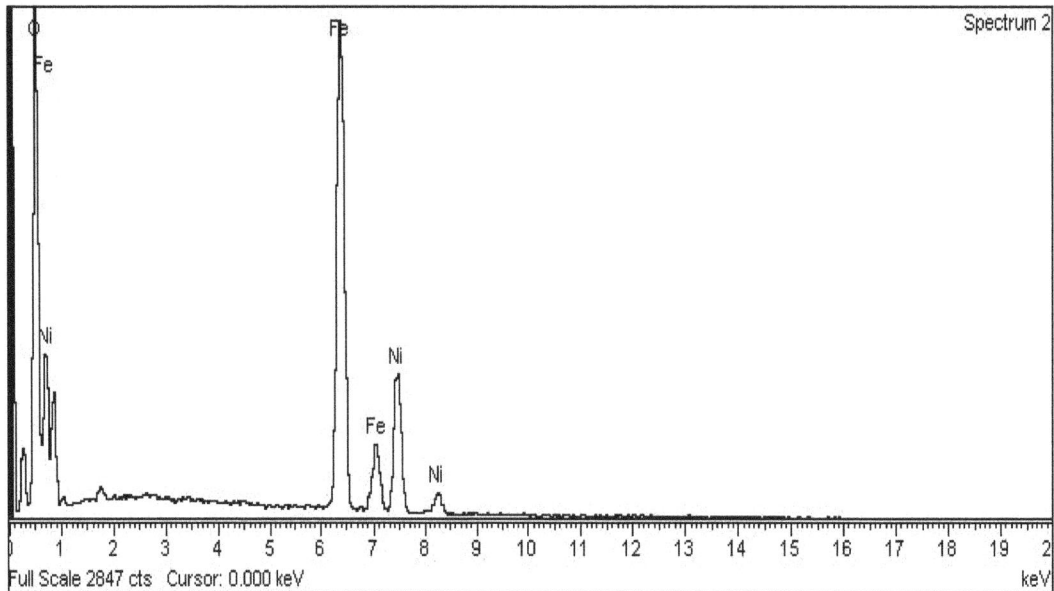

**Figure 5.** EDS of sample which heated at 600°C.

at 400 and 500°C show a superparamagnetic property. FT-IR spectrum also shows that $NiFe_2O_4$ nanoparticle has been prepared properly.

## ACKNOWLEDGEMENT

The work was supported by University of Sistan and Baluchestan of Iran and authors are grateful for the supports from the University of Sistan and Baluchestan.

## REFERENCES

Abdullah CB, Sadan Ozcanb, Nic C, Ismat Shah S (2008). Solid state reaction synthesis of NiFe2O4 nanoparticles. J. Magn. Magn. Mater. 320:857-863.

Billas IML, Chatelain A, de Heer WA (1994). Magnetism from the Atom to the Bulk in Iron, Cobalt and Nickel Clusters. Science, 265: 1682.

Cullity BD, Graham CD (2008). Introduction to Magnetic Materials, Second Edition, John Wiley & Sons, Inc., Hoboken, NJ, USA.

de Paiva JAC, Grac MPF, Monteiro J, Macedo MA, Valente MA (2009). Spectroscopy studies of NiFe2O4 nanosized powders obtained using coconut water. J. Alloys Compd. 485:637-641.

John J, Abdul KM (2010). Investigation of mixed spinel structure of nanostructured nickel ferrite. J. Appl. Phys. 107:114310

Maaz K, Karim K, Mumtaz A, Hasanain SK, Liu J, Duan JL (2009). Synthesis and magnetic characterization of nickel ferrite nanoparticles prepared by co-precipitation route. J. Magn. Magn. Mater. 321:1838-1842.

Martinez B, Obradors X, Balcells Ll, Rouanet A, Monty C (1998). Low Temperature Surface Spin-Glass Transition in γ- Fe2O3 Nanoparticles Phys. Rev. Lett. 80:181.

Misra RDK, Gubbala S, Kale A, Egelhoff Jr. WFA (2004). comparison of the magnetic characteristics of nanocrystalline nickel, zinc, and manganese ferrites synthesized by reverse micelle technique. Mater. Sci. Eng. B 111:164-174

Nathani H, Gubbala S, Misra RDK (2005). Soft magnetic material (NiFe2O4) particles synthesized by solvent co-precipitation method. Sci. Eng. B 121:126.

Santi M, Chivalrat M, Banjong B, Supapan S (2007). A simple route to synthesize nickel ferrite (NiFe2O4) nanoparticles using egg White. Scripta Mater. 56:797-800.

Sellmyer DJ, Skomski R (2006). Introduction to Advanced Magnetic Nanostructures" (2006). Faculty Publications: Materials Research Science and Engineering Center. Paper 28.

# 10

# Magneto hydrodynamic (MHD) squeezing flow of a Casson fluid between parallel disks

Naveed Ahmed[1], Umar Khan[1], Sheikh Irfanullah Khan[1,2], Yang Xiao-Jun[3], Zulfiqar Ali Zaidi[1,2] and Syed Tauseef Mohyud-Din[1]

[1]Department of Mathematics, Faculty of Sciences, HITEC University, Taxila Cantt, Pakistan.
[2]COMSATS Institute of Information Technology, University Road, Abbottabad, Pakistan.
[3]College of Science, China University of Mining and Technology, Xuzhou, Jiangsu, 221008, China.

Squeezing flow of a Casson fluid is considered between two parallel disks. Upper disk is taken to be impermeable but capable of moving towards or away from the lower fixed and porous disk. Governing equations are derived with the help of conservation laws combined with suitable similarity transforms. Homotopy analysis method (HAM) is then been employed to determine the solution to resulting ordinary differential equation. Numerical solution is also obtained using R-K 4 method and comparison shows an excellent agreement between both solutions. Effects of different physical parameters on the flow are also discussed with the help of graphs along with comprehensive discussions.

Key words: Casson fluid, homotopy analysis method (HAM), squeezing flow, parallel disks, magneto hydrodynamic (MHD) flow, numerical solution.

## INTRODUCTION

Squeezing flow between parallel disks has been an active field of research. Its biological and industrial applications have attracted many researchers towards its study. Numbers of efforts have been made to understand such types of flows in more depth. Motion of pistons is vital for running engines and machines. Squeezing flow under the influence of moving disk is also involved in nasogastric tubes and syringes. Better understanding of these flows leads us to more efficient and effective machines which may be used for both industrial and biological purposes.

After the foundational directions provided by Stefen (1874), many researchers investigated the squeezing flow problems (Reynolds, 1886; Archibald, 1956; Grimm, 1976; Wolfe, 1965; Kuzma, 1968; Tichy and Winer, 1970; Jackson, 1962; Hughes and Elco, 1962). As in most of

the cases fluids under consideration are non-Newtonian hence due to complex nature of these fluids different mathematical models are used to study their flow. For blood type fluids Mrill et al. (1965) and McDonald (1974) depicted a most compatible model known as Casson fluid.

Later Domairry and Aziz (2009) considered the flow of an electrically conducting fluid between two parallel disks of which lower disk is permeable and fluid can enter or exit through it during suction or injection process; upper disk is taken to be impermeable and it moves towards the lower disk with a certain time dependent velocity. They applied homotopy perturbation method (HPM) to approximate the solution. Due to inherent nonlinearities in Navier Stokes equations, exact solution in most of the cases is unlikely, therefore, different approximation

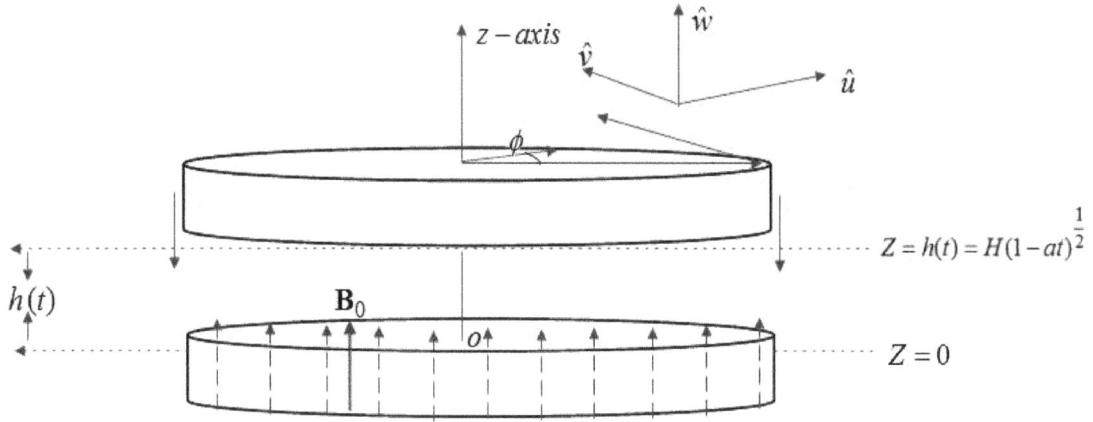

**Figure 1.** Schematic diagram of the problem.

techniques are used to approximate the solution analytically (Abbasbandy, 2007a, b; Abdou and Soliman, 2005a; Noor and Mohyud-Din, 2007; Abdou and Soliman, 2005b; Asadullah et al., 2013; Noor et al., 2008; Mohyud-Din et al., 2009; Nadeem et al., 2012). One of these analytical methods is homotopy analysis method (HAM) that has been effectively applied by different researchers to various nonlinear problems (Liao, 2003; Liao, 2004; Abbasbandy and Zakaria, 2008; Abbasbandy, 2007c; Tan and Abbasbandy, 2008; Hussain et al., 2012; Zeeshan et al., 2012; Hayat et al., 2003; Hayat et al., 2004; Khan et al., 2008; Hayat et al., 2009; Ellahi et al., 2010; Ellahi, 2013; Ellahi, 2012; Ellahi et al., 2012; Hayat et al., 2006).

In this paper, squeezed flow of magneto hydrodynamic (MHD) flow of a non-Newtonian Casson fluid is presented. The governing nonlinear partial differential equations are reduced to a much simpler nonlinear ordinary differential equation by employing a similarity transform. The reduced equation is solved by HAM and effects of emerging parameters are demonstrated graphically coupled with comprehensive discussions. A numerical solution is also carried out by using Runge-Kutta fourth order method to check the validity of analytical solution. An excellent agreement among the solutions is observed.

**MATHEMATICAL ANALYSIS**

Consider MHD incompressible flow of a Casson fluid between parallel infinite disks separated by a distance $h(t) = H(1-at)^{1/2}$. A magnetic field proportional to $B_0(1-at)^{1/2}$ is applied perpendicular to the disks. Based on the assumption of low Reynolds number, the induced magnetic field is neglected. The upper disk at $z = h(t)$ is moving with velocity $\dfrac{aH(1-at)^{-1/2}}{2}$ towards or away from the stationary lower disk at $z = 0$. The physical configuration is presented in Figure 1. Rheological equation of Casson fluid is defined as follows (Nadeem et al., 2012):

$$\tau_{ij} = \left[ \mu_B + \left( \frac{p_y}{2\pi} \right)^{1/n} \right]^n 2e_{ij}, \tag{1}$$

$\mu_B$ is dynamic viscosity of the non-Newtonian fluid, $p_y$ is yield stress of fluid and $\pi$ is the product of component of deformation rate with itself, that is, $\pi = e_{ij}e_{ij}$, where $e_{ij}$ is the $(i, j)$th component of the deformation rate.

We have chosen the cylindrical coordinates system $(r, \phi, z)$. Due to the rotational symmetry of the flow $(\partial/\partial\phi = 0)$, the azimuthal component $v$ of the velocity $V = (u, v, w)$ vanishes identically. Thus, the governing equation for unsteady two-dimensional flow and heat transfer of a Casson fluid are:

$$\frac{\partial \hat{u}}{\partial r} + \frac{\hat{u}}{r} + \frac{\partial \hat{w}}{\partial z} = 0 \tag{2}$$

$$\rho\left( \frac{\partial \hat{u}}{\partial t} + \hat{u}\frac{\partial \hat{u}}{\partial r} + \hat{w}\frac{\partial \hat{u}}{\partial z} \right) = -\frac{\partial \hat{p}}{\partial r} + \mu\left(1 + \frac{1}{\beta}\right)\left( 2\frac{\partial^2 \hat{u}}{\partial r^2} + \frac{2}{r}\frac{\partial \hat{u}}{\partial r} + \frac{\partial^2 \hat{u}}{\partial z^2} + \frac{\partial^2 \hat{u}}{\partial r \partial z} - 2\frac{\hat{u}}{r^2} \right) - \frac{\sigma}{\rho}B^2(t)\hat{u}$$

$$\tag{3}$$

$$\rho\left(\frac{\partial \hat{w}}{\partial t}+\hat{u}\frac{\partial \hat{w}}{\partial r}+\hat{w}\frac{\partial \hat{w}}{\partial z}\right)=-\frac{\partial \hat{p}}{\partial z}+\mu\left(1+\frac{1}{\beta}\right)\left(\frac{\partial^2 \hat{w}}{\partial r^2}+2\frac{\partial^2 \hat{w}}{\partial z^2}+\frac{1}{r}\frac{\partial \hat{w}}{\partial r}+\frac{1}{r}\frac{\partial \hat{u}}{\partial z}+\frac{\partial^2 \hat{u}}{\partial z^2}\right) \tag{4}$$

The boundary conditions are (Domairy and Aziz, 2009):

$$\hat{u}=0, \quad \hat{w}=\frac{dh}{dt} \quad \text{at} \quad z=h(t)$$
$$\hat{u}=0, \quad \hat{w}=-w_0 \quad \text{at} \quad z=0. \tag{5}$$

In the above equations, $\hat{u}$ and $\hat{w}$ are the velocity components in the $r$- and $z$ - directions respectively, $\rho$ is density, $\mu$ dynamic viscosity, $\hat{p}$ pressure, $v$ kinematic viscosity and $w_0$ is the suction/injection velocity.

Substituting the following transformations (Domairy and Aziz, 2009):

$$\hat{u}=\frac{ar}{2(1-at)}f'(\eta), \quad \hat{w}=-\frac{aH}{\sqrt{1-at}}f'(\eta),$$
$$B(t)=\frac{B_0}{\sqrt{1-at}}, \quad \eta=\frac{z}{H\sqrt{1-at}}. \tag{6}$$

into Equations 2 and 3 and eliminating the pressure gradient from the resulting equations, we finally obtain

$$f_{n+1}(\eta)=A+\frac{1}{2}A_1\eta^2+\frac{1}{6}A_2\eta^3$$
$$-\left(\frac{\beta+1}{\beta}\right)\int_0^\eta \frac{(\eta-s)^3}{3!}\left(\begin{array}{c}-S(3f_n'''(s)+sf_n'''(s)-2f_n(s)f_n'''(s))\\-M^2f_n(s)f_n'''(s)\end{array}\right)ds, \tag{7}$$

with the boundary conditions

$$f(0)=A, \quad f'(0)=0,$$
$$f(1)=\frac{1}{2}, \quad f'(1)=0. \tag{8}$$

Where $S$ denotes the squeeze number, $A$ the suction/blowing parameter and $M$ is the Hartman number, defined as:

$$S=\frac{aH^2}{2v}, \quad M^2=\frac{aB_0^2H^2}{v}. \tag{9}$$

## SOLUTION PROCEDURE

Zero order deformation problem

Following the procedure proposed by Liao (2003), it is forthright to choose following initial guess:

$$f_0(\eta)=A+\frac{1}{2}(3-6A)\eta^2+(-1+2A)\eta^3. \tag{10}$$

Linear operator is selected as:

$$L_f=\frac{df^4}{d\eta^4}. \tag{11}$$

Above operator satisfies the following property:

$$L_f(C_1+C_2\eta+C_3\eta^2+C_4\eta^3)=0, \tag{12}$$

where $C_i(i=1-4)$ are the constants. Zero order deformation problem can now be constructed as follows (Noor et al., 2008):

$$(1-q)L_f[\tilde{f}(\eta,q)-f_0(\eta)]=qhN_f[\tilde{f}(\eta,q)], \tag{13}$$

$$\tilde{f}(0,q)=A, \quad \tilde{f}'(0,q)=0, \quad \tilde{f}(1,q)=\frac{1}{2}, \quad \tilde{f}'(1,q)=0, \tag{14}$$

where $q\in[0,1]$ is an embedding parameter and $h$ is nonzero auxiliary parameter. Nonlinear operator is

$$N_f[\tilde{f}(\eta,q)]$$
$$=\frac{\partial^4 \tilde{f}(\eta,q)}{\partial \eta^4}-\left(\frac{\beta}{1+\beta}\right)\left\{S\left(\eta\frac{\partial^3 \tilde{f}(\eta,q)}{\partial \eta^3}+3\frac{\partial^2 \tilde{f}(\eta,q)}{\partial \eta^2}-2\tilde{f}(\eta,q)\frac{\partial^3 \tilde{f}(\eta,q)}{\partial \eta^3}\right)\right.$$
$$\left.-M^2\frac{\partial^2 \tilde{f}(\eta,q)}{\partial \eta^2}\right\}. \tag{15}$$

For $q=0$ and $q=0$ we have

$$\tilde{f}(\eta,0)=f_0(\eta), \qquad \tilde{f}(\eta,1)=f(\eta). \tag{16}$$

$\tilde{f}(\eta,p)$ can be expressed as a Taylor's series in terms of $q$, that is,

$$\tilde{f}(\eta,q)=f_0(\eta)+\sum_{m=1}^{\infty}f_m(\eta)q^m, \quad f_m(\eta)=\frac{1}{m!}\frac{\partial^m f(\eta,q)}{\partial \eta^m}\bigg|_{p=0}. \tag{17}$$

Substituting $q=1$ in above equation we obtain

$$\tilde{f}(\eta,1)=f_0(\eta)+\sum_{m=1}^{\infty}f_m(\eta). \tag{18}$$

$m^{th}$-order deformation problem

$m$ times differentiation of zero order problem depicted by Equation 13 and setting $q=0$ leads to

$$L_f[\tilde{f}_m(\eta)-\chi_m f_{m-1}]=h\Re_m^f(\eta), \tag{19}$$

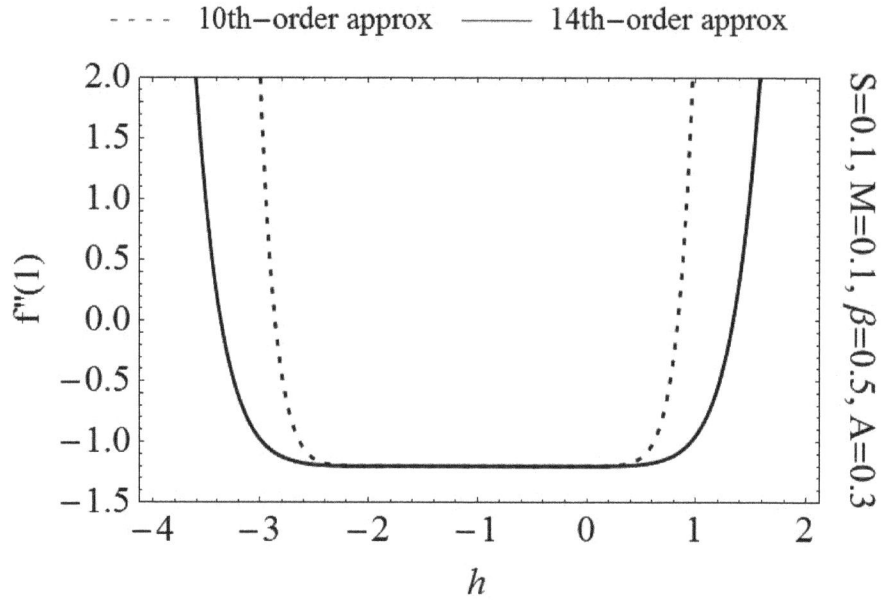

**Figure 2.** *h* curves for the function *f* for different orders of approximations.

Where

$$\Re_m^f(\eta) = f_{m-1}''' - \left(\frac{\beta}{1+\beta}\right)\left\{S\left(\eta f_{m-1}'' + 3f_{m-1}''\right) - M^2 f_{m-1}'' + 2S\sum_{k=0}^{m-1} f_{m-1-k}'' f_k''\right\},$$

(20)

$$\chi_m = \begin{cases} 0, & m \le 1 \\ 1, & m > 1 \end{cases}$$

(21)

The general solution of Equation 19 is

$$f_m(\eta) = f_m^\circ(\eta) + C_1^m + C_2^m \eta + C_3^m \eta^2 + C_4^m \eta^3,$$

(22)

where $f_m^\circ(\eta)$ represents the special solution; also

$$C_1^m = -f_m^\circ(0), \quad C_2^m = -\left.\frac{\partial f_m^\circ(\eta)}{\partial \eta}\right|_{\eta=0}$$

$$C_3^m = -3C_1 - 2C_2 - 3f_m^\circ(1) + \left.\frac{\partial f_m^\circ(\eta)}{\partial \eta}\right|_{\eta=1}$$

(23)

$$C_4^m = 2C_1 + C_2 + 2f_m^\circ(1) - \left.\frac{\partial f_m^\circ(\eta)}{\partial \eta}\right|_{\eta=1}$$

Above higher order solution can be substituted to Equation 18 to obtain the final solution.

### Convergence of the solution

Obtained series solution given by Equation 18 contains an auxiliary parameter $h$. As pointed out by Liao (2003), this parameter is the key to control convergence of series solution. Acceptable range of $h$ can be determined by identifying the line segment of so called $h$-curves which is parallel to $h$ axis. In our particular problem this can be achieved by seeing the range in which $f''(1)$ bears the same magnitude for any value of $h$ within that range. Figure 2 is displayed to demonstrate convergence region for two orders of approximations namely 10th and 14th. It clearly shows that the acceptable region of $h$ is to be between -2.4 and 0.4.

### RESULTS AND DISCUSSION

Acceptable range for auxiliary parameter $h$ has been discussed in previous section. In our analysis and discussions we use $h = -0.9$ as an optimal value of $h$. After ensuring the convergence of series solution our concern now is to see the influences of suction/blowing parameter $A$, squeeze number $S$, Hartmann number $M$ and Casson fluid parameter $\beta$ on velocity is examined. For convenience, we divide our discussions into two parts; one dedicated to investigate the upshots on varying physical parameter for the case suction ($A > 0$) and the other one describes the same effects for the case of blowing ($A < 0$).

### Suction case

Effects of increasing suction at lower disk on both axial and radial velocities are displayed in Figures 3 and 4

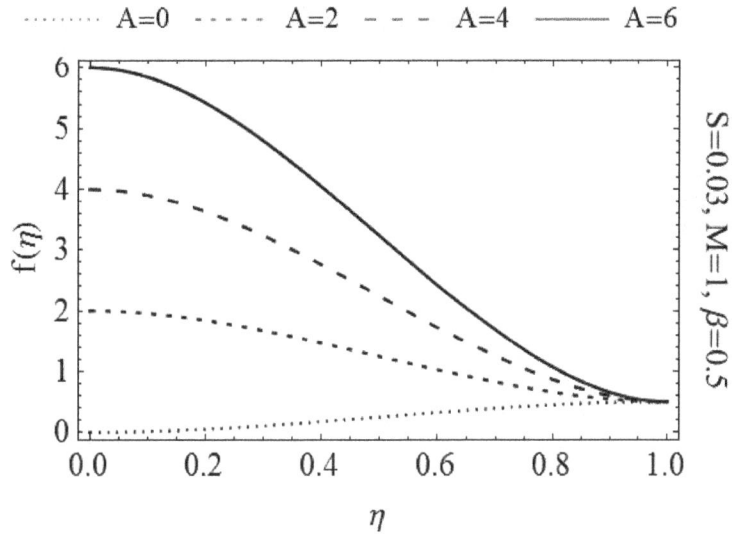

**Figure 3.** Effects of *A* on axial velocity.

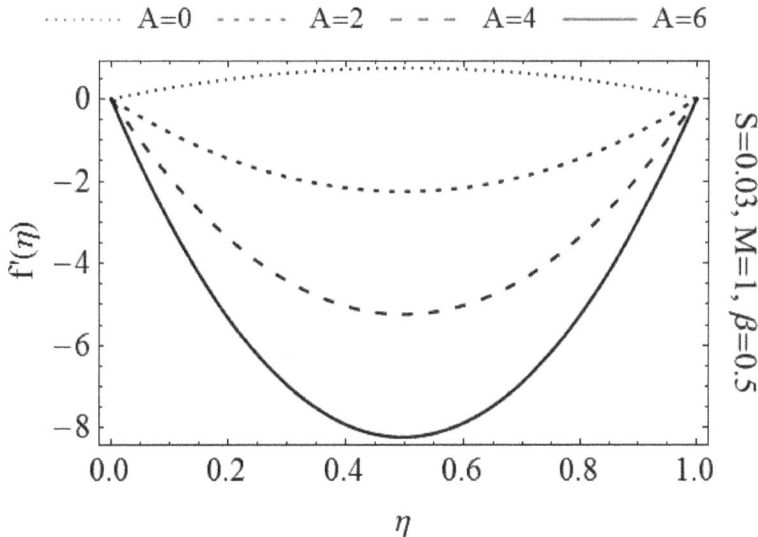

**Figure 4.** Effect of *A* on radial velocity.

respectively. It is evident that increasing value of A results in higher absolute values of both the velocities. As increasing suction allows more fluid to flow near the lower disk therefore a decrease in boundary layer thickness is expected.

Influences of squeeze parameter S on axial and radial velocities are displayed in Figures 4 and 5 respectively. Here S<0 corresponds to the movement of upper disk towards lower one. On the other hand, S>0 describes away movement of the same disk. It can be seen from Figure 5 that for squeezing motion of upper disk

combined with suction axial velocity near the center is increased while for dilating motion a decrease in axial velocity is observed. From Figure 6 one can see the behavior of radial velocity for same variations in S. It is evident for expanding motion; an accelerated radial flow is observed near the upper disk however this trend changes gradually as we move away from it. Somewhere near the center this trend gets converted into an opposite one; that is, from that point to lower disk a delayed motion is observed. For contracting motion of upper disk combined with suction at lower disk effects of increasing

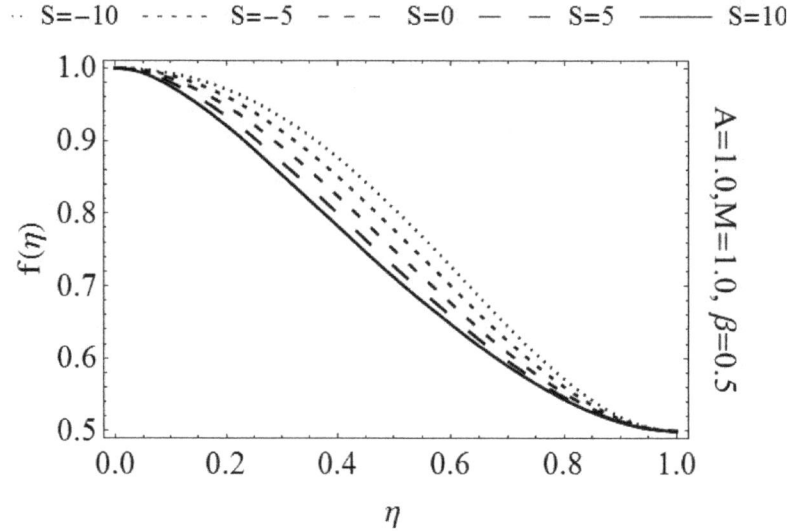

**Figure 5.** Effects of *S* on axial velocity.

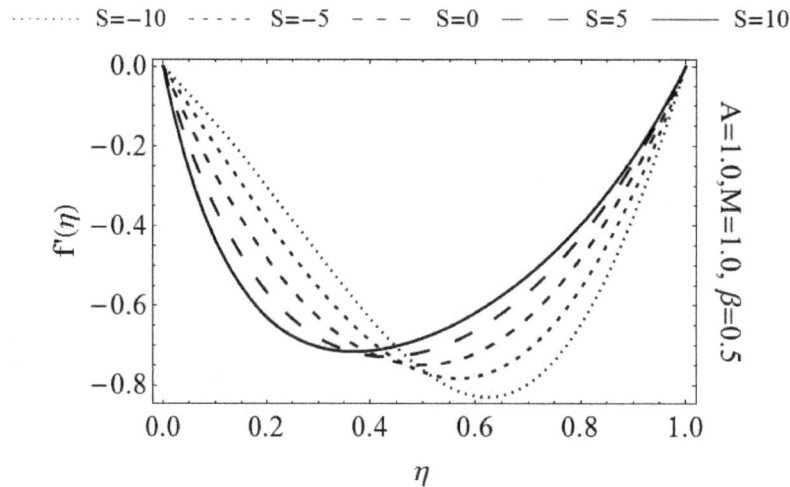

**Figure 6.** Effect of *S* on radial velocity.

absolute values of *S* are quite opposite to the case of expanding motion. In this case, radial velocity near upper disk decreases while near the lower disk an accelerated radial flow is observed.

Graphical results describing flow behavior under increasing Hartmann number *M* are displayed in Figures 7 and 8. From Figure 7, it can be seen that the axial velocity $f(\eta)$ is a decreasing function of M. As apparent from Figure 8, absolute value of radial velocity $f'(\eta)$ increases near the disks while in center, it behaves oppositely. It is also worth mentioning that effect near upper disk is more prominent as compared to lower one. Figures 9 and 10 respectively are dedicated to display

behavior of axial and radial velocity for increasing Casson parameter $\beta$. Axial velocity is a decreasing function of $\beta$ as shown in Figure 9. Effects of $\beta$ on axial velocity are more visible in central region as compared to the area near disks. Furthermore, $\beta \to \infty$ gives us the flow of viscous fluid. Figure 10 shows that the radial velocity near lower disk decreases with rising $\beta$. However, after moving some distance away from the lower disk this behavior changes into an opposite one; that is, after $\eta > 0.4$ we observe an accelerated radial flow. One may also see that effects of $\beta$ near the disks are very slight as compared to the region far from them.

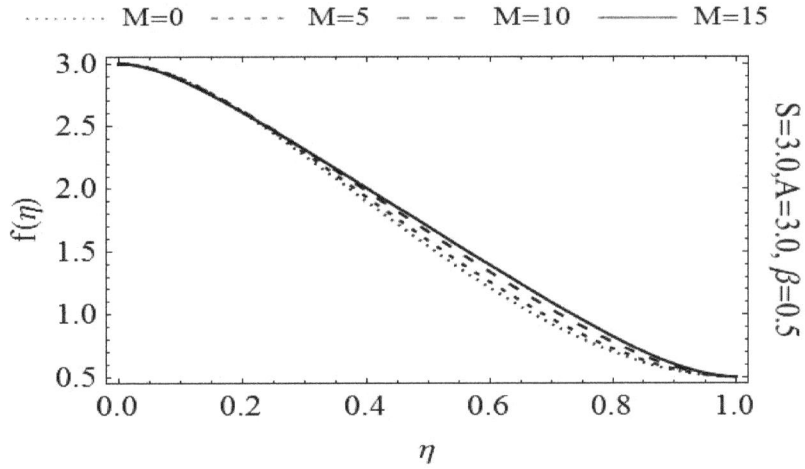

**Figure 7.** Effects of *M* on axial velocity.

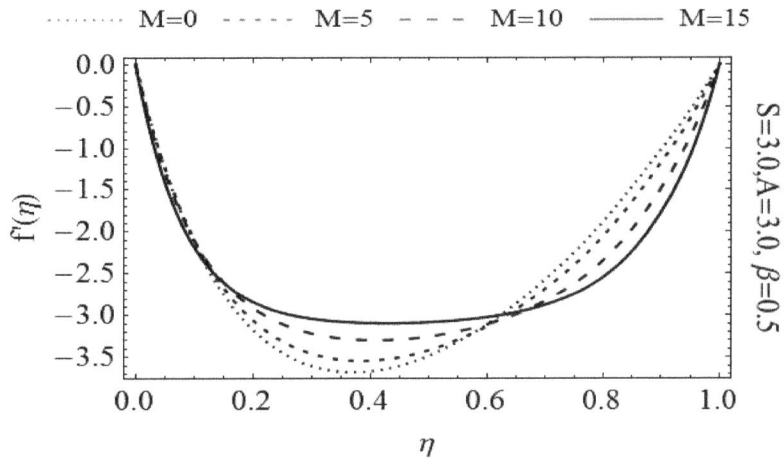

**Figure 8.** Effect of *M* on radial velocity.

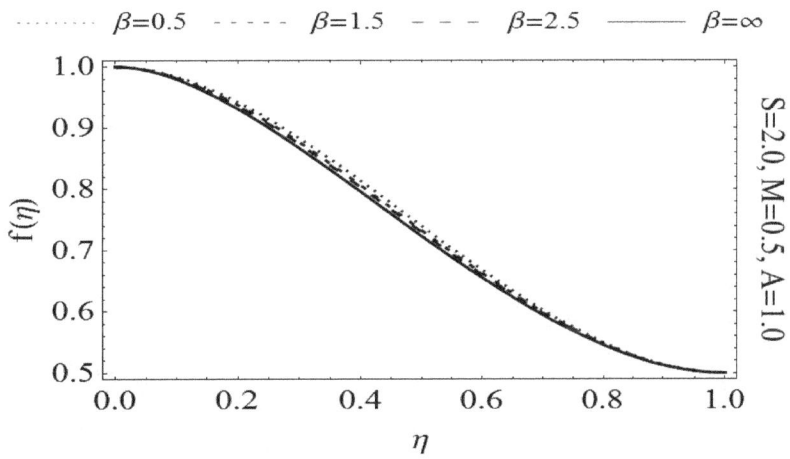

**Figure 9.** Effects of $\beta$ on axial velocity.

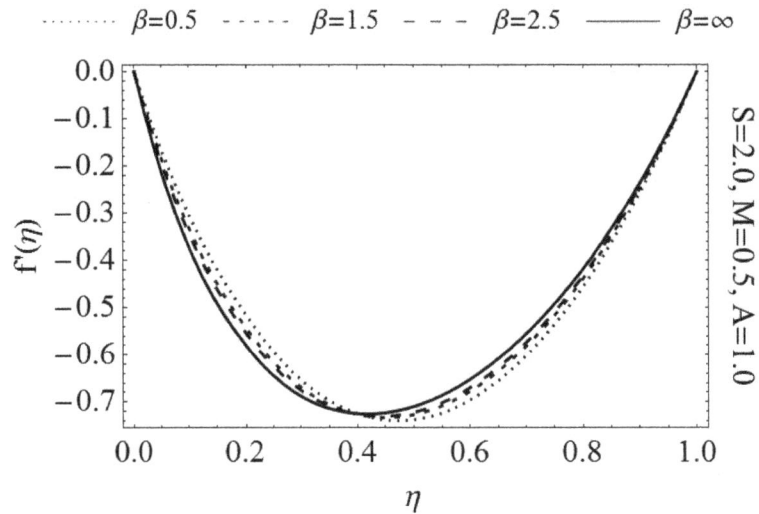

**Figure 10.** Effect of $\beta$ on radial velocity.

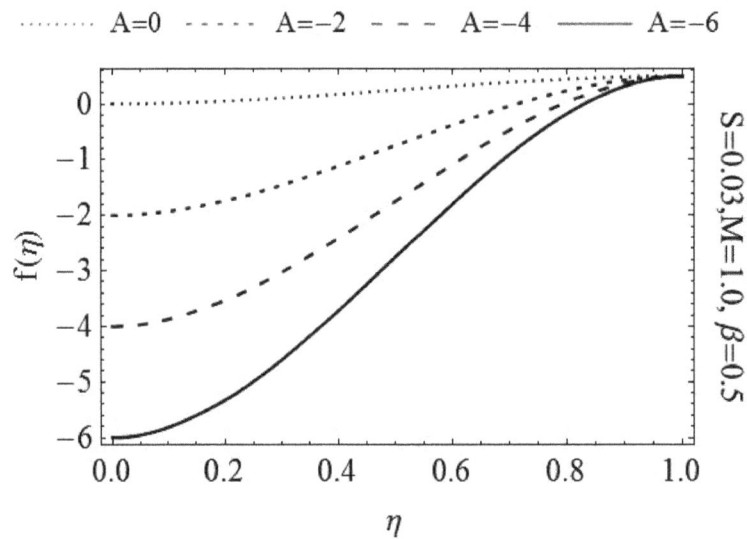

**Figure 11.** Effects of $A$ on axial velocity.

### Blowing case

Here we discus influences of involved physical parameter in the case when blowing occur at the lower disk. Figures 11 and 12 declare that the influence on increasing injection leads to increased absolute values of both axial and redial velocities. Figures 13-18 show that the effects of $S$, $M$ and $\beta$ on both axial and radial flow are opposite in blowing case as compared to the ones obtained for suction case. Same problem is solved numerically by using a well-known RK-4 method. Comparison for is presented in Table 1. It can be observed that both numerical and analytical solutions are in excellent agreement.

### Conclusion

Squeezing flow between parallel disks is presented. Homotopy analysis method (HAM) has been employed to obtain analytical solution to the problem. Influences of emerging flow parameters are discussed in detail with the help of graphs. It is also concluded that the effects of physical parameter on axial and redial velocities are quite

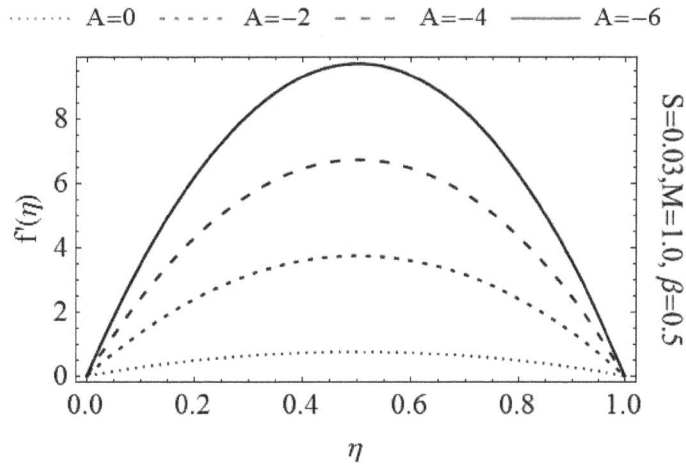

**Figure 12.** Effect of *A* on radial velocity.

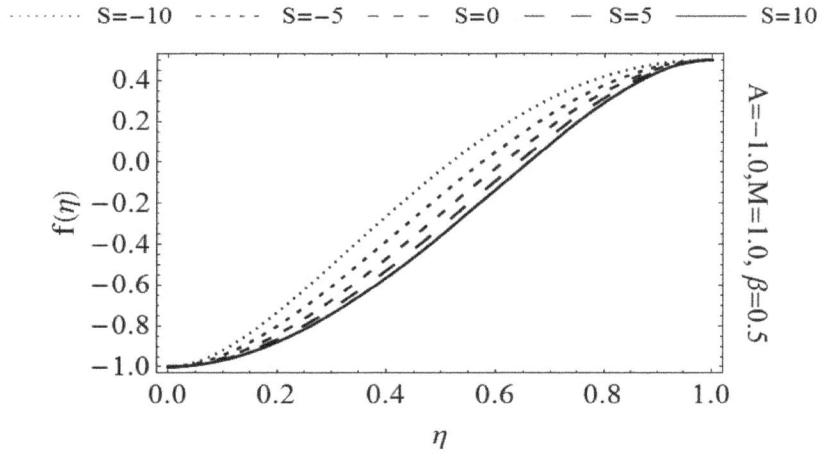

**Figure 13.** Effects of *S* on axial velocity.

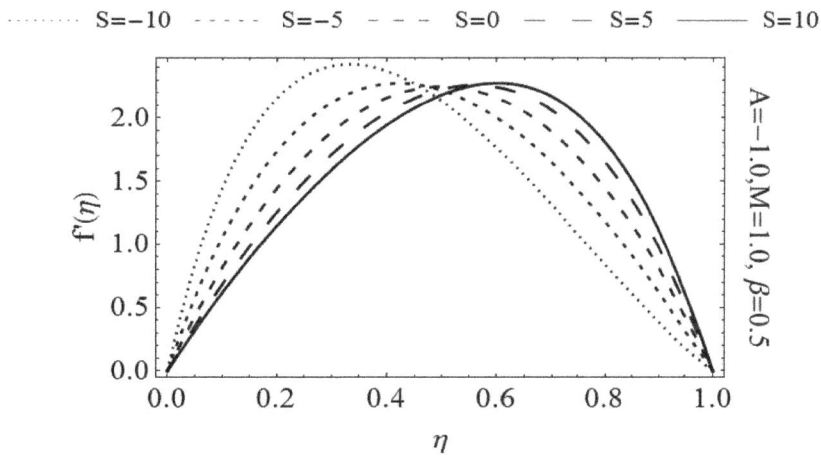

**Figure 14.** Effect of *S* on radial velocity.

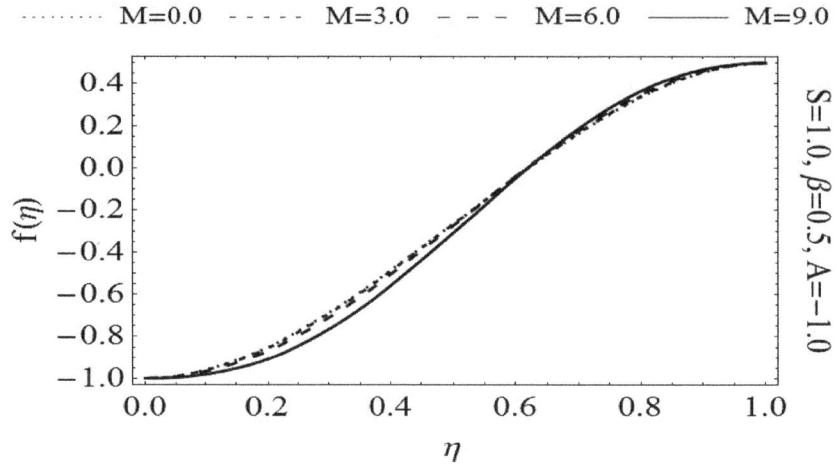

**Figure 15.** Effects of *M* on axial velocity.

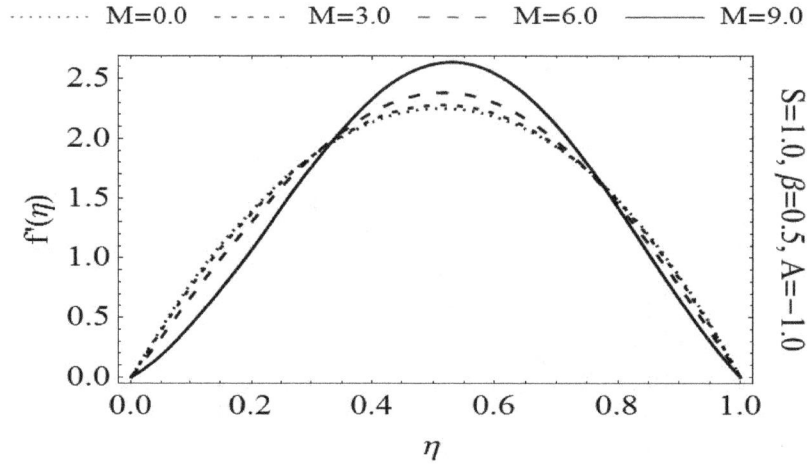

**Figure 16.** Effect of *M* on radial velocity.

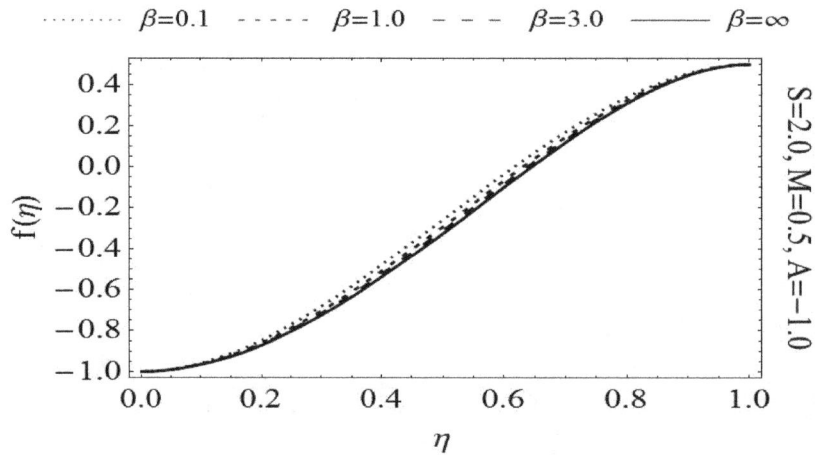

**Figure 17.** Effects of $\beta$ on axial velocity.

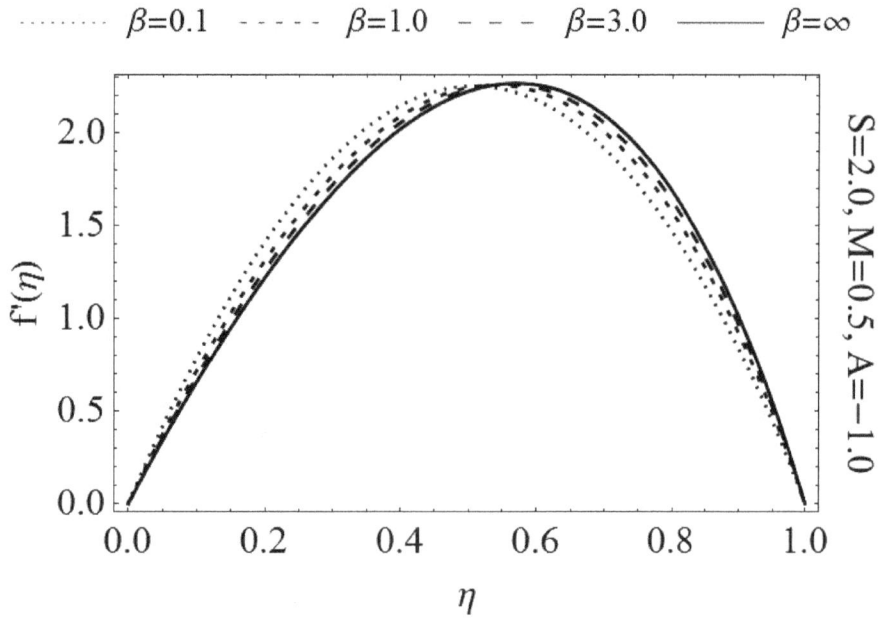

**Figure 18.** Effect of $\beta$ on radial velocity.

**Table 1.** Comparison of HAM and numerical solutions for $S = 2.0, A = 1.0, M = 0.5, h = -0.9, \beta = 0.5.$

| $\eta$ | $f(\eta)$HAM | Numerical | $f'(\eta)$HAM | Numerical |
|--------|--------------|-----------|---------------|-----------|
| 0 | 1.000000 | 1.000000 | 0 | 0 |
| 0.1 | 0.984012 | 0.984012 | -0.303368 | -0.303368 |
| 0.2 | 0.942321 | 0.942321 | -0.516753 | -0.516753 |
| 0.3 | 0.883192 | 0.883192 | -0.654099 | -0.654099 |
| 0.4 | 0.813695 | 0.813695 | -0.725515 | -0.725515 |
| 0.5 | 0.740051 | 0.740051 | -0.737995 | -0.737995 |
| 0.6 | 0.667911 | 0.667911 | -0.695962 | -0.695962 |
| 0.7 | 0.602598 | 0.602598 | -0.601667 | -0.601667 |
| 0.8 | 0.549306 | 0.549306 | -0.455454 | -0.455454 |
| 0.9 | 0.513283 | 0.513283 | -0.255919 | -0.255919 |
| 1.0 | 0.500000 | 0.500000 | 0 | 0 |

opposite in the cases of suction to the blowing. A numerical solution using well known R-K 4 method has also been obtained for the sake of comparison. It is found that the results agree exceptionally well.

## REFERENCES

Abbasbandy S (2007a). The application of homotopy analysis method to solve a generalized Hirota-Satsuma coupled KdV equation, Phys. Lett. A. 361:478–483.

Abbasbandy S (2007b). A new application of He's variational iteration method for quadratic Riccati differential equation by using Adomian's polynomials, J. Computational Appl. Math. 207:59-63.

Abbasbandy S (2007c). Numerical solutions of nonlinear Klein-Gordon equation by variational iteration method, Int. J. Num. Methods Eng. 70:876-881.

Abbasbandy S, Zakaria FS (2008). Soliton solutions for the fifth-order K-dVEquation with the homotopy analysis method, Nonlinear Dynamics, 51:83–87.

Abdou MA, Soliman AA (2005a). New applications of variational iteration method, Physica D. 211(1-2):1-8.

Abdou MA, Soliman AA (2005b). Variational iteration method for solving Burger's and coupled Burger's equations. J.Computational Appl. Math. 181:245-251.

Archibald FR (1956). Load capacity and time relations for squeeze films. J. Lubrication Technol. 78:A231–A245.

Asadullah M, Khan U, Ahmed N, Manzoor R, Mohyud-Din ST (2013). Int. J. Modern Math. Sci. 6:92-106.

Domairy G, Aziz A (2009), Approximate Analysis of MHD Squeeze Flow between Two Parallel Disks with Suction or Injection by Homotopy Perturbation Method, Mathematical Problems in Engineering, article ID/2009/603916.

Ellahi R (2012). A Study on the Convergence of Series Solution of Non-Newtonian Third Grade Fluid with Variable Viscosity: By Means of Homotopy Analysis Method, Adv. Math. Phys. Article ID 634925, 11 pages.

Ellahi R (2013). The effects of MHD and temperature dependent viscosity on the flow of non-Newtonian nanofluid in a pipe: Analytical solutions, Appl. Math. Modeling, 37:1451-1467.

Ellahi R, Hayat T, Mahomed FM, Zeeshan A (2010). analytical solutions for MHD flow in an annulus, Commun. Nonlin. Sci. Numer. Simul. 15:1224-1227.

Ellahi R, Raza M, Vafai K (2012).Series solutions of non-Newtonian nanofluids with Reynolds' model and Vogel's model by means of the homotopy analysis method, Mathe. Computer Modelling, 55:1876-1891.

Grimm RJ (1976). Squeezing flows of Newtonian liquid films: an analysis includes the fluid inertia. Appl. Sci. Res. 32(2):149-166,

Hayat T, Ellahi R, Ariel PD, Asghar S (2006). Homotopy Solution for the Channel Flow of a Third Grade Fluid, Nonlinear Dynamics, 45:55-64.

Hayat T, Ellahi R, Asghar S (2004). Unsteady periodic flows lows of a magnetohydrodynamic fluid due to noncoxial rotations of a porous disk and a fluid at infinity, Math. Computer Modelling, 40:173-179.

Hayat T, Ellahi R, Mahomed FM (2009). The Analytical Solutions for Magnetohydrodynamic Flow of a Third Order Fluid in a Porous Medium, Zeitschriff Naturforsch, 64a:531-539.

Hayat T, Mumtaz S, Ellahi R (2003). MHD unsteady periodic flows due to non-coaxial rotations of a disk and a fluid at infinity, Acta Mathematica Sinica, 19:235-240.

Hughes WF, Elco RA (1962). Magnetohydrodynamic lubrication flow between parallel rotating disks. J. Fluid Mechanics, 13:21-32.

Hussain A, Mohyud-Din ST, Cheema TA (2012). Analytical and Numerical Approaches to Squeezing Flow and Heat Transfer between Two Parallel Disks with Velocity Slip and Temperature Jump, Chinese Phys. Lett. 29:114705.

Jackson JD (1962). A study of squeezing flow, Appl. Sci. Res. A. 11:148-152.

Khan I, Ellahi R, Fetecau C (2008). Some MHD Flows of a Second Grade Fluid through the Porous Medium. J. Porous Media, 11:389-400.

Kuzma DC (1968). Fluid inertia effects in squeeze films. Appl. Sci. Res. 18:15-20.

Liao SJ (2003). Beyond perturbation: introduction to the Homotopy Analysis Method, CRC Press, Boca Raton, Chapman and Hall, 2003.

Liao SJ (2004). On the homotopy analysis method for nonlinear problems, Appl. Math. Computation, 147:499-513.

McDonald DA (1974). Blood Flows in Arteries, 2nd ed. Arnold, London

Mohyud-Din ST, Noor MA, Waheed A (2009). Variation of parameter method for solving sixth-order boundary value problems, Communication Korean Math. Soc. 24:605-615.

Mrill EW, Benis AM, Gilliland ER, Sherwood TK, Salzman EW (1965). Pressure flow relations of human blood hollow fibers at low flow rates. J. Appl. Physiol. 20:954-967.

Nadeem S, Haq UIR, Lee C (2012). MHD flow of a Casson fluid over an exponentially shrinking sheet, Scientia Iranica, 19:1150-1553.

Noor MA, Mohyud-Din ST (2007). Variational iteration technique for solving higher order boundary value problems, Appl. Math. Computation, 189:1929-1942.

Noor MA, Mohyud-Din ST, Waheed A (2008). Variation of parameters method for solving fifth-order boundary value problems. Appl. Math. Infor. Sci. 2:135 -141.

Reynolds O (1886). On the theory of lubrication and its application to Mr. Beauchamp Tower's experiments, including an experimental determination of the viscosity of olive oil, Philosophical Transactions of the Royal Society of London, 177:157-234.

Stefen MJ (1874). Versuch¨ Uber die scheinbare adhesion, Sitzungsberichte der Akademie der Wissenschaften in Wien. Mathematik-Naturwissen, 69:713-721.

Tan Y, Abbasbandy S (2008). Homotopy analysis method for quadratic Riccati differential Equation, Communications Nonlinear Sci. Num. Simulation, 13:539-546.

Tichy JA, Winer WO (1970). Inertial considerations in parallel circular squeeze film bearings, J. Lubrication Technol. 92:588-592.

Wolfe WA (1965). Squeeze film pressures. Appl. Sci. Res. 14:77-90.

Zeeshan A, Ellahi R, Siddiqui AM, Rahman HU (2012). An investigation of porosity and magnetohydrodynamic flow of non-Newtonian nanofluid in coaxial cylinders, Int. J. Phys. Sci. 7(9):1353-1361.

# 11

# Interface energy of solid cadmium (Cd) phase in the Cd-Pb (Cadmium-Lead) binary alloy

## Fatma Meydaneri[1]*, Buket Saatçi[2] and Ahmet Ülgen[3]

[1]Department of Metallurgy and Materials Engineering, Faculty of Engineering, Karabük University, 78050, Karabük, Turkey.
[2]Department of Physics, Faculty of Arts and Sciences, Erciyes University, 38039, Kayseri, Turkey.
[3]Department of Chemistry, Faculty of Arts and Sciences, Erciyes University, 38039, Kayseri, Turkey.

The grain boundary groove shapes for the Cadmium (Cd) phase in equilibrium with the Lead (Pb)-Cd eutectic liquid were obtained with radial heat flow apparatus. From the obtained grain boundary groove shapes, the Gibbs-Thomson coefficient, solid-liquid and grain boundary energy for the Cd phase in equilibrium with the Pb-Cd eutectic liquid were determined to be $(8.35 \pm 0.41) \times 10^{-8}$ Km, $(128.00 \pm 12.8) \times 10^{-3}$ Jm$^{-2}$ and $(239.33 \pm 26.32) \times 10^{-3}$ Jm$^{-2}$, respectively. The thermal conductivities of the solid and liquid phases at the eutectic composition and temperature were also measured by using radial heat flow and Bridgman-type directional growth apparatus, respectively.

**Key words:** Crystal growth, phase equilibria, thermal conductivity, Gibbs-Thomson coefficient, grain boundary energy.

## INTRODUCTION

Despite being hazardous substances cadmium (Cd), Lead (Pb) and their alloys are sometimes used technologically. Cd is a Group IIB element, while Pb is in group IVB of the periodic table. Cd is extensively used in the electroplating and battery industries. Pb is of significant relevance in making cable covering, plumbing, ammunition and as a sound absorber. Eutectic alloys are the basis of most engineering materials, and have relatively low melting points, excellent fluidity, and good mechanical properties. Knowledge of the thermodynamics of materials provides fundamental information about the stability of phases and about the driving forces for chemical reactions and diffusion processes.

Hence, the purpose of this study is to determine some thermodynamic properties such as thermal conductivity coefficients ($\kappa_S$, $\kappa_L$), Gibbs-Thomson coefficient ($\Gamma$), solid-liquid interfacial energy ($\sigma_{SL}$) and the grain boundary energy ($\sigma_{GB}$) of Cd solution in equilibrium with the Pb-Cd eutectic liquid from the observed grain boundary groove shapes. These data are of great importance for the development of electronic, semi-conductor, superconductor materials and interconnection technologies, especially in microelectronics. The solid-liquid surface energy ($\sigma_{SL}$) is defined as the reversible work required to create a unit area of the interface at constant temperature, volume and chemical potentials (Woodruff, 1973) and it plays a critical role in phase transformations. Theoretical and experimental works have been made to determine the values of $\sigma_{SL}$ in various materials by using different methods for the last 50 years. Unfortunately, it is not easy to measure $\sigma_{SL}$ even for a pure material, and very little progress has been made in its measurement for a multi-component system. One of the most common techniques to determine the solid-liquid interface energy is to use the equilibrated grain boundary groove shapes. Observation of grain boundary groove shape in an alloy is obviously very difficult. The technique was used to measure $\sigma_{SL}$ of

---
*Corresponding authors. E-mail: meydaneri@yahoo.com.

**(a)**

**(b)**

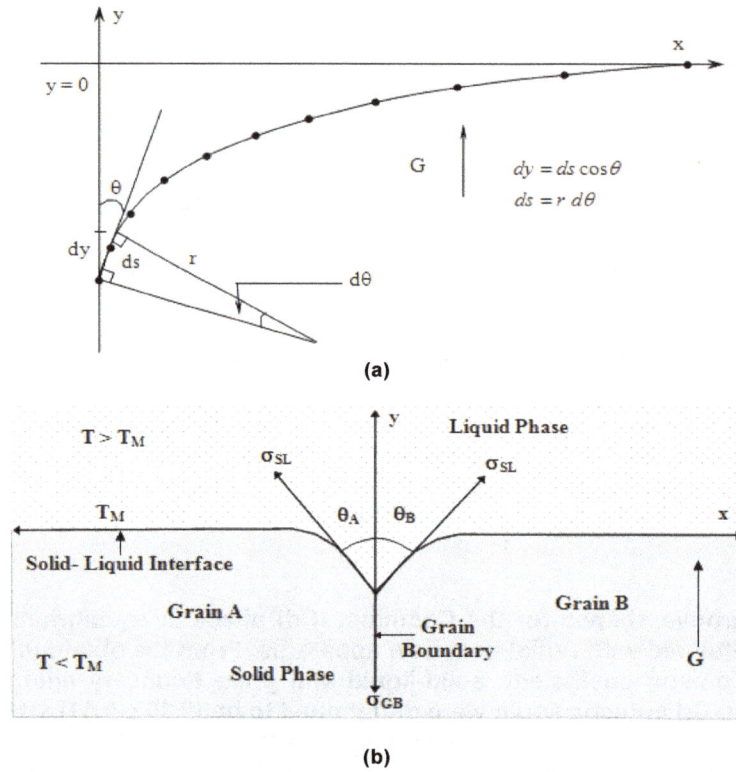

**Figure 1.** (a) Solid-liquid interface in a temperature gradient, showing the x, y coordinates, $\sigma_{SL}$, surface tension, $\sigma_{GB}$, grain boundary tension and θ. (b) Schematic illustration of an equilibrated grain boundary groove.

metallic alloy systems by Gündüz and Hunt, and to equilibrium solid-liquid and solid-solid interface in constant temperature gradient (G). Also, Gündüz and Hunt developed a numeric model to calculate the Gibbs-Thomson coefficient ($\Gamma$) that the distinguishing characteristics of the materials.

The Gibbs-Thomson coefficient ($\Gamma$) is expressed in the form of a change in undercooling $\Delta T_r$ with r the radius of the curvature as:

$$\Delta T_r = \frac{\Gamma}{r} \qquad (1)$$

Equation 1 can be integrated in the y-direction (perpendicular to the interface in 2D) from the flat interface to a point on the curve

$$\int_0^y \Delta T_r \, dy = \Gamma \int_0^y \frac{1}{r} \, dy \qquad (2)$$

The left hand side of Equation 2 may be evaluated if $\Delta T_r$ is known for any y-point. The thermal conductivities of the solid ($\kappa_S$) and liquid ($\kappa_L$) phases are not equal so that the left hand side of Equation 2 can be integrated by

using the numerically calculated $\Delta T_r$ values (Gündüz and Hunt, 1985, 1989). The right hand side of Equation 2 may be evaluated for any shape by defining $ds = r \, d\theta$ (s is the distance along the interface and $\theta$ is the angle of a tangent to the interface with y-axis (as shown in Figure 1), which is obtained by fitting a Taylor expansion to the adjacent points on the curve) giving

$$\Gamma \int_0^y \frac{1}{r} \, dy = \Gamma (1 - \sin\theta) \qquad (3)$$

This allows the Gibbs-Thomson coefficient to be determined for a measured grain boundary groove shape. To obtain accurate values of the Gibbs-Thomson coefficient the shape of the interface, the temperature gradient in the solid ($G_S$) and the conductivities of the solid ($\kappa_S$) and liquid ($\kappa_L$) must be known.

The solid-liquid surface energy is obtained from the thermodynamic definition of the Gibbs-Thomson coefficient which is expressed as:

$$\Gamma = \frac{\sigma_{SL}}{\Delta S^*} \qquad (4)$$

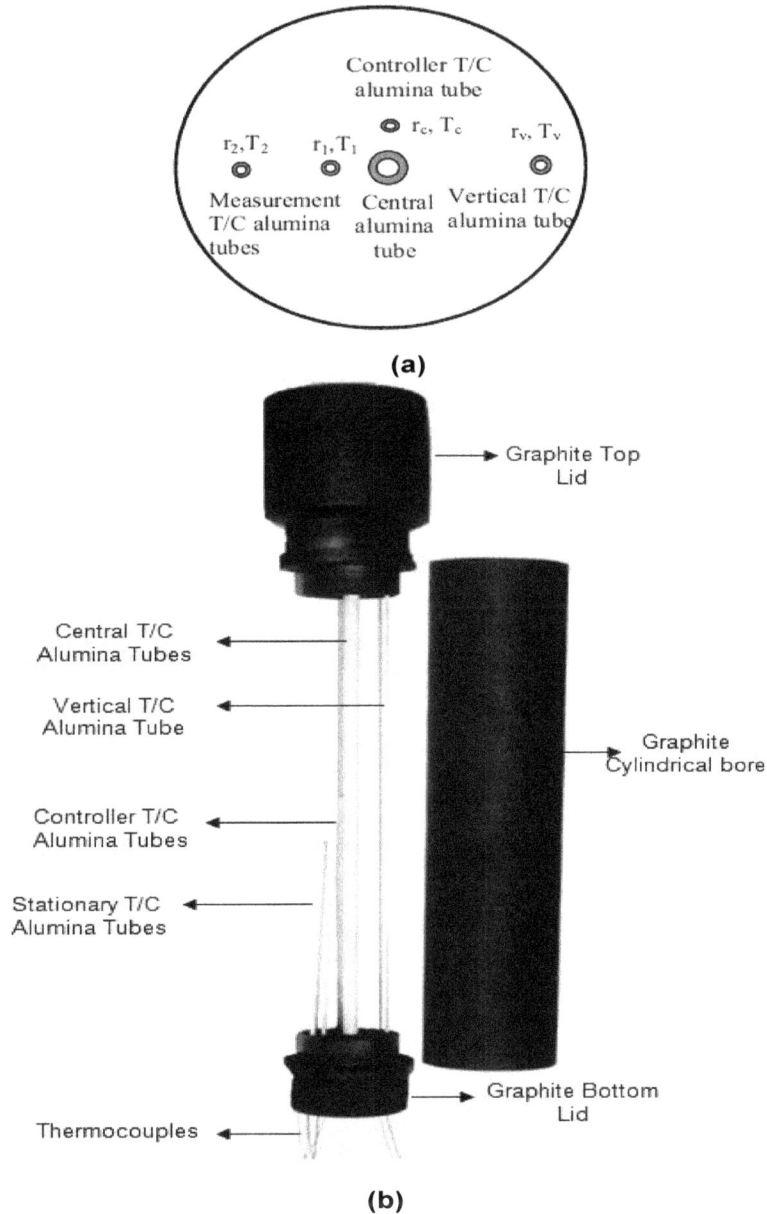

**Figure 2.** (a) Transverse cross-section of specimen. (b) Longitudinal section of alumina tubes and thermocouples placed to graphite crucible.

where $\Delta S^*$ is the effective entropy of melting per unit volume which must be known.

## EXPERIMENTAL PROCEDURE

### The sample crucible

The sample crucible is composed of three parts, a cylindrical bore (30 mm inner diameter (ID), 40 mm outer diameter (OD) and 170 mm in length), and the top and bottom lids which were tightly fitted to the cylindrical part, as shown in Figure 2. Three stationary thermocouples (inserted in a 0.8 mm ID, 1.2 mm OD alumina tube) were fitted on the bottom lid, while two thermocouples ($r_1$ and $r_c$)

were placed 1.0 to 1.5 mm away from the central alumina tube, the other one ($r_2$) was placed about 11 mm away from the central alumina tube. $r_c$ was used for the control unit, and the other two ($r_2 > r_1$) were used for measuring the temperature. A moveable thermocouple (inserted in a 1 mm ID, 2 mm OD alumina tube) was also placed 10 mm away from the center and used for measuring the vertical temperature variation. Also, a thin-walled central alumina tube (2 mm ID, 3 mm OD) was used to isolate the central heating wire (Kanthal A1) on the bottom lid.

### Sample preparation

The phase diagram of Pb-Cd alloy has been examined (Hansen and Anderko, 1985). The composition was determined to be

Cd-20 wt.% Pb to obtain the Cd phase in equilibrium with the eutectic liquid (Cd-17.4 wt.% Pb). (Cd-20 wt.% Pb) alloy was melted in a vacuum furnace by using 4N pure Cd and 4N pure Pb. After several stirrings, the melted alloy was poured into a graphite sample crucible held in a specially constructed casting furnace at approximately 30K above the eutectic temperature, $T_E$, of the alloy (521K). The molten metal was then directionally solidified from bottom to top to ensure that the sample crucible was completely full. The sample was prepared to be placed in the radial heat flow apparatus.

## Radial heat flow apparatus

In order to observe the equilibrated grain boundary groove shapes in eutectic alloy systems, Gündüz and Hunt (1985, 1989) designed a radial heat flow apparatus. Maraşli and Hunt (1996) improved the experimental apparatus for higher temperature. Recently, Maraşli and others studied ternary alloy systems by using the apparatus (Akbulut et al., 2009; Ocak et al., 2010; Aksöz et al., 2011). The details of the apparatus and experimental procedures are given in Maraşli and Hunt (1996), Akbulut et al. (2009), Ocak et al. (2010), Aksöz et al. (2011), Meydaneri et al. (2011, 2012), Saatçi et al. (2007), and Bulla et al. (2007). In the present work, a similar apparatus was used to observe the grain boundary groove shapes of the solid (Cd - 0.25 wt.% Pb) solution in equilibrium with the (Cd-17.4 wt.% Pb) eutectic liquid. A schematic illustration of the radial heat flow apparatus and the block diagram of the experimental system are shown in Figure 3.

## Equilibration of the sample

The central heating wire (Kanthal A1, typically about 1.7 mm diameter and 200 mm in length) was placed inside a thin-walled central alumina tube. The terminal-points of the heating element were threaded and screwed into 7 mm copper rods. The casting sample was placed inside a water cooled jacket, and to increase the experimental sensitivity, the gap between the cooling jacket and the specimen was filled with free running sand. Then, the thermocouples (K type, 0.5 mm thick insulated) were placed into alumina tubes fixed on the bottom lid, and the jacket was placed in the radial heat flow apparatus. The sample was heated from the center by the heating wire and the outside of the sample was kept cool with the water cooling jacket. A thin layer (1 to 3 mm thick) was melted around the central heating wire and the specimen was annealed in a very stable temperature gradient for 18 days to get the equilibrium solid solution (Cd-0.25 wt.% Pb) with the eutectic liquid (Cd-17.4 wt.% Pb). The temperature on the sample was controlled to an accuracy of ±0.01K with a Eurotherm 9706 type controller. The temperature control was procured by measuring the oscillation period (PID) and setting actual time constants. During the annealing period, the temperature in the specimen and the longitudinal temperature variations on the sample were continually recorded by the stationary thermocouples and a moveable thermocouple, respectively, by using a Pico TC-08 data-logger via computer. The potential difference (Vh) between the ends of the central heating element was measured with Hewlett Packard 34401 multimeters. The current (I) passing from the central heating wire was measured with a TES 3012 Digital Clampmeter. The input power was also recorded periodically. The temperature in the sample was stable at about ± 0.01 to 0.02K for hours and ±0.05K for up to 18 days. Effective cooling is very important to obtain a high temperature gradient and a well quenched solid-liquid interface. At the end of the annealing time the specimen was rapidly quenched by turning off the input power.

## Metallography

The equilibrated sample was removed from the furnace and cut transversely into ~20 mm lengths. The transverse sections were ground flat with 800, 1000, and 2400 grid SiC papers, respectively, before mounting. Grinding and polishing were then carried out using the standard techniques. After polishing, the samples were etched with a suitable enchant (5 ml nitric acid + 5 ml acetic acid + 90 ml glycerin) to observe the equilibrated solid-liquid interface.

The equilibrated grain boundary groove shapes which occurred on the solid-liquid interface were carefully photographed with a charge couple device (CCD) digital camera placed on top of an Olympus BH2 light optical microscope using a 40 x objective. A graticule (100 × 0.01 = 1 mm) was also photographed using the same objective. The digital camera had rectangular pixels. Thus, the magnifications in the x and y directions are different. The photographs of the equilibrated grain boundary groove shapes and the graticule in the $x$ and $y$ directions were photographed by using Adobe Photoshop CS2 version software, so that accurate measurements of the groove coordinate points on the groove shapes could be made.

## Geometrical correction for the groove coordinates

In order to obtain accurate $\Gamma$ values by the numerical method, not only the $G$ and $\kappa_S$, values but also the coordinates on the grain boundary grooves must be measured accurately. The coordinates of the cusp, $x$, $y$ should be measured using the coordinates $x$, $y$, $z$ where the $x$-axis is parallel to the solid-liquid interface, the $y$ axis is normal to the solid-liquid interface and the $z$ axis lies at the base of the grain boundary groove as shown in Figure 4a. The coordinates of the cusp $x'$, $y'$ from the metallographic section must be transformed to the $x$, $y$ coordinates. Maraşli and Hunt devised a geometrical method to make appropriate corrections to the groove shapes and the details of the geometrical method are given in Maraşli and Hunt (1996). The relation between $x$ and $x'$ can be expressed as:

$$x = x'\cos\alpha = x'\frac{\sqrt{a^2+d^2}}{\sqrt{a^2+b^2+d^2}} \qquad (5)$$

and the relation between $y'$ and $y$ can be expressed as:

$$y = y'\cos\beta = y'\frac{d}{\sqrt{a^2+d^2}} \qquad (6)$$

where $d$ is the distance between the first and second plane along the $z'$ axis, $b$ is the displacement of the grain boundary position along the $x'$ axis, $a$ is the displacement of the solid-liquid interface along the $y'$ axis, $\alpha$ is the angle between the $x'$ axis and $x$ axis, and $\beta$ is the angle between the $y'$ axis and $y$ axis as shown in Figure 4. In this work, the values of $a$, $b$ and $d$ were measured in order to transform the cusp coordinates $x'$, $y'$ into the $x$, $y$ coordinates as follows.

Two perpendicular reference lines (approximately 0.1 mm thick and 0.1 mm deep) were marked near the grain boundary groove on the polished surface of the sample as shown in Figure 4c. The samples were then polished and the grain boundary groove shapes were photographed. The thickness of the sample $d_1$ was measured with a digital micrometer (resolution of 1 μm) at several points of the sample to obtain the average value. After the thickness measurements had been made the sample was again polished to remove a thin layer (approximately 40 to 50 μm) from the sample surface. The same grain boundary groove shapes were again photographed and the thickness of the sample $d_2$ was measured

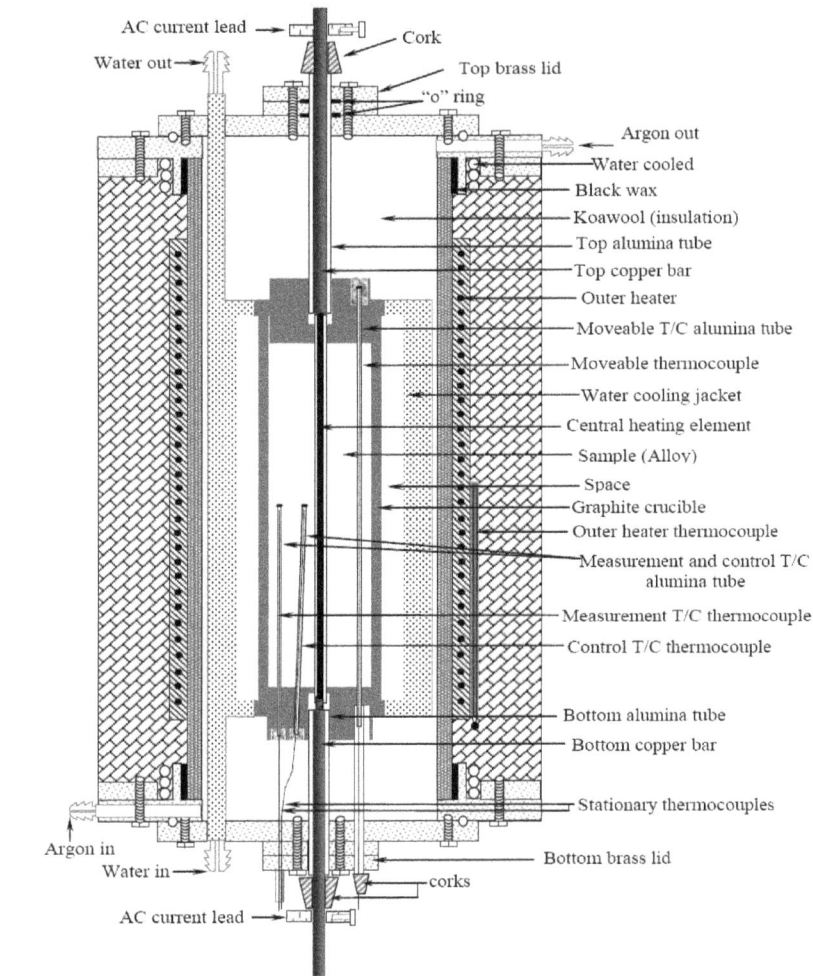

**(a)**

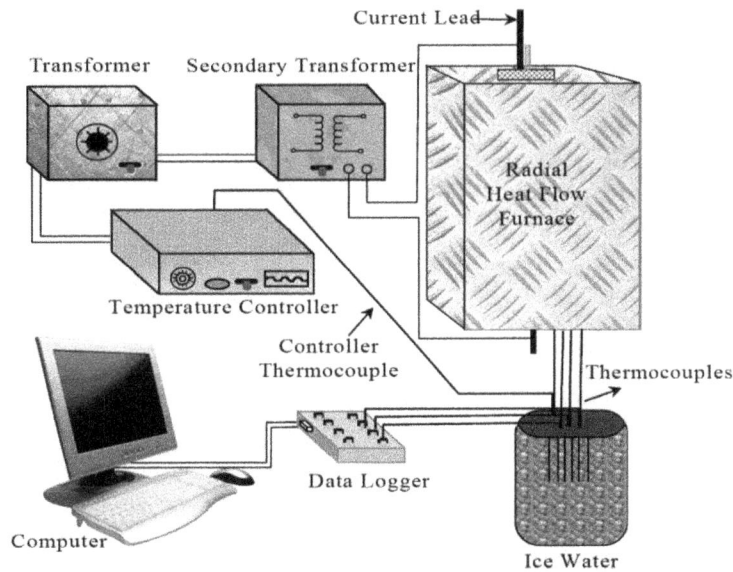

**(b)**

**Figure 3.** (a) Schematic illustration of radial heat flow apparatus. (b) The block diagram of the experimental system.

(a)

(b)

(c)

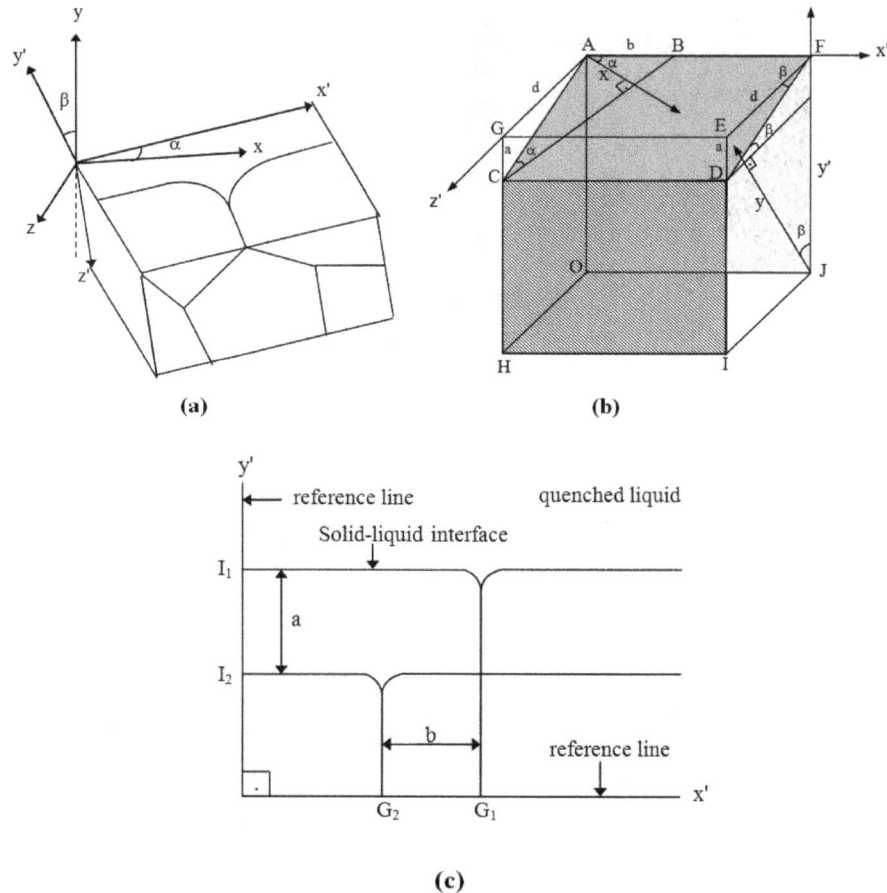

**Figure 4.** (a) Schematic illustration of the relationship between the actual coordinates, $x$, $y$ and the measured coordinates, $x'$, $y'$ of the groove shape. (b) Schematic illustration for the metallic examination of the sample: where B is the location of the grain boundary groove shape onto first plane OJFA, C is the location of the grain boundary groove shape onto second plane HIDC, AB = b, CG = ED = a and AG = d. (c) Schematic illustration of the displacement of the grain boundary groove shape position along the $x'$ and $y'$ axis (Maraşli and Hunt, 1996).

with the same micrometer. The difference between the thickness of the sample, $d = d_1 - d_2$, gave the layer removed from the sample surface. The photographs of the grain boundary groove shapes were mounted on one another using *Adobe Photoshop CS2* version software to measure the displacement of the solid-liquid interface along the $y'$ axis and the displacement of the grain boundary groove position along the $x'$ axis (Figure 4b).

Thus, the required $a$, $b$ and $d$ measurements were made so that appropriate corrections to the shape of the grooves could be deduced. As shown in Equations 5 and 6, if the values of $a$, $b$ and $d$ are measured, then the groove coordinates, $x'$, $y'$ can be transformed into $x$, $y$ coordinates.

**The thermal conductivity of the solid and liquid phases**

The radial heat flow method is the most appropriate technique for measuring the conductivities in the solid phase. The thermal conductivity of the solid Cd solution is also needed to calculate the temperature gradient on the solid phases. The sample crucible was made from graphite to measure the thermal conductivity of the sample, as shown in Figure 2.

The sample was heated up in steps of 50K up to 513K (8K below $T_E$). First, isotherms macroscopically parallel to the axial center of the sample were obtained for the expected temperature, by moving the central heater wire up and down, and the sample was kept at this steady state condition for about 2 h. Then the total input power $Q$ and $T_1$, $T_2$ temperatures were recorded with a *Hewlett Packard 34401* type multimeter, *TES 3012 Digital Clampmeter* and a *Pico TC-08* data-logger at this condition. Finally, the sample was left to cool to room temperature. The cooled sample was moved from the radial heat flow apparatus and was cut transversely near to the measurement points.

The transversely cut samples were ground and polished for $r_1$ and $r_2$ measurements. The distances were measured with an *Olympus BH2* light optical microscope to an accuracy of ±0.01 mm. The transverse and longitudinal sections of the sample were examined for porosity, cracks and casting defects to make sure that these would not introduce any errors to the measurements.

The temperature gradient in the cylindrical specimen is given by Fourier's law

$$G_S = \left(\frac{dT}{dr}\right) = -\frac{Q}{A\kappa_S} \qquad (7)$$

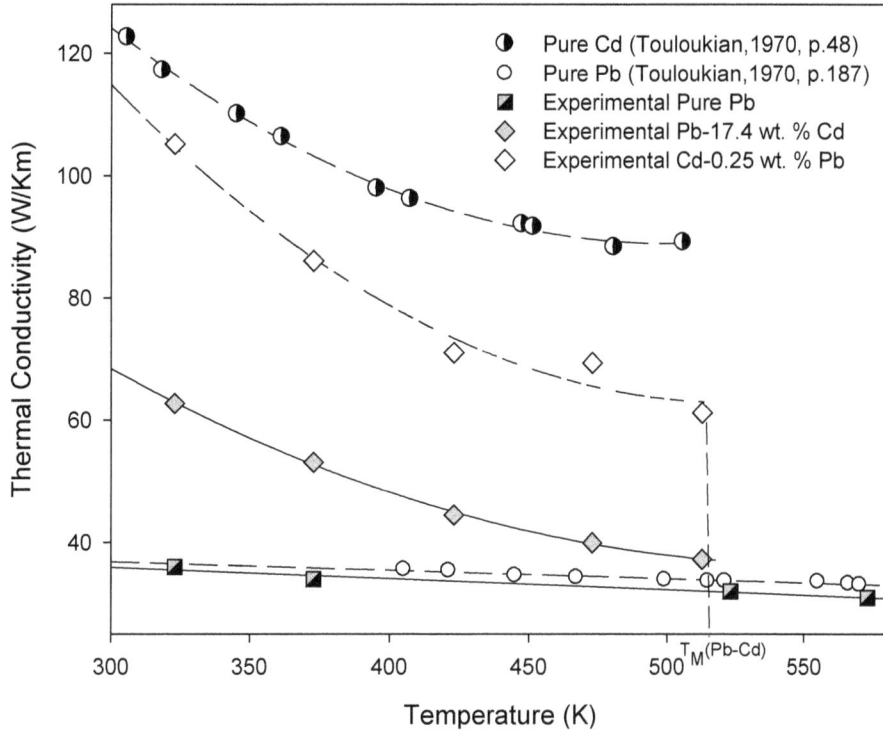

**Figure 5.** Thermal conductivities of solid phases versus temperature for pure Cd (Touloukian et al., 1970, p.48), pure Pb (Touloukian et al., 1970, p. 187), experimental pure Pb, experimental Cd-17.4 wt.% Pb and experimental Cd-0.25 wt.% Pb alloys.

where $Q$ is the total input power from the center of the specimen, $A$ is the surface area of the specimen and $\kappa_S$ is the thermal conductivity of the solid phase. Integration of Equation 7 gives

$$\kappa_S = \frac{1}{2\pi\ell} ln(\frac{r_2}{r_1}) \frac{Q}{(T_1 - T_2)} \qquad (8)$$

where $\ell$ is the length of the heating element, $r_1$ and $r_2$ are the fixed distances from the center of the sample, and $T_1$ and $T_2$ are the temperatures at the fixed positions ($r_2 > r_1$) $r_1$ and $r_2$, respectively. If $Q$, $r_1$, $r_2$, $T_1$ and $T_2$ can be accurately measured for a well-characterized sample, then reliable $\kappa_S$ values can be determined (Erol et al., 2005). The $\kappa_S$ values at the melting temperatures of the materials were obtained by extrapolating the thermal conductivity curves to the melting temperature. The thermal conductivity values of the eutectic solid solution (Cd-17.4 wt.% Pb) and solid solution (Cd-0.25 wt.% Pb), $\kappa_S$, were obtained as 35.94 and 62.60 W/Km, respectively. The thermal conductivity variation of the solid phases for the materials with temperature is shown in Figure 5. The thermal conductivity ratio of the equilibrated eutectic liquid phase (Cd-17.4 wt.% Pb) to the solid solution (Cd-0.25 wt. % Pb), $R = \kappa_L (liquid(Cd - 17.4\,wt.\%\,Pb\,solution))/\kappa_S(\,solid\,(Cd - 0.25\,wt.\%\,Pb))$ must be calculated to obtain the Gibbs-Thomson coefficients with the numerical method. The thermal conductivity ratio can be obtained by a Bridgman-type growth apparatus. The heat flow away from the interface through the solid phase must balance that of the heat flow through the liquid phase plus the latent heat generated at the interface, that is, (Porter and Easterling, 1991)

$$VL = \kappa_S G_S - \kappa_L G_L \qquad (9)$$

where $V$ is the growth rate, $L$ is the latent heat, and $G_S$ and $G_L$ are the temperature gradients in the solid and liquid phases, respectively, and $\kappa_S$ and $\kappa_L$ are the thermal conductivities of the solid and liquid phases, respectively. For low growth rates, $VL \ll \kappa_S G_S$, so that the conductivity ratio, $R$, is given by

$$\kappa_L G_L \cong \kappa_S G_S, \quad R = \frac{\kappa_L}{\kappa_S} = \frac{G_S}{G_L} \qquad (10)$$

The directional growth apparatus, first constructed by McCartney (1981), was used to find the thermal conductivity ratio, $R = \kappa_L / \kappa_S$. A thin-walled graphite crucible, 6.3 mm OD × 4 mm ID × 180 mm length, was used to minimize convection in the liquid phase. Molten eutectic alloy (Cd-17.4 wt.% Pb) was poured into the thin-walled graphite tube and the molten alloy was then directionally frozen from bottom to top to ensure that the crucible was completely full.

The specimen was then placed in the Bridgman directional growth apparatus. The specimen was heated to 30K over the melting temperature of the alloy. The specimen was then left to reach thermal equilibrium for at least 2 h. The temperature in the specimen was measured with insulated K type thermocouples. In the present work, a 1.2 mm OD × 0.8 mm ID alumina tube was used to insulate the thermocouple from the molten alloy, and the thermocouples were placed vertical to the heat flow direction. At the end of equilibration, the temperature in the specimen was stable to ±0.5K for the short term period and to ±1K for the long term period. When the specimen temperature stabilized, the directional growth was begun by turning the motor on. The cooling rate was recorded with a data-logger via computer.

The velocity of the motor used in the present work was

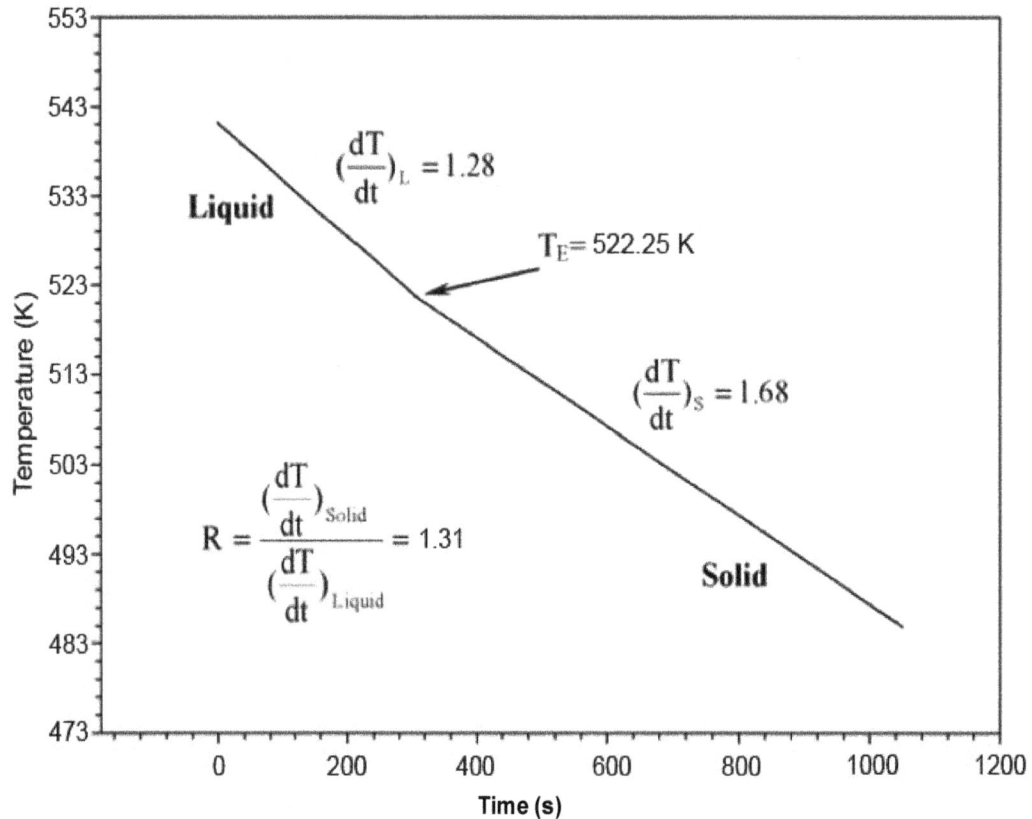

**Figure 6.** Cooling curve for the eutectic Cd-17.4 wt.% Pb alloy.

5 mm/min, while the solid-liquid interface passed the thermocouples, the slope of the cooling rate for the liquid and solid phases was observed. When the thermocouple reading reached approximately 40 to 50K below that of the melting temperature, the growth was stopped by turning the motor off. The thermal conductivity ratio was obtained from the cooling rate ratio of the liquid phase to the solid phase. The cooling rate of the liquid and solid phases is given by:

$$\left(\frac{dT}{dt}\right)_L = \left(\frac{dT}{dx}\right)_L \left(\frac{dx}{dt}\right)_L = G_L V \qquad (11)$$

and

$$\left(\frac{dT}{dt}\right)_s = \left(\frac{dT}{dx}\right)_s \left(\frac{dx}{dt}\right)_s = G_S V \qquad (12)$$

From Equations 11 and 12, the thermal conductivity ratio can be written as:

$$R = \frac{\kappa_L}{\kappa_S} = \frac{G_S}{G_L} = \frac{\left(\frac{dT}{dt}\right)_S}{\left(\frac{dT}{dt}\right)_L} \qquad (13)$$

where $\left(\frac{dT}{dt}\right)_s$ and $\left(\frac{dT}{dt}\right)_L$ values were directly measured from the temperature versus time, and the cooling curve is shown in Figure 6. The thermal conductivity ratio of the eutectic liquid to the eutectic

solid, $R = \kappa_L \left(eutectic\right)/\kappa_S \left(eutectic\right)$ was found to be 1.31. The thermal conductivity value of the eutectic solid solution (Cd-17.4 wt.% Pb), $\kappa_S$, was obtained as 35.94 W/Km. Therefore, the thermal conductivity value of the (Cd-17.4 wt.% Pb) eutectic liquid solution, $\kappa_L$, was obtained as 47.08 W/Km. According to the Pb-Cd phase diagram, the solid solubility of Pb in Cd is very restricted and suggested a solubility (about 0.25 wt.)% Pb at the eutectic temperature of 521K. The thermal conductivity of the $\kappa_S$ (solid Cd-0.25 wt. % Pb) was found to be 62.60 W/Km. The value of $R = \kappa_L \left(liquid(Cd - 17.4 \, wt. \% \, Pb \, solution)\right)/\kappa_S \left(solid \, (Cd - 0.25 \, wt. \% \, Pb)\right)$ was found to be 0.76. The values of thermal conductivity used in the calculations are given in Table 1.

**Measurement of temperature gradient in the solid phase**

At the steady-state the temperature gradient at radius $r$ is given by

$$G_S = \frac{dT}{dr} = -\frac{Q}{2\pi r \ell \kappa_S} \qquad (14)$$

where $Q$ is the input power, $\ell$ is the length of the heating wire, $r$ is the distance to the solid-liquid interface from the center of the sample and $\kappa_S$ is the thermal conductivity of the solid phase. The average temperature gradient of the solid phase must be determined for each grain boundary groove shape. The temperature gradient of the Cd solid phase was calculated by using the measured values in Equation 14 for each grain boundary groove shape (Table 2).

**Table 1.** Thermal conductivity of solid and liquid phases in Cd-Pb alloy system.

| Alloy | Phase | Melting temperature (K) | $\kappa$ (W/Km) | $R = \kappa_L/\kappa_S$ |
|---|---|---|---|---|
| Pb-Cd (Eutectic Composition) | (Solid Phase) Cd-17.4 wt.% Pb | 521 | 35.94 | 1.31 |
|  | (Liquid Phase) Cd-17.4 wt. % Pb |  | 47.08 |  |
| Pb-Cd | (Solid Phase) Cd-0.25 wt. % Pb | 521 | 62.60 | 0.76 |
|  | (Liquid Phase) Cd-17.4 wt. % Pb |  | 47.08 |  |

**Table 2.** Gibbs-Thomson coefficients for the solid Cd phase in equilibrium with the eutectic liquid.

| Groove No. | $\alpha^o$ | $\beta^o$ | $G_S \times 10^2$ (K/m) | Gibbs-Thomson coefficient | |
|---|---|---|---|---|---|
|  |  |  |  | $\Gamma_{LHS} \times 10^{-8}$ (Km) | $\Gamma_{RHS} \times 10^{-8}$ (Km) |
| 1 | 17.06 | 10.87 | 14.98 | 8.55 | 8.69 |
| 2 | 27.07 | 29.88 | 16.31 | 8.51 | 8.73 |
| 3 | 14.95 | 6.10 | 16.08 | 8.15 | 8.85 |
| 4 | 17.23 | 6.05 | 19.45 | 8.49 | 8.64 |
| 5 | 9.17 | 9.47 | 17.17 | 7.56 | 8.56 |
| 6 | 36.33 | 23.41 | 16.19 | 8.76 | 8.65 |
| 7 | 9.74 | 3.68 | 19.38 | 8.87 | 7.21 |
| 8 | 16.42 | 2.50 | 19.53 | 8.11 | 8.87 |
| 9 | 5.54 | 11.41 | 19.28 | 7.50 | 7.95 |
| 10 | 5.27 | 16.94 | 16.21 | 8.37 | 7.99 |

$\overline{\Gamma} = (8.35 \pm 0.41) \times 10^{-8}$ Km.

## RESULTS AND DISCUSSION

### The Gibbs-Thomson coefficient

When the thermal conductivity of the solid phase, $\kappa_S$, and the thermal conductivity of the liquid phase, $\kappa_L$, are not equal, $(\kappa_S \neq \kappa_L)$, the curvature under cooling, $\Delta T_r$, is no longer a linear function of the distance. If the thermal conductivity ratio $(R = \kappa_L/\kappa_S)$ of the equilibrated liquid phase to solid phase, the coordinates of the grain boundary groove shapes, geometric correction factors and the temperature gradients for each grain boundary groove shape are known, the Gibbs-Thomson coefficients $(\Gamma)$ can be obtained using the numerical method. The numerical method is described in detail in Gündüz and Hunt (1985, 1989).

In this study, the values of $\Gamma$ for the solid Cd solution in equilibrium with the eutectic liquid (Cd-17.4 wt.% Pb) were determined by this numerical method using ten equilibrated grain boundary groove shapes, and the results are given in Table 2. Typical grain boundary groove shapes for the solid Cd phase in equilibrium with the eutectic liquid are shown in Figure 7. As can be seen from Figure 7, the grains and interfaces of the system are very clear. The average value of $\Gamma$ is found to be $(8.35 \pm 0.41) \times 10^{-8}$ km for solid Cd (Table 2).

### The effective entropy change

It is also necessary to know the entropy of fusion per unit volume, $\Delta S^*$, for the solid phase to determine the solid-liquid interfacial energy and the entropy change per unit volume for an alloy is given by Gündüz and Hunt (1985):

$$\Delta S^* = \frac{(1-C_S)(S_A^L - S_A^S) + C_S(S_B^L - S_B^S)}{V_S} \quad (15)$$

where $S_A^L$, $S_A^S$, $S_B^L$ and $S_B^S$ are the partial molar entropies for $A$ and $B$ materials and $C_S$ is the solid composition. Since the entropy terms are generally not available, for convenience, the undercooling at constant composition may be explained by the change in composition at constant temperature.

For a spherical solid (Christian, 1975)

$$\Delta C_r = \frac{2\sigma_{SL}V_S(1-C_L)C_L}{rRT_M(C_S - C_L)} \quad (16)$$

where $R$ is the gas constant, $T_M$ is the melting temperature, $C_S$ and $C_L$ are the compositions of the equilibrated solid and liquid phases and $V_S$ is the molar volume of the solid phase. For small changes

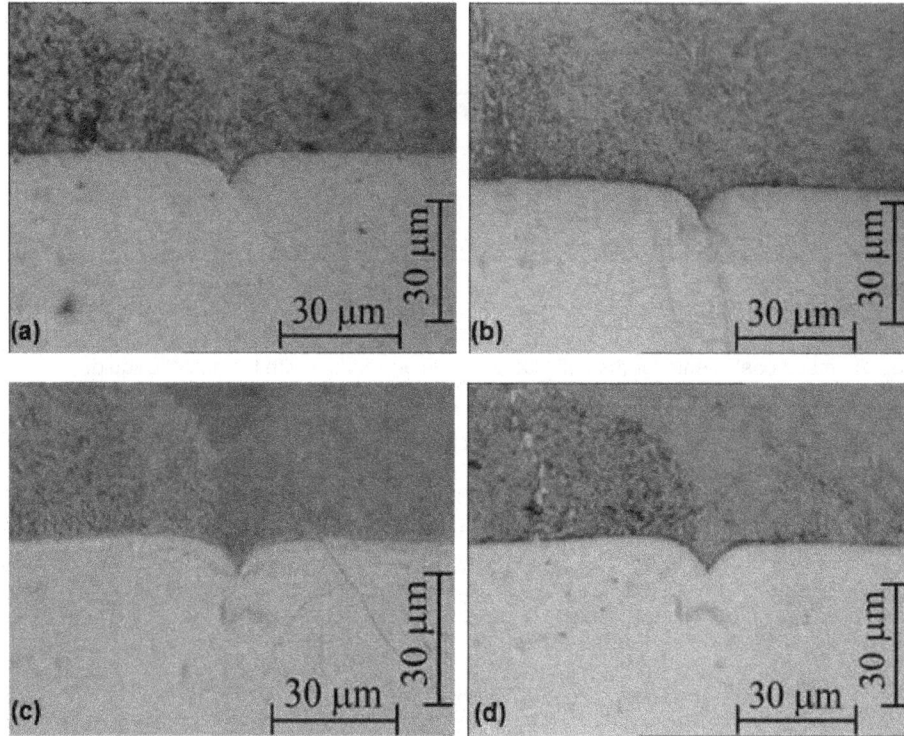

**Figure 7.** Typical grain boundary groove shapes for the solid Cd phase in equilibrium with the eutectic liquid.

$$\Delta T_r = m_L \Delta C_r = \frac{2 m_L \sigma_{SL} V_S (1 - C_L) C_L}{r R T_M (C_S - C_L)} \qquad (17)$$

where $m_L$ is the slope of liquidus. For a spherical solid $r_1 = r_2 = r$ and the curvature undercooling is written by

$$\Delta T_r = \frac{2 \sigma_{SL}}{r \Delta S^*} \qquad (18)$$

From Equations 17 and 18, the entropy change for an alloy is written as:

$$\Delta S^* = \frac{R T_M (C_S - C_L)}{m_L V_S (1 - C_L) C_L} \qquad (19)$$

The molar volume, $V_S$ is expressed as:

$$V_S = V_c N_a \frac{1}{n} \qquad (20)$$

where $V_c$ is the volume of the unit cell, $N_a$ is the Avogadro's number and $n$ is the number of molecules per unit cell. The values of the relevant constant used in the determination of entropy change per unit volume are given in Table 3. The entropy of fusion per unit volume for the solid Cd phase is found to be $(1.533 \pm 0.076) \times 10^6$ J/K m$^3$.

## The solid-liquid interface energy ($\sigma_{SL}$)

The solid-liquid interface energy ($\sigma_{SL}$) can be obtained from the Gibbs-Thomson equation for isotropic interfacial energy (Gündüz and Hunt, 1985, 1989; Maraşli and Hunt, 1996; Akbulut et al., 2009; Ocak et al., 2010; Meydaneri et al., 2011, 2012) and it is expressed as:

$$\Gamma = \frac{\sigma_{SL}}{\Delta S^*} \qquad (21)$$

The solid-liquid interface energy for the solid Cd solution in equilibrium with the eutectic liquid solution was obtained using the values of the Gibbs-Thomson coefficient and the entropy of fusion per unit volume. The value of the solid-liquid interface energy was found to be $(128.00 \pm 12.8) \times 10^{-3}$ Jm$^{-2}$. A comparison with previous works is also shown in Table 4. The value of $\sigma_{SL}$ is in good agreement with previous theoretically and experimentally calculated values of $\sigma_{SL}$ for solid Cd.

## Grain boundary energy

According to force balance at the grain boundary groove, if the solid-liquid interface energy is calculated, it is possible to determine the grain boundary energy. When

**Table 3.** Some physical properties of solid Cd phase in the Cd-Pb alloy system.

| System | For solid Cd phase in the Cd-Pb alloy |
|---|---|
| Composition of solid phase, Cs | Solid Cd (solid Cd-0.25 wt.% Pb) (Hansen and Anderko, 1985) |
| Composition of quenched liquid phase, $C_L$ | Eutectic Liquid (Cd-17.4 wt.% Pb) (Hansen and Anderko, 1985) |
| The value of f(C) for solid Cd | -3.56 (Hansen and Anderko, 1985) |
| Melting Temperature, $T_M$ (K) | 521 (Hansen and Anderko, 1985) |
| Liquidus slope of solid Cd, $m_L$ (K/at.fr.) | -774 (Hansen and Anderko, 1985) |
| Crystal structure of solid Cd | Hexagonal A3 |
| Lattice parameters of Cd (Å) | a = 2.978, c = 5.617 |
| n | 6 |
| Molar volume of solid Cd, $V_S \times 10^{-6}$ ($m^3$) | 12.99 |
| Molar mass of Cd (g) | 112.40 |
| Density (g/cm$^3$) | 8.65 |
| Entropy change of fusion for solid Cd, $\Delta S^*$ (J/Km$^3$) | $1.533 \pm 0.076 \times 10^6$ |

$$f(C) = \frac{C_S - C_L}{(1 - C_L)C_L}$$

**Table 4.** A comparison of the $\sigma_{SL}$, $\sigma_{GB}$ and $\Gamma$ values for solid Cd phase obtained in the present work with the values of $\sigma_{SL}$, $\sigma_{GB}$ and $\Gamma$ obtained in previous works.

| System | Liquid phase (C_L) | Solid phase (C_S) | Temperature (K) | $\Gamma \times 10^{-8}$ (Km) | $\sigma_{SL} \times 10^{-3}$ (J m$^{-2}$) | $\sigma_{GB} \times 10^{-3}$ (J m$^{-2}$) |
|---|---|---|---|---|---|---|
| Bi-Cd (Keşlioğlu et al., 2004) | Cd-0.03 at.% Bi | Bi-54.6 at.% Cd | 418.7 | $8.28 \pm 0.33$ | $81.22 \pm 7.31$ | $154.32 \pm 18.52$ |
| Cd-Zn (Saatçi and Pamuk, 2006) | Cd-5.0 at.% Zn | Cd-26.5 at. % Zn | 539 | $8.16 \pm 0.65$ | $121.46 \pm 0.97$ | $242.38 \pm 1.93$ |
| Cd-Pb [PW] | Cd-17.4 wt.% Pb | Cd-0.25 wt. % Pb | 521 | $8.35 \pm 0.41$ | $128.00 \pm 12.8$ | $239.33 \pm 26.32$ |

[PW]: Present work.

the interface energy is isotropic, the force balance can be expressed as:

$$\sigma_{SS} = \sigma_{SL}^A \cos \theta_A + \sigma_{SL}^B \cos \theta_B \qquad (22)$$

where $\theta_A$ and $\theta_B$ are the angles that the solid-liquid interfaces make with the y- axis. If the grains on either side of the grain boundary are approximately the same, the grain boundary energy can be expressed by:

$$\sigma_{GB} = 2\sigma_{SL} \cos \theta \qquad (23)$$

where $\theta = \frac{\theta_A + \theta_B}{2}$ is the average angle value that the solid-liquid interfaces make with the y-axis. As shown in Equation 23, $\sigma_{GB}$ is not sensitive to the error in $\theta$ for small $\theta$ values. According to Equation 23, the value of $\sigma_{GB}$ should be smaller or equal to twice that of the solid-liquid interface energy, that is, $\sigma_{GB} \leq 2\sigma_{SL}$.

The grain boundary energy was calculated from Equation 23 using the related $\sigma_{SL}$ and $\theta$ for groove shapes. The average grain boundary energy value for the ten grain boundary groove shapes was found to be $\sigma_{GB}$ = (239.33 ± 26.32) ×10$^{-3}$ J.m$^{-2}$.

A comparison of the values of the solid Cd solution obtained in the present work with the values determined in previous works is given in Table 4.

Interfacial free energy and its anisotropy are considered to play a critical role in many phase transformations. The determination of the effects of anisotropy on the interfacial energy is very difficult. In the present work, the interfacial energy of the solid Cd solution in equilibrium with the eutectic liquid solution was assumed to be isotropic.

### The experimental error ratio in the present work

The coordinates of the equilibrated grain boundary groove shapes were measured with an optical microscope to an accuracy of ±10 μm.

The thickness of the sample for geometrical correction was measured with a digital micrometer with ±1 μm resolution. Thus, the error ratio in the measurements of the equilibrated grain boundary coordinates was less than 0.2%.

The experimental error ratio in the measurement of $\kappa_S$ is the sum of the fractional uncertainty of the measurements of power, temperature differences, length of heating wire and position of thermocouples which can be expressed as:

$$\left|\frac{\Delta\kappa_S}{\kappa_S}\right| = \left|\frac{\Delta Q}{Q}\right| + \left|\frac{\Delta T_1}{T_1}\right| + \left|\frac{\Delta T_2}{T_2}\right| + \left|\frac{\Delta\ell}{\ell}\right| + \left|\frac{\Delta r_1}{r_1}\right| + \left|\frac{\Delta r_2}{r_2}\right| \qquad (24)$$

The experimental error ratio in the thermal conductivity measurements is found to be about 5%.

The experimental error ratio in the measurement of the temperature gradient of the solid phase is the sum of the fractional uncertainty of the power measurement, length of heating wire, thermal conductivity and thermocouples' positions which can be expressed as:

$$\left|\frac{\Delta G_S}{G_S}\right| = \left|\frac{\Delta Q}{Q}\right| + \left|\frac{\Delta\ell}{\ell}\right| + \left|\frac{\Delta r}{r}\right| + \left|\frac{\Delta\kappa_S}{\kappa_S}\right| \qquad (25)$$

If Equation 25 is compared with Equation 24, the experimental errors from the measurements of $Q$, $\ell$, $r$, $\Delta T$ in Equation 25 already exist in the fractional uncertainties at Equation 24. Thus, the total experimental error in the temperature gradient measurements is equal to the experimental error in the thermal conductivity measurements and is about 5%.

The experimental error ratio in the determination of the Gibbs-Thomson coefficient is the sum of the experimental error ratios in the measurements of the temperature gradient, thermal conductivity and groove coordinates. Thus, the total error ratio in the determination of the Gibbs-Thomson coefficient is about 5%. The error ratio in the determined entropy change of fusion per unit volume is estimated to be about 5% (Tassa and Hunt, 1976).

The experimental error ratio in the determination of solid-liquid interfacial energy is the sum of the experimental error ratios of the Gibbs-Thomson coefficient and the entropy change of fusion per unit volume. Thus, the total experimental error ratio of the solid-liquid interfacial energy obtained in the present work is about 10%. The experimental error ratio in the determination of $\theta$ angles was found to be 1%. Thus, the total experimental error ratio in the resulting grain boundary energy is about 11%.

## Conclusion

(1) The equilibrated grain boundary groove shapes were obtained for the solid Cd solution in equilibrium with the eutectic liquid by using a radial heat flow apparatus, and the Gibbs-Thomson coefficient, $\Gamma$, was calculated to be $\Gamma = (8.35 \pm 0.41) \times 10^{-8}$ Km by using the groove shapes.

(2) The effective entropy change per unit volume, $\Delta S^*$, was calculated to be $(1.533 \pm 0.076) \times 10^6$ J/Km$^3$ by using the phase diagrams and related parameters.

(3) The solid-liquid interface energy, $\sigma_{SL}$, was determined to be $\sigma_{SL} = (128.00 \pm 12.8) \times 10^{-3}$ Jm$^{-2}$ from the Gibbs-Thomson equation by using the Gibbs-Thomson coefficient, $\Gamma$, and the effective entropy change, $\Delta S^*$.

(4) The grain boundary energy, $\sigma_{GB}$, was determined to be $\sigma_{GB} = (239.33 \pm 26.32) \times 10^{-3}$ Jm$^{-2}$ by using the angle $\theta$ and the related $\sigma_{SL}$.

## ACKNOWLEDGEMENTS

This work was financially supported by the Erciyes University Research Foundation under Contract No: FBD-09-846. The authors are grateful to the foundation for its financial support.

## REFERENCES

Akbulut S, Ocak Y, Maraşli N, Keşlioğlu K, Kaya H, Çadirli E (2009). Determination of interfacial energies of solid Sn solution in the In-Bi-Sn ternary alloy. Mater. Charact. 60:183-192.

Aksöz S, Ocak Y, Maraşli N, Keşlioğlu K (2011). Determination of thermal conductivity and interfacial energy of solid Zn solution in the Zn-Al-Bi eutectic system. Exp. Therm Fluid Sci. 35:395-404.

Bulla A, Carreno-Bodensiek C, Pustal B, Berger R, Bührig-Polaczek A, Ludwig A (2007). Determination of the solid-liquid interface energy in the Al-Cu-Ag system. Metall. Mater. Trans. A 38:1956-1964.

Christian JW (1975). The theory of transformations in metals and alloys. part I, 2nd ed., Oxford, Pergamon, P. 169.

Erol M, Keşlioğlu K, Şahingöz R, Maraşli N (2005). Experimental determination of thermal conductivity of solid and liquid phases in Bi-Sn and Zn-Mg binary eutectic alloys. Met. Mater. Int. 11:421-428.

Gündüz M, Hunt JD (1985). The measurement of solid-liquid surface energies in the Al-Cu, Al-Si and Pb-Sn systems. Acta. Metall. 33:1651-1672.

Gündüz M, Hunt JD (1989). Solid-liquid surface-energy in the Al-Mg system. Acta. Metall. 37:1839-1845.

Hansen M, Anderko K (1985). Constitutions of binary alloys. 2nd ed, New York: McGraw-Hill. P. 432.

Keşlioğlu K, Erol M, Maraşli N, Gündüz M (2004). Experimental determination of solid-liquid interfacial energy for Cd in Bi-Cd liquid solutions. J. Alloy. Compd. 385:207-213.

Maraşli N, Hunt JD (1996). Solid-liquid surface energies in the Al-CuAl$_2$, Al-NiAl$_3$ and Al-Ti systems. Acta. Mater. 44:1085-1096.

McCartney DG (1981). D.Phil. Thesis. University of Oxford, UK, pp. 85-175.

Meydaneri F, Payveren M, Saatçi B, Özdemir M, Maraşli N (2012). Experimental determination of interfacial energy for solid Zn solution in the Sn-Zn eutectic system. Met. Mater. Int.18:95-104.

Meydaneri F, Saatçi B, Özdemir M (2011). Experimental determination of solid-liquid interfacial energy for solid Sn in the Sn-Mg system. Appl. Surf. Sci. 257:6518-6526.

Ocak Y, Akbulut S, Maraşli N, Keşlioğlu K (2010). Interfacial energies of solid Sn solution in the Sn-Ag-In ternary alloy. Chem. Phys. Lett. 496:263-269.

Ocak Y, Aksöz S, Maraşli N, Keşlioğlu K (2010). Experimental determination of thermal conductivity and solid-liquid interfacial energy of solid Ag$_3$Sn in the Ag-Sn-In ternary alloy. Intermetallic 18:2250-2258.

Porter DA, Easterling KE (1991). Phase transformations in metals and alloys. Van Nostrnad Reinhold Co. Ltd. UK, P. 204.

Saatçi B, Çimen S, Pamuk H, Gündüz M (2007). The interfacial free energy of solid Sn on the boundary interface with liquid Cd-Sn eutectic solution. J. Phys. Condens. Matter. 19:326219-326230.

Saatçi B, Pamuk H (2006). Experimental determination of solid-liquid surface energy for Cd solid solution in Cd-Zn liquid solutions. J. Phys: Condens. Mater. 18:10143-10155.

Tassa M, Hunt JD (1976). The measurement of Al-Cu dendrite tip and eutectic interface temperatures and their use for predicting the extent of the eutectic range. J. Cryst. Growth 34:38-48.

Touloukian YS, Powell RW, Ho CY, Klemens PG (1970). Thermal conductivity metallic elements and alloys. New York: Plenum 1:48, 187.

Woodruff DP (1973). The Solid-Liquid Interface. vol. 36, Cambridge University Press, Cambridge, pp. 1-2.

# 12

# A planar microstrip metamaterial resonator using split ring dual at Ku-Band

Nitin Kumar[1] and S. C. Gupta[2]

[1]ECE Department, Uttarakhand Technical University, Dehradun, Uttarakhand, India.
[2]ECE Department, DIT University, Dehradun, Uttarkhand, India.

**This paper introduces a new planar microstrip metamaterial resonator, the novelty of this paper lays in its unit cell design. The unit cell is formed by connecting metallic traces of two edge coupled split ring resonators to form the infinity symbol on one side of the substrate, and an array of conducting wires on the other. An RLC equivalent model of the structure is also proposed, it can be advantageous to use this model to identify the resonant frequency along with the root of the negative permeability and negative permittivity. The model shows resonance at 17 GHz. The structure was designed and simulated using EM solver Ansys HFSS, the extracted s-parameter matrix was analyzed to determine the effective permittivity, permeability and index of refraction. The structure shows negative values for effective $\varepsilon$, $\mu$ at resonant frequency 16.5 GHz. At frequencies where both the recovered real parts of $\varepsilon$ and $\mu$ are simultaneously negative, the real part of the index of refraction is also found to be negative.**

**Key words:** Microstrip metamaterial, negative refraction, permittivity, permeability, RLC circuit.

## INTRODUCTION

Metamaterials, first named and theoretically discussed by Veselago (1968), are studied widely throughout the world. These are an artificially engineered material showing electromagnetic properties not readily found in naturally occurring material, such as, property of negative refractive index and artificial magnetism (Sabah, 2010; Mahmood, 2004; Sulaiman et al., 2010; Smith et al., 2001; Sharma et al., 2011). Recently work is done in direction of making a perfect lens using metamaterials (He-Xiu Xu et al., 2013a, b).

Metamaterials are often characterized in terms of their effective material properties, such as effective electric permittivity and effective magnetic permeability. Any one of these parameters, or even both of them may be simultaneously negative. The former is known as single negative material (SNG), if only effective permittivity is negative it is called Epsilon negative material (ENG),

whereas if only effective permeability is negative it is called as Mu-negative material (MNG). The latter is referred to as left-handed metamaterials (LHM), double negative (DNG), or negative refractive index material (NRIM).

Artificial plasmas show negative effective permittivity for all frequencies smaller than plasma frequency of the Plasmon medium (Pendry et al., 1996). Effective negative permeability can be obtained in the well known Split-ring-resonator structure, but only for a narrow magnetic resonant frequency band (Pendry et al., 1999). In past few years, metamaterials has been a naive topic of interest among the research fraternity. Over these years various innovative structures have been reported.

This paper presents design and simulation of a new planar microstrip metamaterial resonator, exhibiting negative index of refraction. In comparison to the papers

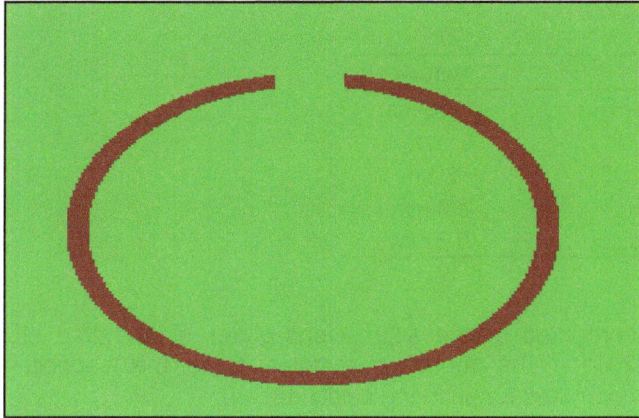

**Figure 1.** A Circular Split-Ring Resonator Structure (SRR).

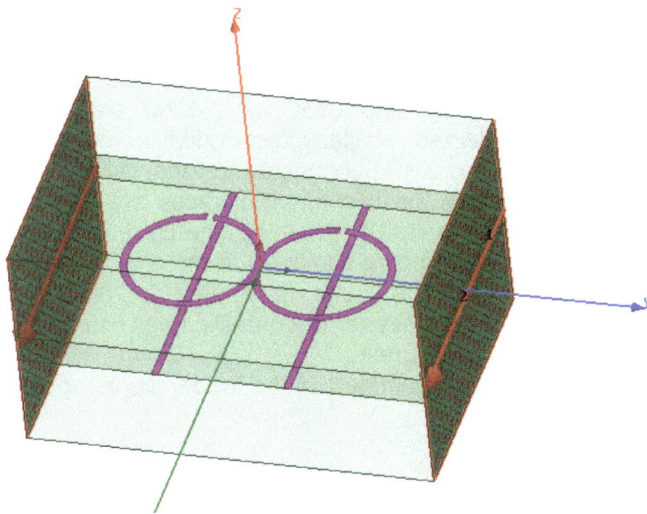

**Figure 2a.** Structure of Infinity Shaped Metamaterial (ISM) kept in a waveguide with wave-ports.

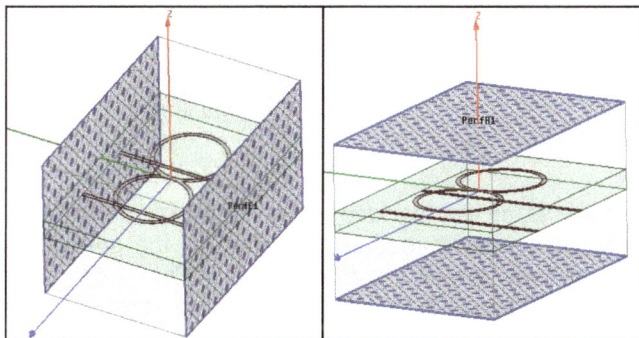

**Figure 2b.** Schematic diagram of ISM, showing the boundary conditions: PEC and PMC boundaries respectively.

cited here, the novelty of this paper lies in the unit cell design, where two edge coupled circular split-ring resonators are connected to form the infinity symbol and an array of straight wire conductors is also used. The overall size of the metamaterial unit is very small 6 mm × 8 mm. Since the geometry of the structure resembles the shape of mathematical symbol 'infinity' it will be referred to as Infinity Shaped Metamaterial (ISM) in rest of the paper. The structure was simulated using Ansys HFSS and the extracted s-parameter values ($S_{11}$ and $S_{21}$) was analyzed to calculate index of refraction. The results promises a bandwidth of around 8 GHz (~8.5 to 16.5 GHz).

Many researchers have done the analysis of SRR unit cells and SRR arrays, and shown that SRRs behave as LC resonators that can be excited by external magnetic flux. The analysis of SRRs by accurate circuit models can be effectively used to estimate the behavior of SRR structures in a simple, efficient manner. Also, explicit relationships between electrical parameters, dimensions of the SRR structure and its frequency dependent transmission/reflection behavior may be found.

Another method called NRW technique (Suganthi et al., 2012) was also used to calculate effective permittivity and effective permeability from s-parameters, using which refractive index can be calculated. The results obtained from all techniques were compared and found in satisfactory agreement with each other.

## DESIGN SETUP

Figure 1 above shows a single Circular Split-Ring Resonator, and Figure 2a shows the unit cell of Infinity Shaped Metamaterial (ISM).

It has been shown in various papers that a single SRR provides magnetic resonance and supports negative effective permeability (Pendry et al., 1999), in this paper it can be seen from Figure 2a that, ISM can be formed by connecting traces of two SRRs in edge coupled configuration in the shape of mathematical symbol 'infinity'. This structure behaves as 2 SRR's connected in series. Figure 2a also shows the ground plane, which is composed of an array of straight wire conductors instead of a continuous sheet of copper. These straight wire conductors are placed directly beneath the slit of the SRRs lying on other side of the substrate; they will provide a virtual path for the currents to continue flowing in the split rings.

As suggested by Pendry et al. (1996), the electric field should be parallel to the wire while the magnetic field should be perpendicular to the SRR. To retrieve the scattering parameters the radiation setup of the structure is done in an air filled waveguide. The electric (PEC) and magnetic (PMC) fields are defined over the walls of the waveguide in such a manner that the aforementioned conditions are satisfied, and is shown in Figure 2b. The structure is fed RF signals ranging from 15 to 18 GHz,

**Table 1.** Parameter table.

| Parameters | Values |
|---|---|
| Substrate (Duroid (tm)) with Thickness | 0.786 mm |
| Relative dielectric constant | 1.1 |
| Radius of outer circle of the ring | 2 mm |
| Radius of inner circle of the ring | 1.8 mm |
| Width of split/Width of wire conductor | 0.2 mm |

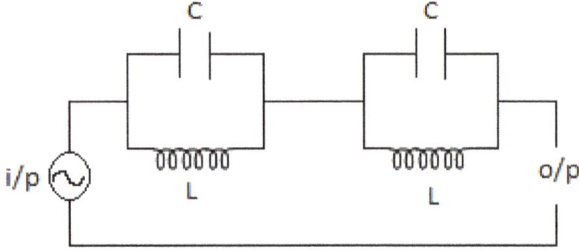

**Figure 3.** Simplified equivalent circuit of ISM unit cell.

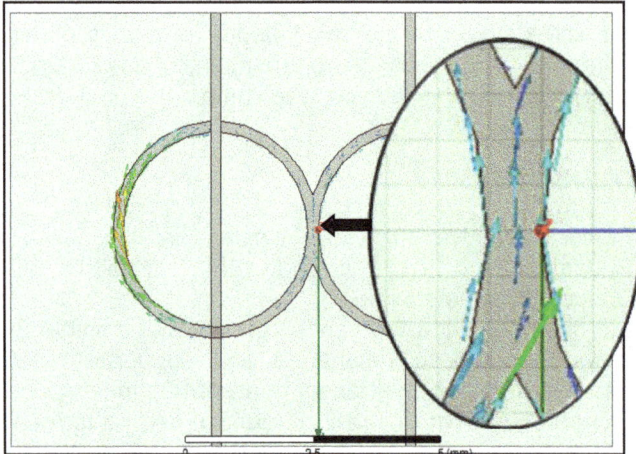

**Figure 4.** Distribution of current over the metallic rings, also showing the addition of current in the area common to both rings.

with the help of wave-ports (Figure 2a). The physical parameters of the structure are mentioned in Table 1.

Thus, the proposed structure ISM is a composite of split-rings and array of wires; both components are required to obtain negative effective permittivity and negative effective permeability in a single structure.

## Two port equivalent circuit model of ISM

The simplified two-port equivalent circuit representation suggested for ISM unit cell is shown in Figure 3, where $L$ is the self-inductance of the metal loop, which can be

computed by the expressions given in (Mondher et al., 2011). The model parameter $C$ is the capacitance computed for split ring calculated as:

$$C = C_{pp} + C_s$$

where $C_{pp}$ and $C_s$ are parallel plate and surface capacitances, respectively. The resonant frequency of the structure can be calculated by $f = 1/2\pi LC$. For simplicity of design and calculations, the effect of coupling between strip used as ground and metallic SRRs is neglected; similarly the mutual coupling effect between the two SRRs connected to form 'infinity' is also neglected. Since the two metallic traces are connected together in series w.r.t the feeding, the current flowing in two rings must combine additively in the area common to both the rings. The same is verified by plotting the distribution of current over the metallic rings using HFSS, and is shown in Figure 4.

The electrical parameters of the model are computed as:

$$L = \mu_0 r \left[ log\left(\frac{2r}{g}\right) + 0.9 + 0.2\left(\frac{g}{2r}\right)^2 \right] \qquad (1)$$

where, L represents the inductance of SRR, g represents the width of the split, and r is the average or mean radius. To calculate the total capacitance, a simple analytical approximate expression may be used. First, the surface capacitance is determined analytically by using analytical expressions for the electric field of a split-ring, and is given by (Mondher et al., 2011):

$$C_s = \frac{2\varepsilon_0(t+w)}{\pi} log \frac{4r_i}{g} \qquad (2)$$

where, $C_s$ is the surface capacitance, $\varepsilon_0$ permittivity of free space, w represents the width of the metallic split ring, t thickness of the metal used for split ring, $r_i$ inner radius of the split ring, and g width of the split.

Secondly, the gap capacitance or parallel plate capacitance of the split is computed as:

$$C_{pp} = \frac{\varepsilon_0 \varepsilon_r A}{g} \qquad (3)$$

where, $C_{pp}$ is the parallel plate capacitance (of the split), $\varepsilon_r$ relative permittivity, A is area of the plate of capacitor

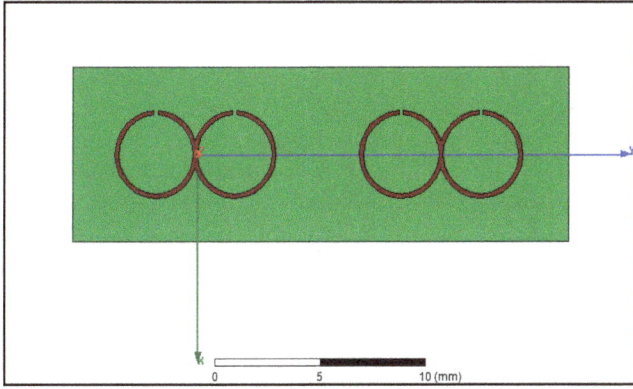

**Figure 5.** Array of two ISM elements.

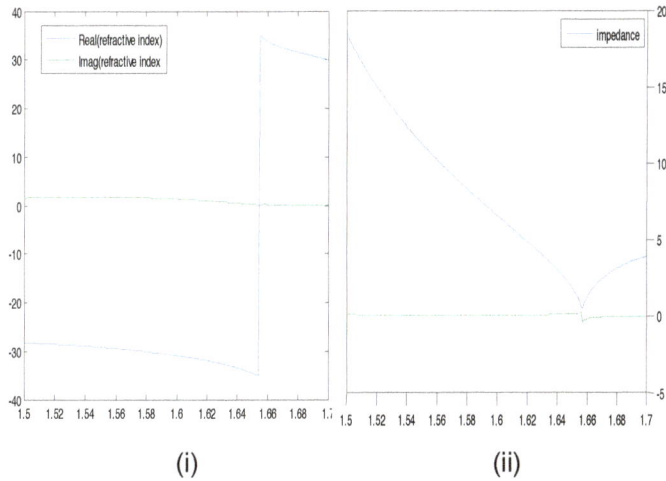

(i)                                    (ii)

**Figure 6.** (i) Refractive Index (ii) Wave impedance versus Frequency.

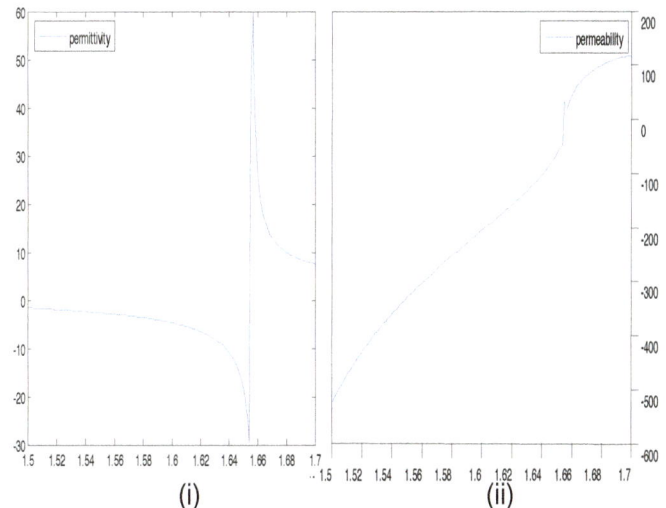

(i)                                    (ii)

**Figure 7.** (i) Effective permittivity, (ii) Effective permeability versus Frequency in GHz.

(here $A = t * w$ ).Using the above formulae, the value of inductance and capacitance was calculated, and the resonant frequency using these values comes out to be ~17 GHz.

A linear array of two elements (shown in Figure 5) was then developed and analyzed, the coupling effects between two ISM unit cells can be described by a parallel RC circuit in the shunt branch. This coupling equivalent circuit is connected in series between two ISM blocks. The coupling parameters $C_m$ and $L_m$ can be computed by similar approaches used for the computation of the parameters $C$ and $L$ mentioned previously. The effect of mutual coupling between two ISM elements was found to be too small, and had a very little effect on the resonant frequency. Hence, it was not considered here.

## SIMULATION AND RESULTS

The ISM structure was designed and simulated using EM solver Ansys HFSS. With extracted s-parameter matrix, value of refractive index $n$ and wave impedance $z$ was calculated using the following equations (Sabah, 2010; Smith et al., 2001).

$$n = \frac{1}{kd} \cos^{-1}\left\{ \frac{1 - s_{11}^2 + s_{21}^2}{2 s_{21}} \right\} \qquad (4)$$

$$z = \pm \sqrt{\frac{(1 + s_{11})^2 - s_{21}^2}{(1 - s_{11})^2 - s_{21}^2}} \qquad (5)$$

The value of effective permittivity ε and effective permeability μ then may be computed as $\varepsilon_{eff} = n/z$ and $\mu_{eff} = n * z$.

The condition Im{n}≥0 fix the choice for sign of 'n'. Similarly the condition Re{z}≥0 fixes the choice for sign of 'z'. An improved parameter retrieval method given in (Liu and Wang, 2012) is as follows:

$$n = \frac{ln\left( \frac{s_{21}}{1 - s_{11} \frac{z-1}{z+1}} \right)}{ikd} \qquad (6)$$

where, $k$ is wave number, $d$ is thickness of ISM unit cell.

We calculate $z$ first, and then $n$ can be calculated from Equation (6). All of the above formulae were programmed in MATLAB 2009a to obtain the required plots. Refractive index versus frequency curve using Equation (6) and wave impedance using Equation (5) are shown in Figures 6 and 7:

After calculating $n$ and $z$, the value of effective permittivity and permeability was computed and the graphs versus frequency are shown below:

The graphs above suggests metamaterial behavior of ISM at ~16.5 GHz. Although the results above are calculated using well known techniques, one more

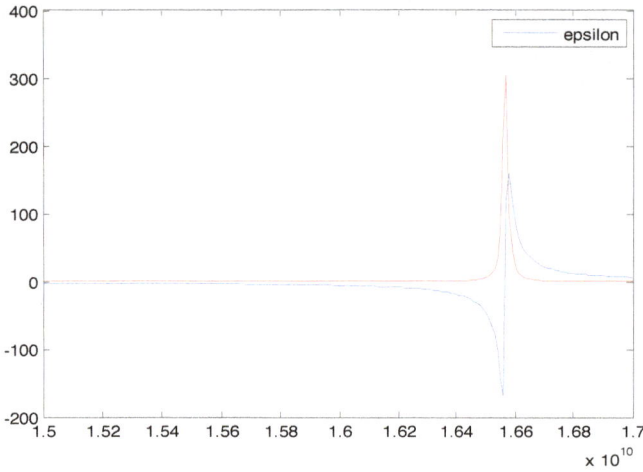

**Figure 8.** Effective permittivity versus Frequency in GHz.

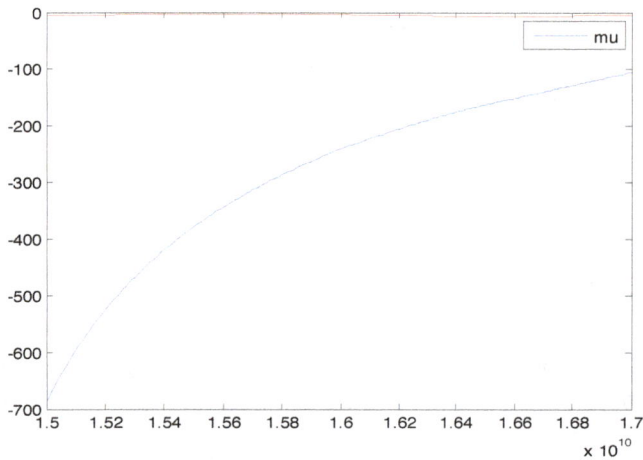

**Figure 9.** Effective permeability versus Frequency in GHz.

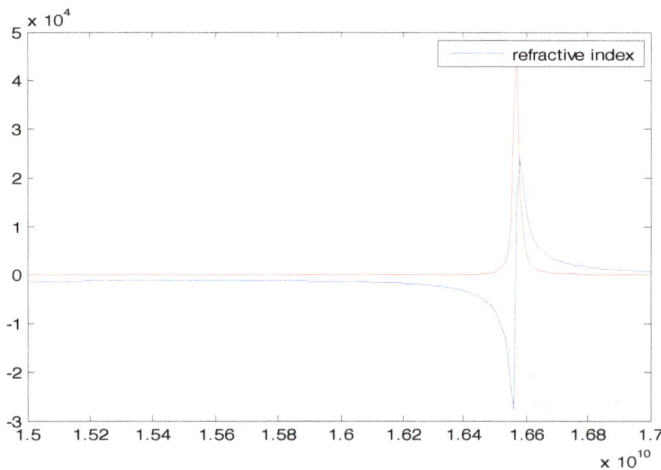

**Figure 10.** Refractive Index versus Frequency in GHz.

technique, the NRW parameter retrieval approach is also used in this paper to reinforce the results already obtained. A separate MATLAB code was developed based on NRW approach to find the medium properties using extracted S11 and S21 parameters. The results obtained using NRW approach shown in Figure 8 to 10 are in satisfactory agreement with those produced from Equation (4) to (6) (shown in Figures 6 and 7). The $\varepsilon_{eff}$ and $\mu_{eff}$ of the medium are related to $S$-parameters by the Equations (7) and (8) below (Suganthi et al., 2012):

$$\varepsilon_{eff} = \frac{2}{jk_0 d}\frac{1-V_1}{1+V_1} \tag{7}$$

$$\mu_{eff} = \frac{2}{jk_0 d}\frac{1-V_2}{1+V_2} \tag{8}$$

where $k_0$ is a wave number equivalent to $2\pi/\lambda_0$, $d$ is the thickness of the substrate and $V1$ and $V2$.

$$V_1 = S_{21} + S_{11} \tag{9}$$

$$V_2 = S_{21} - S_{11} \tag{10}$$

After calculating $\varepsilon_{eff}$ and $\mu_{eff}$ using above equations, refractive index '$n$' can be computed using:

$$n = \pm\sqrt{\varepsilon_{eff} \times \mu_{eff}} \tag{11}$$

Using MATLAB, graphs for effective permittivity, effective permeability, refractive index versus frequency are plotted and are shown in Figures 8 to 10:

Figure 10 shows negative value of refractive index below ~16.5 GHz. The results obtained from HFSS for ISM unit cell and for linear array of 1×2 ISM elements are shown in Figures 11 and 12.

In Figures 11 and 12 the dip in value of $S_{11}$(dB), shows the resonant frequency of ISM unit cell and array of 1×2 ISM elements, respectively. The resonant frequency in both cases is 16.58 and 16.5 GHz, approximately same. The graph (Figure 13(i)) shows the phase of $S_{11}$ and $S_{21}$(Radians) for ISM unit cell and its array. The phase of $S_{11}$ and $S_{21}$ crosses each other and shows zero crossing at resonant frequency, which suggests the presence of metamaterial property. Also, the metamaterial property was preserved in case of linear array of two or more elements.

For further analysis a linear array of 10 elements was prepared to observe any deviation in resonant frequency, Figure 14 shows the structure and results for a linear array of 10 elements. From the results (Figure 13 (ii)) it may be observed that the shift in resonant frequency is too small to be considered.

The results obtained for a single ISM element, array of 1×2 ISM elements, and array of 1×10 ISM elements, all

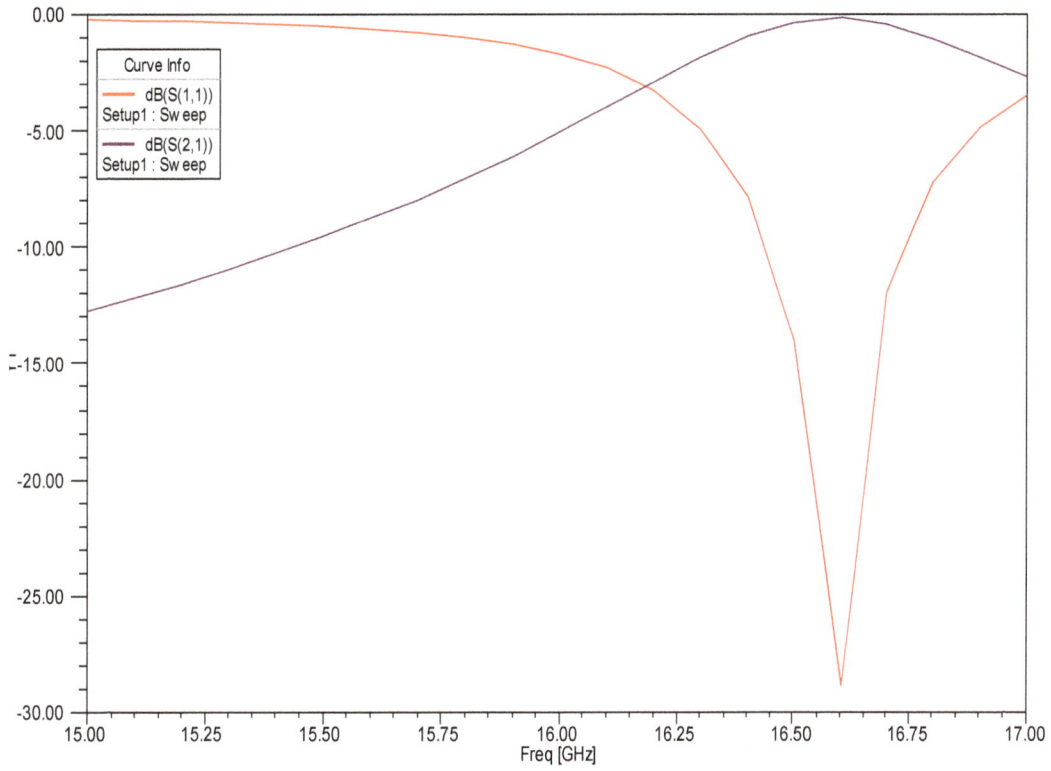

**Figure 11.** $S_{11}$(Red), $S_{21}$(Brown) in dB versus Frequency for ISM unit cell.

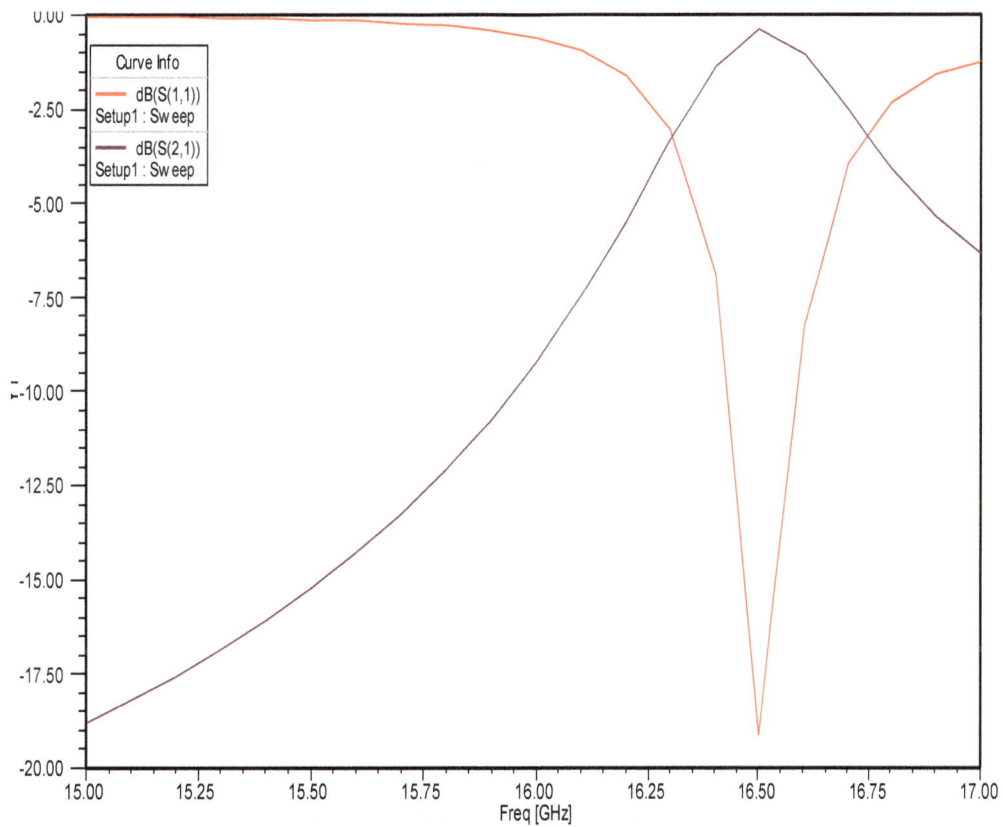

**Figure 12.** $S_{11}$(Red), $S_{21}$(Brown) in dB versus Frequency for 1×2 array ISM.

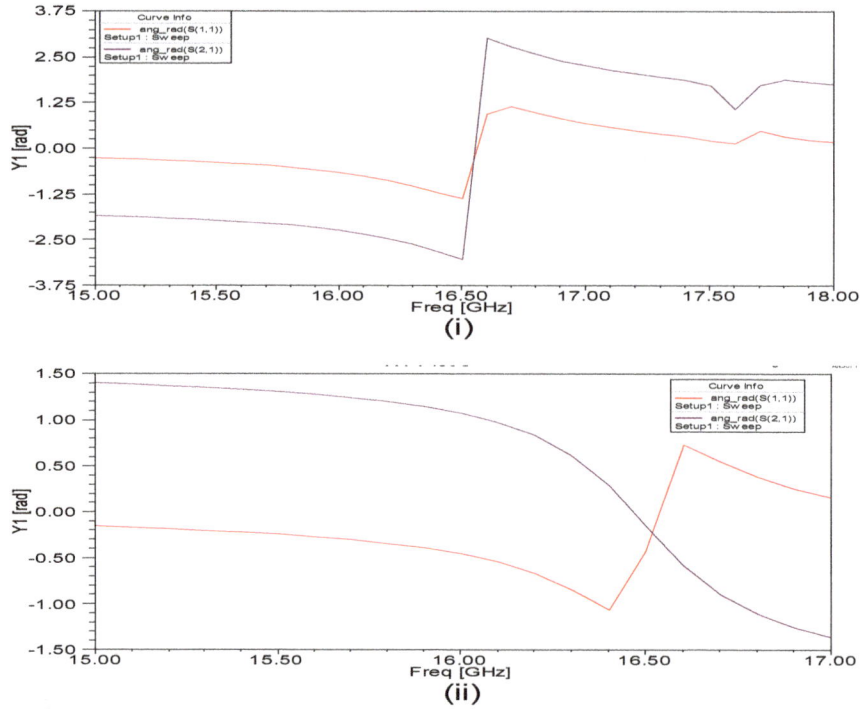

**Figure 13.** Phase of S11 (red), S21(brown) in radians, (i) ISM unit cell, (ii) Array of 1x2 ISM elements versus Frequency in GHz.

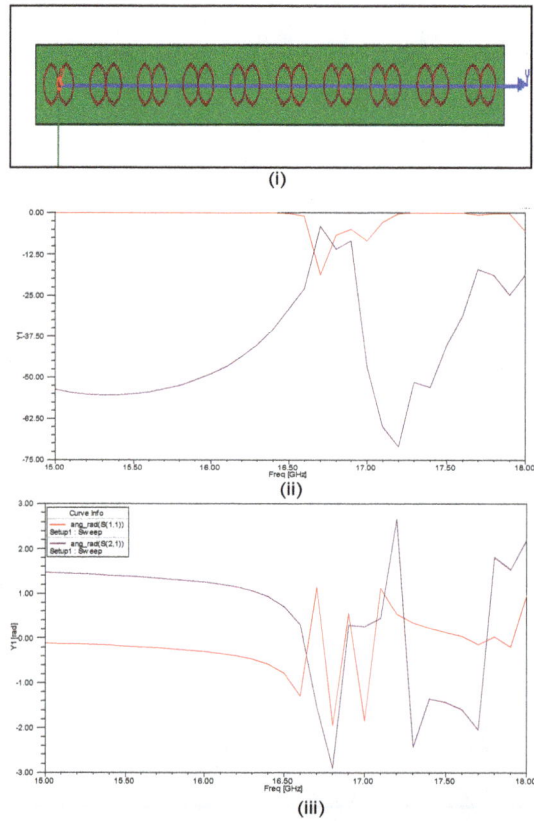

**Figure 14.** Linear array 10 elements (i) Structure, (ii) S11(red) and S21(brown) in dB, (iii) Phase of S11(red), S21(brown) in radians.

are in good agreement with each other and shows resonance around 16.5 GHz.

## Conclusion

A new planar microstrip metamaterial resonator using circular split ring dual, connected in the shape of 'infinity' and array of straight wire conductors is presented, it exhibits the property of negative index of refraction at Ku-band. Results obtained using HFSS are verified by coding formulae for refractive index, effective permittivity, and effective permeability in MATLAB and plotting curves versus frequency. The results obtained from all the techniques mentioned in this paper are found in satisfactory agreement. In future, the presented ISM resonator can be incorporated with microstrip antennas to get highly directional beam patterns either by using it as a substrate or by using it as a metamaterial cover kept in front of the antenna; also a physical model of ISM resonator may be fabricated.

## ACKNOWLEDGEMENTS

The authors are thankful to DEAL (Defense Electronics Applications Laboratory), Dehradun for extending their laboratory to use Ansoft HFSS.

## REFERENCES

Mahmood SF (2004). A new miniaturized annular ring patch resonator partially loaded by a metamaterial ring with negative permeability and permittivity.IEEE Antenna Wireless Propagation Letters, 3:19-22.

Mondher L, Jamel BT, Fethi C (2011). A New Proposed Analytical Model of Circular Split Ring Resonator. Journal of Materials Science and Engineering B1, pp. 696-701.

Pendry JB, Holden AJ, Robbins DJ, Stewart WJ (1999). Magnetism from Conductors and EnhanceNonlinear Phenomena. IEEE Transactions On Microwave Theory And Techniques, 47(11):2075-2084.

PendryJB, Holden AJ, Stewart WJ, Youngs I (1996). Extremely Low Frequency Plasmons in Metallic Mesostructures.Physical Rev. Lett. 76(25);4773-4776.

Sabah C (2010). Tunable metamaterial design composed of triangular split ring resonator and wire strip for S- and C- microwave bands.Progress In Electromagnetic Research, B. 22:341-357.

Sharma V, Pattanik SS, Garg T, Devi S (2011). A microstrip metamaterial split ring resonator.Int. J. Phys. Sci. 6(4):660-663,

Smith DR, Schultz S, Markoš P, Soukoulis CM (2001). Determination of Effective Permittivity and Permeability of Metamaterials from Reflection and Transmission Coefficients.PACS Nos. 41.20.Jb, 42.25.Bs, 73.20.Mf.

Suganthi S, Raghavan S, Kumar D., Hosimin Thilagar S (2012).A Compact Hilbert Curve Fractal Antenna on Metamaterial Using CSRR.PIERS Proceedings, Kuala Lumpur, Malaysia, pp.136-140.

Sulaiman AA, Othman A, Jusoh MH, Baba NH, Awang RA, Ain MF (2010). Small Patch Antenna on Omega structure metamaterials. Europ J. Sci. Res. 43(4):527-537.

Veselago VG (1968). The Electrodynamics Of Substances With Simultaneously Negative Values Of ε And μ.Soviet Physics Uspekhi, Volume 10(4):509-514.

He-Xiu Xu, Wang, Guang-Ming Qing Qi, Mei Lv, Yuan-Yuan, Gao Xi (2013 a). Three-Dimensional Super Lens Composed of Fractal Left-Handed Materials. Advanced Optical Materials, 1(7):495–502.

He-Xiu Xu Wang, Guang-Ming Qing Qi, Mei Lv, Yuan-Yuan Gao, Xi (2013 b). Metamaterial lens made of fully printed resonant-type negative-refractive index transmission lines. Appl. Phys. Lett. 102(19):193502.

He-Xiu Xu Wang, Guang-Ming Qing Qi, Mei (2013 c). Hilbert-Shaped Magnetic Waveguided Metamaterials for Electromagnetic Coupling Reduction of Microstrip Antenna Array. IEEE Transactions On Magnetics, 49(4):1526-1529.

Liu Z , Wang P (2012). A C-Band High Gain Microstrip Antennausing Negative Permeability Metamaterial on Low Temperature Co-Fire Ceramic Substrate.Proceedings of ISAP 2012, Nagoya, Japan, pp. 878-881.

# Eco-partitioning and indices of heavy metal accumulation in sediment and *Tilapia zillii* fish in water catchment of River Niger at Ajaokuta, North Central Nigeria

Olatunde Stephen Olatunji[1]  and Oladele Osibanjo[2]

[1]Department of Chemistry, Faculty of Applied Sciences, Cape Peninsula University of Technology Bellville Western Cape, South Africa.
[2]Basel Regional Coordination Centre, Faculty of Science, University of Ibadan, Ibadan, Oyo State Nigeria.

In this study the distribution and accumulation indices of some heavy metals in sediments and *Tilapia zillii* fish in freshwater catchment of River Niger by Ajaokuta Steel Company (ASC), North Central Nigeria were investigated. Water, bottom sediments and *Tilapia zillii* fish samples were collected upstream and downstream of the drainage column by ASC, Ajaokuta. The sample were digested according to standard methods and analysed for Cd, Mn, Cr, Ni, Cu, Zn and Pb using flame atomic absorption spectrometer. Accumulation indices or factor (AI or AF) of the investigated heavy metals were defined using the ratio of mean concentration $C_o$ in component/organism and that in the surrounding water $C_w$ at steady state (AI/AF = $C_o/C_w$). Sediment accumulation indices (AI) of the metals were: Cd, 5.4; Mn, 3.4; Cr, 1.6; Ni, 12.5; Cu, 1.6; Zn, 25.9 and Pb, 411.6, while the AI of the metals in fillets of *T. zillii* were Cd, 3.0; Mn, 2.1; Cr, 1.6; Ni, 5.1; Cu, 4.6; Zn, 3.2 and Pb, 14.0. Seasonal climate changes induces little marginal or no changes in the AIs of the metals except for Pb (841.5, 411.6) and Cd (11.6, 5.4) in sediment, and Pb (32, 14) and Cd (5.3, 3.0) ($p<0.05$) in fish fillets. Thus significant changes in metal AIs may be the consequence of their concentration levels in the aquatic ecosystem. Therefore, accumulation indices or factors may be an estimate of ecosystem status, and may be a useful tool for monitoring and predictive risk assessment (MPRA) purposes.

Key words: Partitioning, accumulation indices, heavy metals, fish fillet, bottom sediment.

## INTRODUCTION

The sustenance of aquatic ecosystem is a function of the quality and health of the water catchment, the physico-chemical composition and the balance in the dynamics of the ecosystem structure (Waldichuk, 1977). The quality of water in streams, rivers and lakes, as well those of coastal and marine water is hampered by the presence of contaminant substances such as heavy metals, when present in concentrations beyond natural levels. The

notoriety of heavy metals arises from its non degradable character which results in its persistence. The tendency for the accumulation of heavy metals beyond tolerable concentration in biotic and non-biotic functions, gives rise to its bio-toxic tendencies (Sieckhaus, 2009).

The efficiency of the dynamics of material circulation relies on the type, nature and sources of input from the different release points. Orlob (1975) reported that circulation was an important determinant of ecosystem response to material input into aquatic environment. Partitioning and accumulation of heavy metals in the different compartments of an aquatic environment is therefore a function of the ecosystem response to contaminant loads (Zhou et al., 2008). The response mechanism of aquatic ecosystem to input of contaminants is influenced such factors as, temperature zoning, vertical stratification, quality of inflow from the arterial networks/tributaries, ecosystem flora and fauna population, primary productivity, turbidity, light penetration, environment/habitat loss or creation, and hydraulic residence time variation resulting from sedimentation and the underlying water geology (Okabe et al., 1993; Leclerc et al., 1995; Shen et al., 1995; Bailey and Hamilton, 1997; Hamilton and Schladow, 1997; Schladow and Hamilton, 1997; Cook and Burkhard, 1998).

The partitioning of heavy metal in water follows series of complicated pathways including microbial, flora and fauna uptake, siltation/sedimentation etc., and this may lead to accumulation in aquatic biota and in different compartments of the water catchment (Moustafa, 2000). Heavy metals accumulation depends on metals availability and concentration, the rate of metabolism of bioaccumulative chemicals in biotic function, and or their stabilization in non-biotic function such as sediments (Mitsch et al., 1989). The accumulation of heavy metal pollutants in fish and other aquatic foods was reported to be a potential route of human exposure (ATSDR, 1999). Thus, high concentration of heavy metals in fresh water rivers present a major safety and health risk to humans and higher heterotroughs, who depend on such system for food resources.

The accumulation of chemical substances is an intrinsic property, and this determines their potential environmental hazard (Schlechtriem et al., 2012). Assessing the accumulation and or bioaccumulation potential of heavy metals is crucial to environmental and human risk assessment, because it is one of the main features required for environmental monitoring. According to Adolfsson et al. (2012), modern chemical legislation requires measurement of bioaccumulation factor (index) of large number of chemicals in fish and other edible aquatic organism. Data on accumulation indices needed for quantitative measurement of heavy metal partitioning between components of freshwater aquatic ecosystems, in order to define their exposure risk, as well as ecosystem health and status, especially that of River Niger are very limited. The indices of accumulation of contaminants such as heavy metals, may be leading in deriving pollution predictive indices (PPI) which can defines contaminant status of and aquatic ecosystem. Guidelines for the determination of bioaccumulation factor (index of bioaccumulation) are listed in the Organization for Economic Cooperation and Development (OECD) technical guideline (TG) 305.

In this study the distribution and accumulation indices (factors) of some heavy metals in sediments and *Tilapia zillii* fish in fresh water catchment of River Niger ecosystem by Ajaokuta Steel Company, North Central Nigeria were investigated. This is in order to define the dynamics of heavy metals partitioning and accumulation in the column segment, and the likely ecosystem response to input of heavy metals from the activities of Ajaokuta Steel Company.

## MATERIALS AND METHODS

### Description of study area

Ajaokuta is situated along the lower River Niger drainage at about N 07.508848°, E 006.692085° (Figure 1). Lower River Niger stretches from below the Niger-Benue confluence at Lokoja North Central Nigeria flowing between the geo-reference coordinate N 07.747450°, E 006.756990$^V$ northern latitude, down toward the equatorial zone N 04.394008°, E 007.085884°, down to the Niger Delta where it delivers its content via Bonny estuary into the Atlantic Ocean. The inter-annual mean discharge of 5589 m$^3$ s$^{-1}$ was recorded between, 1946 to 1992. Historically, the minimum and maximum drainage flows ever measured in River Niger were 500 and 27,600 m$^3$ s$^{-1}$ respectively (Hubert, 2000; Abrate et al., 2013). A number of tributary rivers contribute to drainage volume of the Lower River Niger, with the largest average contribution of 3500 m$^3$ s$^{-1}$ from River Benue. The mean water residence time in the reservoir is around 40 days while the water temperature in the reservoir generally ranges between 27 and 34°C (Hubert, 2000). A number of riverside communities are found along the lower River Niger drain, whose occupations include fishing, animal farming and other agriculture activities. A power generation plant and a steel production plant are sited along the bank of the Lower River Niger at Geregu and Ajaokuta respectively.

### Sample collection

Samples were collected up-stream and down-stream, along River Niger column by Ajaokuta Steel Company Complex located on N 07.508448°, E 006.692500° (altitude range 74 to 187 m) at Ajaokuta, North Central Nigeria. Three sample types consisting of one hundred and sixty samples each of water samples, bottom river sediment and *T. zillii* (Redbelly Tilapia) fish species were collected over 24 months period spanning between period 2004 and 2005.

### Sample preparation

Sediment samples were sorted to eliminate pebbles and coarsy materials, and air-dried under ambient conditions inside the laboratory for seven-two hours. Fish samples that is, *T. zillii* fish samples were rinsed with distilled water and dried in hot air oven at temperature of 80°C to constant weight (Zheng et al., 2007). The dried soil and sediment samples were pulverized sieved through a nylon sieve of 2 mm mesh size. The dried fish were air cooled to room temperature, and sorted to remove bones from the fish muscles (fillets). The fish fillets were blended using National blender with stainless steel cutters to fine particle sizes suitable for digestion.

### Samples digestion

#### Water samples

The method prescribed by the American Public Health Association

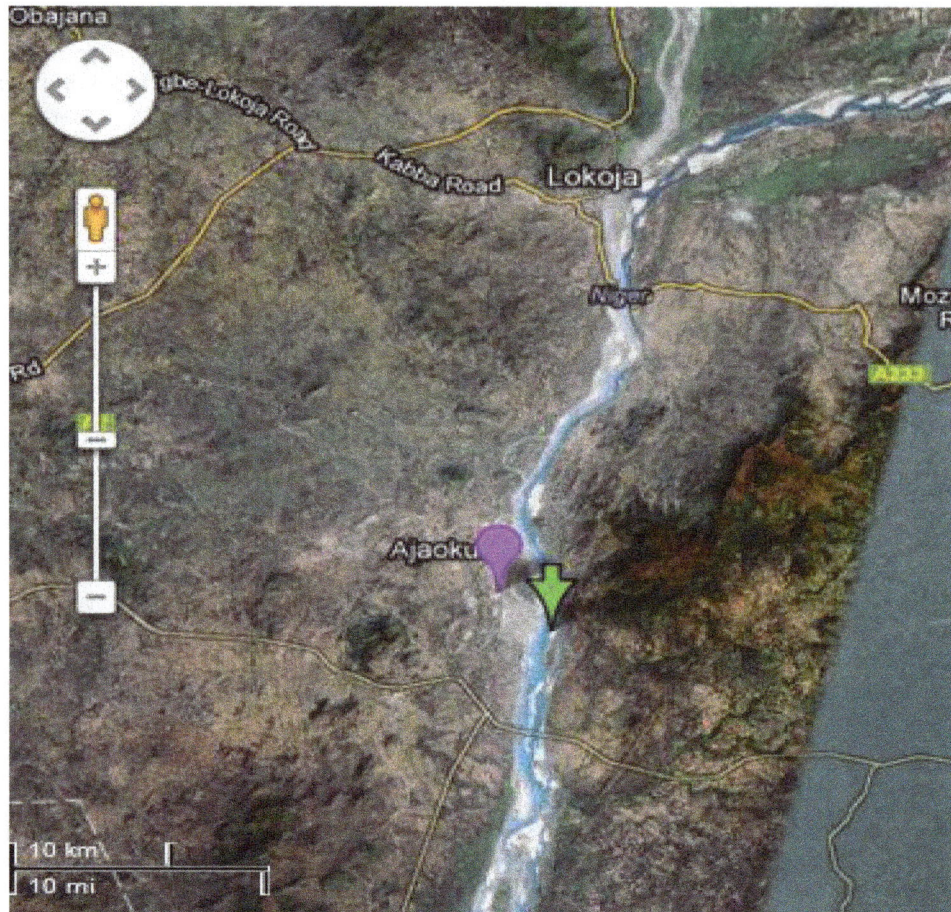

**Figure 1.** Map showing the Lower River Niger Drainage flow from below Lokoja through Ajaokuta down to the massive Niger Delta (Source: Google Maps).

(APHA, 1998) was used. The water samples were mixed from which about 100 ml each of the water samples was measured into separate clean 250 ml beakers. The samples were digested by heating on hot plate to volume of about 20 ml. The solutions of digests were filtered into 100 ml volumetric flasks, and made up mark with distilled water.

**Sediment samples**

Five gram each of sediment samples was weighed into 250 ml nitric acid pretreated teflons beakers. 50 ml and 2 M nitric acid analar grade reagent was added to each beaker and heated in a water bath for two hours (Oniawa, 2000). The resulting sample digests was filtered into 100 ml volumetric flasks and made up to 100 cm$^3$ mark with distilled water.

**Fish samples**

One  gram dry weight of each fish samples muscles (fillets) were weighed into 100 ml kjeldahl flasks into which 1 ml concentrated nitric acid analar grade reagent was added. These were heated in water bath at 80°C for 3 h to ensure complete digestion (UNEP, 1984). The resulting sample solutions were filtered into 100 ml volumetric flasks and made up to 100 cm$^3$ mark with distilled water.

## RESULTS

### Concentrations of heavy metals in water, sediment and fish fillet of *T. zillii*

The concentration levels (mg/L) of heavy metals in water samples collected from River Niger were ranged; Mn, 1.74 to 8.37 (3.85 ± 0.93); Zn, 0.98 to 4.82 (2.72 ± 0.57); Cu, 0.58 to 4.50 (2.17 ± 0.73); Cr, 0.53 to 4.09 (2.08 ± 1.27); Ni, 0.48-1.12 (0.78 ± 0.12); Cd, 0.02-0.13 (0.05 ± 0.02); and Pb, 0.01 to 0.16 (0.03 ± 0.02) (Table 1). The results showed fairly stable metal concentration in water, with distribution sequence in the order Mn > Zn > Cu > Cr > Ni > Cd > Pb. The concentration of metals were within the fresh water guideline requirement for aquatic life recommended by the Canadian Council of Ministers of Environment (CCME, 1999).

The concentrations (mg/kg) of heavy metals in sediments from River Niger were: Zn, 36.64 – 96.23 (70.70 ± 10.68); Mn, 4.97 to 21.77 (13.24 ± 2.04); Pb, 8.84 to 17.52 (12.35 ± 1.14); Ni, 2.65 to 18.61 (9.67 ± 2.91); Cu, 0.89 to 8.21 (3.58 ± 1.32); Cr, 0.48 to 13.08

**Table 1.** Mean concentration levels of heavy metals detected in water, sediments and fish samples.

| Sample type | Cd | Mn | Cr | Ni | Cu | Zn | Pb |
|---|---|---|---|---|---|---|---|
| **Water** | | | | | | | |
| Range | 0.02 – 0.13 | 1.74 – 8.37 | 0.53 – 4.09 | 0.48 – 1.12 | 0.58 – 4.50 | 0.98 – 4.82 | 0.01 – 0.16 |
| Mean concentration ± standard deviation (mg/L) | 0.05±0.02 | 3.85±0.93 | 2.08±1.27 | 0.78±0.12 | 2.17±0.73 | 2.72±0.57 | 0.03±0.02 |
| Fresh water guideline (CCME, 1999) | 0.06 | 4.0 | 0.01 – 5.00 | 1.4 | 0.05 – 2.00 | 5.0 – 15.0 | 0.05 – 0.10 |
| **Sediment** | | | | | | | |
| Range | 0.07 – 0.62 | 4.97 – 21.77 | 0.48 – 13.08 | 2.65 – 18.61 | 0.89 – 8.21 | 36.64 – 96.23 | 8.84 – 17.52 |
| Mean concentration ± standard deviation (mg/kg) | 0.27±0.07 | 13.24±2.04 | 3.38±0.76 | 9.76±2.91 | 3.58±1.32 | 70.70±10.68 | 12.35±1.14 |
| GESAMP guideline (1984) | 0.11 | 770.00 | - | - | 33.00 | 95.00 | 19.00 |
| **Fillets of *T. zillii* fish** | | | | | | | |
| Range | 0.06 – 0.25 | 1.32 – 17.59 | 1.78 – 5.37 | 1.36 – 6.20 | 4.78 – 19.34 | 4.09 – 8.93 | 0.21 – 0.68 |
| Mean concentration ± standard deviation (mg/kg) | 0.15±0.03 | 8.29±3.60 | 3.38±0.65 | 4.03±0.94 | 10.10±2.95 | 8.95±1.43 | 0.42±0.10 |
| FAO/WHO Guideline (1983) | 2.00 | 1250 | 50 | - | 30.00 | 100.00 | 2.00 |

(3.38 ± 0.76) and Cd, 0.07 to 0.62 (0.27 ± 0.07). The distribution of the investigated metals in sediment were in the order Zn > Mn > Pb > Cu > Cr > Ni > Cd. These concentrations are within typical levels for uncontaminated sediments as suggested by the Joint Group of Experts on the Scientific Aspect of Marine Pollution (GESAMP, 1984).

*T. zillii* fish fillets heavy metals concentrations (mg/kg) in fish harvested from River Niger at Ajaokuta, were as follows: Cu, 4.78 to 19.34 10.10 ± 2.; Zn, 4.09 to 8.93 8.95 ± 1.43; Mn, 1.32 to 17.59 8.29 ± 3.60; Ni, 1.36 to 6.20 4.03 ± 0.94; Cr, 1.78 to 5.37 3.38 ± 0.65; Pb, 0.21 to 0.68 0.42 ± 0.10; Cd, 0.06 to 0.25 0.15 ± 0.03. Therefore, *T. zillii* contains safe levels of heavy metals, and this is indicative of the heavy metal levels of their habitat in River Niger at Ajaokuta. The concentration distribution sequence of the investigated metals in sediment were in the order Cu > Zn > Mn > Ni > Cr > Pb > Cd.

although this depends on the magnitude of impact on the physico-chemical and biological conditions. Some heavy metals are taken up and bio-accumulate in higher concentrations, thus producing higher concentration of the substances in the organism than in its environment. This may lead to significant change in the ecosystem structure, where surviving species may have their biomass contaminated, which may result in composition change. Most of these changes may be the result of the accumulation of heavy metals beyond tolerable or health limits.

Early signs of aquatic ecosystem degradation resulting from the accumulation of some heavy metals are usually obvious in biological functions. This may include fish kill, fish/fauna migration, antagonistic or synergistic effects on primary productivity and the depletion of other biological functions. Scott et al. (1986) reported that environmental perturbations including habitat alteration, and the degradation of the intra-gravel

# DISCUSSION

## Concentration of heavy metals in river Niger

Substantial amount of contaminants including heavy metals from anthropogenic sources are deposited into surface water via the atmosphere and seasonal runoffs from over land (George et al., 2004). Water in streams and rivers having contaminants levels surpassing the natural or allowable threshold concentrations were reported to impact poor or fair biological condition on the streams (Boward et al., 1999). Aquatic and terrestrial organisms may be exposed to these substances through aquatic food consumption or from other use of such contaminated streams/rivers.

Aquatic ecosystems with high ecological sensitivity may respond in such a way that there may be sharp changes in ecological functions and or disruption which may lead to habitat loss,

**Table 2.** Accumulation factor of heavy metals concentrations in sediment and fish fillets of *Tilapia zillii* with respect to heavy metals concentrations in water of River Niger at Ajaokuta

| Environment | Accumulation medium | Metal | Accumulation factor | |
|---|---|---|---|---|
| | | | Wet season | Dry season |
| River Niger | Sediment | Cd | 5.4 | 11.6 |
| | Sediment | Mn | 3.4 | 2.7 |
| | Sediment | Cr | 1.6 | 5.4 |
| | Sediment | Ni | 12.5 | 9.3 |
| | Sediment | Cu | 1.6 | 1.6 |
| | Sediment | Zn | 25.9 | 24.8 |
| | Sediment | Pb | 411.6 | 841.5 |
| River Niger | *T. zillii* Fish | Cd | 3.0 | 5.3 |
| | *T. zillii* Fish | Mn | 2.1 | 1.4 |
| | *T. zillii* Fish | Cr | 1.6 | 3.4 |
| | *T. zillii* Fish | Ni | 5.1 | 3.8 |
| | *T. zillii* Fish | Cu | 4.6 | 3.3 |
| | *T. zillii* Fish | Zn | 3.2 | 4.0 |
| | *T. zillii* Fish | Pb | 14.0 | 32 |

environment appeared to have a greater impact on fish species than on the non biotic function of the aquatic ecosystem.

**Accumulation of heavy metals *T. zillii* fish fillets and bottom sediment**

The indices of accumulation (AI) or accumulation factor (AF) of contaminants may be used to define the health, safety and potential risk of exposure of aquatic ecosystems resources and components. Derivations of AIs for different contaminants facilitate the possibility of predicting the status of an environment based on comparison of monitoring indices with baseline index. Water quality initiative (WQI) of U.S. EPA illustrates the importance of the linkage between sediments and the water column and its influence on exposure of all aquatic biota (Cook and Burkhard, 1998). Heavy metals do not remain for long in water, hence they re-distributed by partitioning between water, biotic and non-biotic functions of the aquatic ecosystem, especially sediments, fish, microphytes etc. The distribution of metals between sediment and water column and between biological species such as *T. zillii* and water column can be characterized with sediment – water or fish – water quotient index (accumulation factor). Accumulation index or factor (AI or AF) is defined as the ratio of contaminants mean concentration $C_o$ in component/organism and that in the surrounding water $C_w$ at steady state, via all routes of exposure (AF = $C_o/C_w$) (Walker et al., 2003).

The concentrations of heavy metals in sediments of River Niger at Ajaokuta were found to be higher than in its water. Sediment heavy metals index ratio (AI) or accumulation factor (AF) with respect to heavy metals concentrations in water samples (Table 2) were: Cd, 5.4; Mn, 3.4; Cr, 1.6; Ni, 12.5; Cu, 1.6; Zn, 25.9 and Pb, 411.6, and Cd, 11.6; Mn, 2.7; Cr, 5.3; Ni, 9.3; Cu, 1.6; Zn, 24.8 and Pb, 841.5 during wet and dry season respectively. This showed that sediment concentrate Pb several folds, in the order of 411.7 to 841.5 more than Pb levels in water. Zinc, Ni, Cd, Cr and Mn also accumulated 24.8 to 25.9; 9.3 to 12.5; 5.4 to 11.6 1.6 to 5.3; and 2.7 to 3.4 folds respectively in sediments, compared to their levels in water from the same river, while Cu in sediment was accumulated at approximately two fold of its concentrations in water. This implies that, the concentration levels of heavy metals found in sediments were higher than in water (sediment > water) in the same hydrological environment as expected, that is, in River Niger.

*T. zillii* fish also accumulated several folds concentrations of heavy metals in their muscles (fillets) compared with metals concentration levels in water. The accumulation factor of heavy metals of *T. zillii* fish in Rivers Niger were: Cd, 3.0 to 5.3; Mn, 1.4 to 2.1; Cr, 1.6 to 3.4; Ni, 3.8 to 5.1; Cu, 3.3 to 4.6; Zn, 3.2 to 4.0; Pb, 14.0 to 32.0. There was also several fold accumulation of Pb in *Tilapia zillii* fish, 14.0 to 32.0 folds higher compared with Pb levels in water of River Niger at Ajaokuta. Nickel, Cu, Zn, Cd, and Mn accumulated 5.1; 4.6; 3.2; 3.0 and 2.1 folds respectively than in water in River Niger, with Cu at 1.6, approximately one and a half fold its levels.

Indices of heavy metal accumulation may be generic and species specific as well as environment specific. This takes cognizance of the organism uptake/intake, metabolism and chemical composition. Cook and Burkhard (1998) reported that the degree to which quantitative difference in accumulation of heavy metals may be attributed to a particular ecosystem conditions, apart from the influence of organic carbon which determines the bioavailability variables. Also the degrees of depuration of heavy metals in different fish and organisms, affects their retention capacity. Variation in metabolism and mechanism of depuration of different fish accounts for the species differences in retention or holding of heavy metals.

In general, the AIs of heavy metals in sediment were higher than AIs of heavy metals in fish, while metal levels in the water column were the least in water of River Niger.

## Seasonal influence on heavy metal accumulation in aquatic ecosystem

The effect of seasonal climate change was evaluated by comparing the observed indices of heavy metal accumulation in the bottom sediment and fillets of *T. zillii* fish samples during dry and wet season. The accumulation indices of the concentration levels of the heavy metal in fillets of *T. zillii* and bottom sediment of River Niger were slightly higher during dry season water than in wet seasons except for Pb and Cd. This could be the result of stability in flow conditions and less water turbulence associated dry season low water volume in the river drainage.

Deus et al. (2013) reported that water quality is seasonally variable especially with properties including water temperature, dissolved oxygen, and contaminants load. This is because of the variations and distributions of the frequency and the percentage of extreme precipitation to the daily precipitation data of annual rainfall (Zhang et al., 2008). Wang et al. (2008) however, noted that annual extreme precipitation changes and stream flow processes resulted in little changes in terms of various indices. Aside from the sediment AI for Pb (841.5) and Cd (11.6) during dry season, which was twice (p<0.05) the AI (411.6) and (5.4) respectively calculated for wet season, the calculated AIs for the investigated metal were not significantly different (p>0.05) during dry and wet season. Similarly, the AI for Pb (32) and Cd (5.3) in fish fillets during dry season was more than twice the Pb AI (14) and Cd (3.0) respectively calculated for wet season. Mitsch et al. (1989) reported that marginal changes can affects seasonal patterns of nutrient uptake and release especially during the growth season of some aquatic biota, where uptake and immobilization by microflora, microfauna and macrophytes retain contaminants. However the dieback of aquatic plants releases contaminant back to the water column through decomposition. However, the relationship between contaminants discharged in water and aquatic ecosystem is however complex.

## Uncertainty in accumulation indices

Although study results showed that there was no significant change in AI of the investigated metals except for Pb and Cd, the use of the accumulation indices for monitoring and predictive risk assessment (MPRA) purposes may be limited by some degree of uncertainty based on the degree of sensitivity of the aquatic ecosystem. For instance, the uncertainty associated with the use of fish AI may arise is as a result of their migratory nature, and their environment and site specific variation in metabolism and depuration. This may leaves the AI of fish with some uncertainty especially in its use for MPRA. Similarly changes in natural composition of sediment (e.g. reduction in sediment clay mineral and organic carbon) resulting in weak metal stabilization. This may lead to changes in AIs.

## Conclusion

The result showed that the accumulation indices of the investigated heavy metals in River Niger by Ajaokuta Steel Company, Ajaokuta were nearly consistent through the study period, except those for Pb and Cd which showed significant seasonal variation. Although the status an aquatic ecosystem can be determined by compliance monitoring, the result suggest the use of accumulation indices and or factors as a comprehensive status and risk based tool, required for MPRA purposes, especially when ASC comes into operation. This is because of the comprehensive inclusion of contaminant exposure route vis-a-vis metal partitioning into different compartments and biological assimilation. Accumulation of heavy metals in aquatic ecosystem depends on the concentration levels, and the partitioning of the metals in water.

## REFERENCES

Abrate T, Hubert P, Sighomnou D, (2013). A study on the hydrological series of the River Niger. Hydrol. Sci. J. 58(2):271–279.

Adolfsson-Eric M, Akerman G, McLachlan MS (2012). Measuring bio-concentration factors in fish using exposure to multiple chemicals and internal benchmarking to correct for growth. Environ. Toxicol. Chem. 31(8):1853-1860.

APHA (1998). Standard methods for the Examination of Water and Wastewater 20th Ed. American Public Health, Association (APHA), American Water Works Association (AWWA), Water Pollution Control Federation (WPCF), Washington DC, USA, P. 1268.

ATSDR (1999). Potential for Human Exposure. Agency for Toxic Substances and Disease Registry, U. S. Public Health Services "Toxicology Profile for Mercury" March and April 1999 Media Advisory, New MRLS for toxic substances, Retrieved April 21, 2006 from http://atsdr.cdc.gov/toxprofiles/tp46-c5.pdf.

Bailey MC, Hamilton DP (1997). Wind induced sediment re-suspension: A lake wide model, Ecol. Model. 99(2/3):217–228

Boward D, Kayzak P, Stranko S, Hurd M, Prochaska T, (1999). From the mountains to the sea: The state of Maryland's freshwater streams. EPA 903-R-99-023. Maryland Department of Natural Resources, Monitoring and Non-tidal Assessment Division, Annapolis, Maryland.

CCME (1999). Water Quality guidelines for the protection of aquatic life in Chapter 6, Canadian Environmental Quality Guidelines. Canadian Council of Ministers of the Environment, CCME Task Group on Water Quality Guidelines, Guideline Division, Winnpeg and Ottawa, Canada.

Cook PM, Burkhard LP (1998). Development of bioaccumulation factors for protection of fish and wildlife in the Great Lakes. Bethesda, MD, Sept. 11-13, 1996. EPA 823-R-98-002. U.S. EPA Office of Water, pp. 3-19; 3-27.

Deus R, Brito D, Kenov IA, Limac M, Costa V, Medeiros A, Neves R, Alvesa CN (2013). Three-dimensional model for analysis of spatial and temporal patterns of phytoplankton in Tucuruí Reservoir, Pará, Brazil. Ecol. Model. 253:28–43.

FAO/WHO (1983). Compilation of legal limits for hazardous substances in fish and fishery products. FAO Fish Circ. 464:5-100.

GESAMP (1984). The Health of the Ocean: Review of potentially harmful substances – cadmium, lead, and tin. Repository Studies of the Joint Group of Experts on the Scientific Aspect of Marine Pollution. GESAMP 22:1 -114.

Hamilton DP, Schladow SG (1997). Prediction of water quality in lakes and reservoirs: Part I. Model description. Ecol. Model. 96(1–3):91–110.

Hubert P (2000). The segmentation procedure as a tool for discrete modelling of hydrometeorological regimes. Stochastic Environ. Res. Risk Assess. (SERRA) 14:379–304.

Leclerc M, Boudreault A, Bechara JA, Genevieve C (1995). Two-dimensional hydrodynamic modeling: A neglected tool in the instream flow incremental methodology. Trans. Am. Fish. Soc. 124:645–662.

Mitsch WJ, Reeder BC, Klarer DM (1989). The role of wetlands in the control of nutrients with a case study of western Lake Erie. In: Ecological Engineering: An Introduction to Ecotechnology. (WJ Mitsch and SE Jorgensen, eds.). Wiley, New York, pp. 129-158.

Moustafa MZ (2000). Do wetlands behave like shallow lakes in terms of phosphorus dynamics? J. Am. Water Resour. Assoc. 36(1):43-54.

Okabe T, Amou S, Ishigaki M (1993). A simulation model for sedimentation process in gorge-type reservoirs. In: Hadley RF, Mizuyama T (Eds.), Sediment Problems: Strategies for Monitoring, Predicting and Control IAHS Publ. No 217. International Association of Hydrological Sciences, Wallingford, Oxfordshire, UK, pp. 119–126.

Oniawa PC (2000). Roadside topsoil concentration of lead and other heavy metals in Ibadan, Nigeria. Soil Sedim. Contam. 10(6):577-591

Orlob G, (1975). Present problems and future prospects of ecological modelling. In: Russell CS (Ed.), Ecological modeling in a resource management framework. Resources for the future, Inc., Washington DC, USA. pp. 283–312.

Schladow SG, Hamilton DP (1997). Prediction of water quality in lakes and reservoirs: Part II. Model calibration, sensitivity analysis, and application. Ecol. Model. 96(1–3):111–123.

Schlechtriem C, Fliedner A, Schäfers C (2012). Determination of lipid content in fish samples from bioaccumulation studies: Contributions to the Revision of Guideline OECD 305. Environ. Sci. Eur. 24:13-20.

Scott J, Steward C, Stober Q (1986). Effects of urban development on fish population dynamics in Kelsey Creek, Washington. Trans. Am. Fish. Soc. 115:555-567.

Shen H, Tsanis IK, D'Andrea M (1995). A three-dimensional nested hydrodynamic/pollutant transport simulation model for the near shore areas of Lake Ontario. J. Great Lakes Res. 21(2):161–171.

Sieckhaus JF (2009). Chemical, human health and the environment. A guide to the development and control of chemical and energy, p. 271.

UNEP (1984). Sampling of selected marine organisms and sample preparation for trace metal analysis. Reference Methods for Marine Studies No. 7 Rev. 2 UNEP, Geneva.

Waldichuk M (1977). Global Marine Pollution: An Overview. Intergovernmental Oceanographic Commission Technical Series 18, UNESCO, P. 96.

Walker DJ, Clemente R, Roig A, Bernal MP, (2003). The effects of soil amendments on heavy metal bioavailability in two contaminated Mediterranean soils. Environ. Pollut. 122:303-312.

Wang W, Chen X, Shi P, van Gelder PHAJM (2008). Detecting changes in extreme precipitation and extreme streamflow in the Dongjiang River Basin in Southern China. Hydrol. Earth Syst. Sci. 12:207–221.

Zhang DQ, Feng GL, Hu JG (2008). Trend of extreme precipitation events over China in last 40 years. Chin. Phys. B. 17:736–742.

Zheng Z, He L, Li J, Wu Zb (2007). Analysis of heavy metals of muscles intestine tissue in fish – in Banan section of Chingqing from Gorges Reservoir, China. Pol. J. Environ. Stud. 16(6):949 – 958.

Zhou Q, Zhang J, Fu J, Shi J, Jiang G (2008). Biomonitoring: An appealing tool for assessment of metal pollution in the aquatic ecosystem. Analytica Chimica Acta 606:135–150.

# 14

# Enhancement of cutting tool surface coating quality using ionized gaseous medium (IGM)

**S. O. Yakubu**

Department of Mechanical Engineering, Nigerian Defence Academy, Kaduna.

**The use of ionized gaseous medium (IGM) to prepare a hard alloy material surface by grinding and its effect on the coating quality was investigated. During grinding, IGM was fed to the cutting area by different methods namely: clockwise (longitudinal), anticlockwise (opposed) and transverse (Cross) feeding, respectively. Thereafter, the samples were coated with a titanium nitride on a modern vacuum, ionizing apparatus HHB-6.6-u1 by physical vapour deposition (PVD) method, known as condensation and ionized bombardment (CIB). The analysis of the results and tests carried out revealed that IGM improves the quality of coating, especially when IGM was fed anticlockwise and when the corona discharged current ($\tau_k$) was equal to 50 μA. The lowest component forces were also gotten by anticlockwise feeding for example, the component forces ($P_y$ $P_z$) were 2 times and 1.5 times lower compared to longitudinal and cross feeding of IGM, respectively. The micro hardness of inserts ground with IGM was about 10% higher than those ground with other types of fluids. The micro photograph of inserts structure revealed a distinct and better coated layer for the inserts surfaces prepared prior to coating with IGM. Whereas inserts prepared with compressed air and without cutting fluids showed very blur and indistinct coated layer. It was established that inserts whose surfaces were prepared with IGM and then coated with the titanium nitride (TiN) showed tool-life of about 4 times greater than others.**

**Key words:** Ionized gaseous medium, hand alloy tool inserts, coating, condensation and ionized bombardment, titanium nitride.

## INTRODUCTION

The high rate of cutting tools wear during machining (that is, during metal cutting operations, especially during rough machining). It has led to a constant high demand for tools with high wear resistance. This is more pronounced in railway industries (companies) where large quantity of metal layers is removed from the rail wheel-pairs surface everyday. There are many ways of increasing the cutting tool war-resistance (tool life). Some of these ways are the use of cutting fluids, coating the tools surfaces, optimizing the methods of re-sharpening the tool cutting edge(s), etc. However, coating enhances the tool life more than any of the other method (Anikeev et al., 1980).

Irrespective of the type of coating employed, the quality of the coating (that is, adhesiveness of the coating on the surface matrix of the material being coated) depends largely on the quality of the material surface. Therefore, the hard alloy material surface preparation prior to coating deposition is vital.

IGM works on a simple basic principle of drawing atmospheric air through a compressor and the compressed air is passed through a dry electrostatic cutting tool cooling apparatus "Varkash" where it is ionized. The ionized air is then passed to the cutting area (Yakubu and Popov, 2003).

There are two main groups of coating methods namely;

the chemical vapour deposition (CVD) and the physical vapour deposition (PVD). There are varieties of CVD methods for example, in Russia, there is the thermo-gaseous technology (GT). In some countries like Germany, USA, Sweden, Australia, Japan, etc. 60-80% of their hard alloy tools are coated with CVD method (Samoilov et al., 1988).

Major world producers of had alloy tools like Kennametal-Hertel in Germany, Corboloi (USA), Sandvic-coromant in Sweden, Planzee in Australia, Widia (Germany) and Mitsubishi Carbide (Japan), etc. have designed a new technology of coating based on CVD principle. The temperature range for CVD method is 1000-1100 (Lenskaya et al., 1982).

The principle of PVD is based on vapourization of cathode substances in a vacuum of coating deposition apparatus with a simultaneous feeding of reactive gas for example, $N_3$, $O_2$, $CH_4$, etc. There are different types of PVD methods:

1. Methods based on condensation of substances from plasma phase in a vacuum with ionized bombardment for example, CIB in Russia and ion bond in USA.
2. Magneto-electronic ionized spurting (MIS) in Switzerland and its Russian variation, Magnetic Reactive Spurting (MARS).
3. Ionized cladding in Russia.

The ability to control the temperature of the coating zone in PVD makes it possible to coat not only hard alloy materials, but also high speed steels. PVD is a more common and universal method of coating than CVD especially for single layer, multilayer and composition coatings based on nitrides, carbides, carbonitrides, oxides, boridize of metals in groups IV-VI of the periodic table.

CIB's principle of operation is based on the generation of coating substances with cathode spot (flux) using a highly accurate low voltage vacuum arc discharge exclusively in the cathode material vapour. The feeding of the vacuum camera of the coating apparatus with reactive gases for example, nitrogen, methane, etc. in an ionized bombardment environment, facilitates a smooth plasma-chemical reaction which brings about condensation of coating on the cutting surface of a cutting tool (Andrei, 1980). The evaporating plasma-chemical reactions, ionized bombardment and condensation processes take places in the vacuum camera whose metallic body serves as anode.

This research is aimed at preparing a very good surface for a quality coating using unconventional cutting fluids which is ecologically pure that is, human and environmental friendly.

## EXPERIMENTAL PROCEDURES

### Hard alloy inserts surface preparation prior to coating

The preparation of the hard alloy tool inserts surface prior to coating

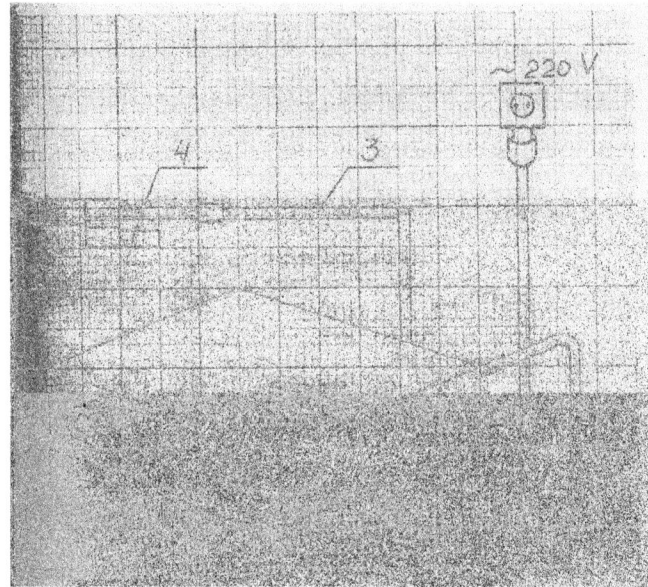

**Figure 1.** The general view of Varkash. 1- power source, 2-nozzle, 3- high voltage connecting cable, 4- "shutser".

was conducted on a flat surface grinding machine using IGM as the cutting fluid. The feeding of IGM to the grinding zone was done with the aid of dry electrostatic cooling apparatus known as "Varkash" (Figure 1). The ionization of the compressed air coming through the compressor was done in the Varkash nozzle with the help of corona discharged current. The hard alloy tool inserts used were from the following grades of hard alloys; T14K8 and T5K10.

In order to improve the efficiency of IGM, it was optimized by feeding it to the grinding area through three different directions namely; longitudinal (clockwise), cross (traverse) and anticlockwise feeding and then, the most effective feeding direction was determined. The values of the corona discharge current were varied from 25 to 100 µA. The grinding was carried out under the following conditions (Regime): Grind speed = 30 m/s, work piece speed (that is, cross feeding) = 2.5 mm/stroke, longitudinal feed = 10 m/min and compressed air pressure = 0.3 MPa.

### Coating deposition process with titanium nitride

Wear resistance coating was carried out on the surface of hard alloy tool inserts after they were been grounded using IGM. The type of coating deposition employed was PVD using a plasma condensation and ionized bombardment method (CIB). CIB comprises of two sequential processes that is, ionized bombardment and condensation of coating. CIB principle of operation is based on the generation of coating substance with cathode flux in a high accurate low voltage cathode material vapour (5-7).

The titanium Nitride (TiN) coating was conducted on a modern ionized vacuum equipment HHB-6.6-U1 (Figure 2). This apparatus can be used to coat cutting tools surfaces with diameter up to 200 mm and length of 250 mm for both hard alloy and HSS materials.

The cathode material used was a titanium evaporator BT1-00-GOST19867-74 by the Russian National Standard. Before the titanium nitride coating started, the hard alloy inserts (samples) underwent pneumatic treatment and degreasing (extraction of grease). This operation was done on a vibrating apparatus called "Vibrint", for 2.5 min with a pressure of 0.1 – 0.2 MPa until tool nose radius of 0.03 – 0.04 mm was achieved, then the samples were put

**Figure 2.** Ionized vacuum coating apparatus HBB-6.6-U1.

into the vacuum camera of ionizing apparatus where preliminary ionized cleaning of the samples within temperature of up to 700°C was conducted. After the ionized cleaning of the samples, the evaporators were put off and allowed to cool to the desired condensation temperature and then put on again to do coating condensation on the samples surfaces. After the coating condensation process, the evaporators were off again and reactive gas ($N_2$) was fed into the vacuum camera with residual pressure of $133 \times 10^5$ Pa.

The residual pressure in the camera was created and controlled from the control board of the coating apparatus. Though the hard alloy tool inserts temperature, the amount of gas fed in and the management of the whole process was done from the control board.The samples surface coating with TiN was executed under the following conditions:

| | | |
|---|---|---|
| The evaporators' arc current | - | 100 A |
| The pressure of nitrogen | - | 0.2 Pa |
| The tool voltage | - | 200 volts |
| The hard alloy inserts tension | - | 100 volts |
| Coating time | - | 20 min |

The chemical composition of the TiN coating was varied by regulating the      arc current of the titanium evaporator, pressure of nitrogen and voltage on the hard alloy tool inserts. The coating quality analysis on the hard alloy tool inserts surface was carried out using an integral evaluation of quality parameters (Vereschaka and Tretyakov, 1986). The use of this method allows in totality

to define and evaluate the main parameters of coating such as adhesive strength of the coating to surface matrix, brittle strength of coating, surface roughness, micro hardness and coating thickness. The quality of coating and its adhesiveness to the surface matrix of the hard alloy inserts surface was determined by the formation of "brittle" cracks after the application of a diamond indenter with a force of 600 N on Rockwell hardness tester TK-2. The test was carried out on three different points on each of the coated insert. The micro hardness of the coated hard alloy inserts were defined using an ultra micro hardness tester "micro Duromat 400" made by Reichert Jung, USA. A diamond indenter with a load of 100 g was used. The test was carried out on seven different points on the coated hard alloy insert surface, other various tests were carried out on the coated inserts such as, microphotograph of their structures, influence of IGM feeding method on the component forces of grinding, influence of IGM on the wear resistance (that is, tool life) of the hard alloy materials. The last tests was done on a universal lathe machine by turning a low carbon steel where the coated hard alloy inserts served as the cutting tool.

## RESULTS AND DISCUSSION

### The effect of IGM on the quality of coating

The result of the test carried out by the method which

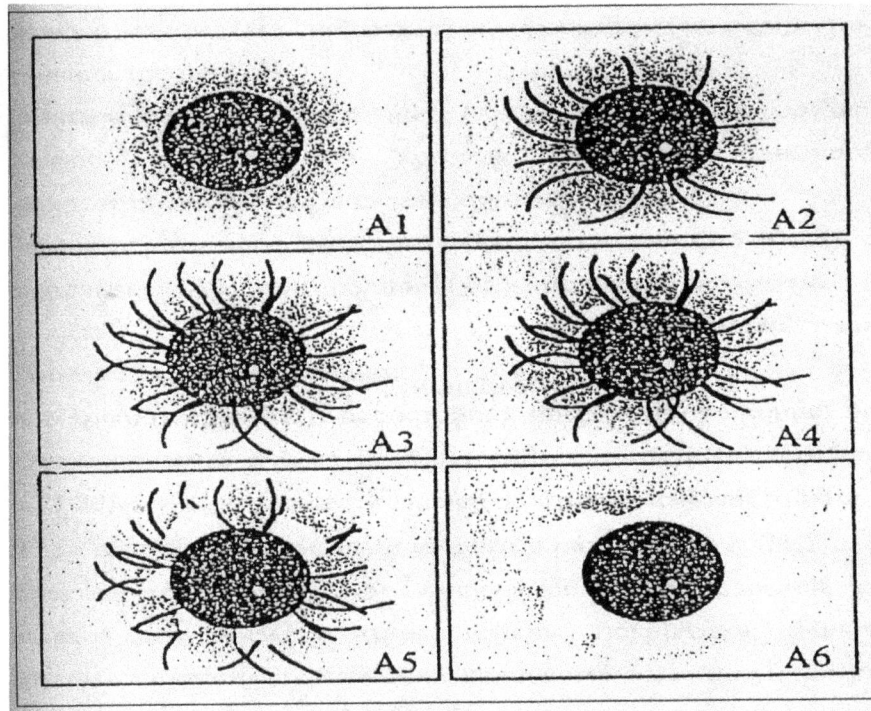

**Figure 3.** Evaluation scale of surface coating quality of hard alloy tools

was earlier on described in the experimental process is presented in Figure 3. The black circle is the imprint left by the diamond indenter on the coated hard alloy material surface. The lines are micro-cracks resulting from the impact force of the indenter.

Diagram $A_1$ and $A_2$ are samples (Figure 3) whose surface were grounded with anticlockwise feeding of IGM. $A_3$ and $A_4$ are inserts whose surfaces were grounded with longitudinal (clockwise) feeding of IGM and lastly $A_5$ and $A_6$ are inserts whose surface were grounded with traverse feeding of IGM.

Samples $A_1$ showed a very high coating quality and there was no sign of exfoliation or formation of micro cracks around the indenter's imprint. Samples $A_2$ have good coating quality because there was no observation of coating exfoliation, but there were very slight traces of micro cracks. In samples $A_3$, an increase in micro racks was observed but no coating exfoliation was found in the samples. The quality of coating in samples $A_3$ is considered normal. In samples $A_4$, there were a lot of micro cracks and a little coating exfoliation. The coating quality of simple $A_4$ is considered fairly okay (satisfactory).

However, the samples $A_5$ and $A_6$ are of poor quality since there was high level of micro cracks and exfoliation of deposited coating. There was a total coating exfoliation in sample $A_6$. Judging from these results, one can conclude that the efficiency of IGM in enhancing the coating quality to a large extent depends on the methods of its feeding during grinding prior t o coating. In other

words, IGM's cooling, lubricating and penetrating properties were more effective during its anticlockwise feeding. Thus, the best quality surface for coating was obtained by anticlockwise feeding of IGM.

**Effect of IGM on micro hardness of the coated inserts**

The results of the micro hardness of two different grades of hard alloys T14K8 and T5K10 showed that all the coated hard alloys inserts had higher micro hardness than those which were neither coated nor ground (that is, master samples). It is worthy to note that among the inserts coated, the micro hardness of those whose surfaces were grounded using IGM was greater than those surfaces that were not grounded.

It was also observed that the micro hardness of hard alloys inserts, whose surfaces were grounded using IGM were generally higher than those grounded with other types of cutting fluids (Figure 4). This high increase in micro hardness of inserts ground with IGM may be associated with the formation of new phase. This shows the importance of preparing (grinding) the insert surface before coating (with wear resistance coating).

Furthermore, the micro hardness of the ground surfaces of coated insert T14K8 was 14% higher than its ungrounded surface. Whereas the micro hardness of the grounded surface of alloy hard insert T5K10 was 10% higher than its unground surface. It was established that micro hardness of hard alloy tips (inserts) got from hard

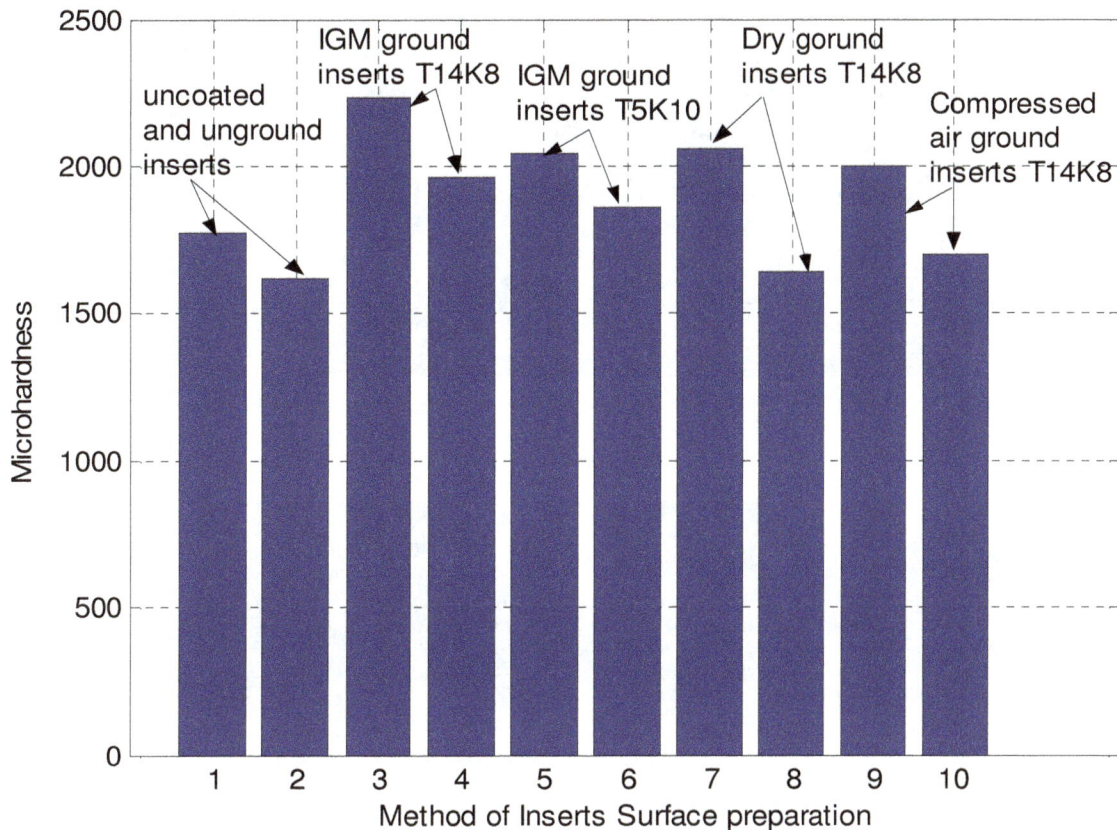

**Figure 4.** The influence of the surface preparation methods on the coated inserts' micro-hardness (1,2 – master samples; 3,4 - ground and unground surfaces T14K8; 5,6 – ground and unground surfaces T5K10; 7,8 – coated and uncoated inserts; 9, 10 – coated and uncoated).

alloy grade T14K8 is generally higher than those got from T5K10.

## Microstructure of coated insert from hard alloy grade T14K8

The micro photograph of the structure of the hard alloy insert which is coated with TiN is presented in Figure 5. In the Figure 5A, is a sample of hard alloy insert ground without the application of cutting fluids; 5B is a microphotograph of hard alloy insert ground with IGM, while 5C is an insert ground with pressurized air. It can be seen from the graph that the three samples have different micro photographic structures. There are three distinct areas in each of the photograph: black, coated layer and base metal. The black part is an empty space, the coated layer indicates the quality of coating and base metal. It was noted that coated layer of the inserts ground with IGM was the most distinct out of the three inserts. This is represented by a very bright white layer between the black part and base metal.

In sample A, it is difficult to differentiate between the base metal and the coated layer. In "C", coated layer is a

bit better, but yet very blur. Thus it is obvious that IGM gives a better coating quality.

## Effect of IGM on the hard alloy inserts tool life (wear resistance)

The efficiency of a cutting tool can be defined through many factors. But the most important is the ability of the cutting tool to resist wearing or failure when it comes in contact with work piece (Yakubu, 2000).

The wearing of a cutting tool is a gradual process subjected to the simultaneous mutual interactions between abrasive, adhesive-fatigue, chemical-oxidizer and diffusion processes (Elgomayel et al., 1979; Gyrevich, 1979). Its effect on tool life and machined surface is negative. Therefore efforts are always made to reduce the tool wearing. The result of the test executed on a TiN coated hard alloy inserts T5K10 was represented graphically in Figure 6, where:

1. Uncoated hard alloy insert from T5K10 used to turn low carbon steel with application of cutting fluid (CF).
2. The same uncoated hard alloy inserts but with the

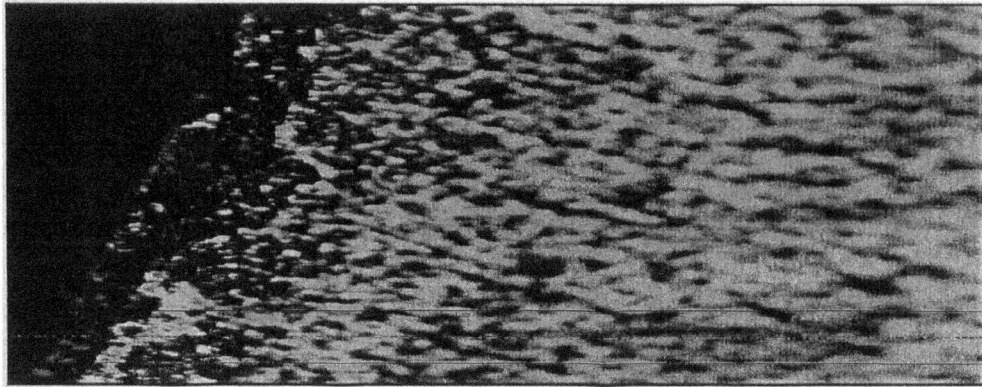

**A**

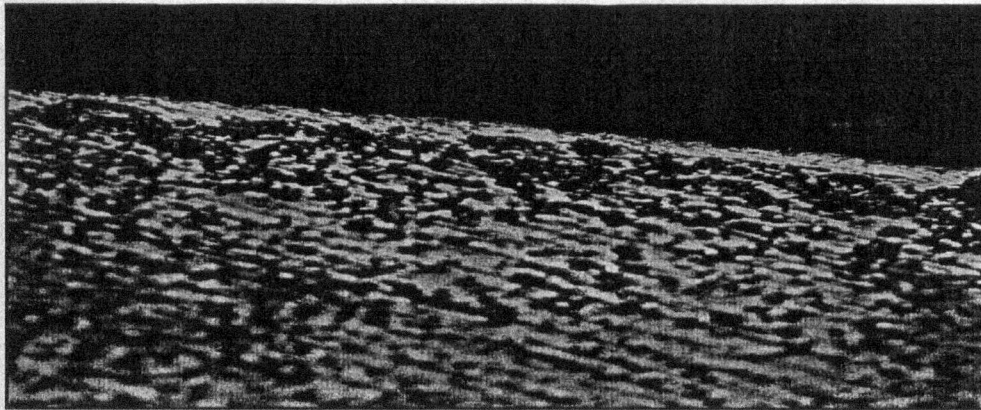

**B**

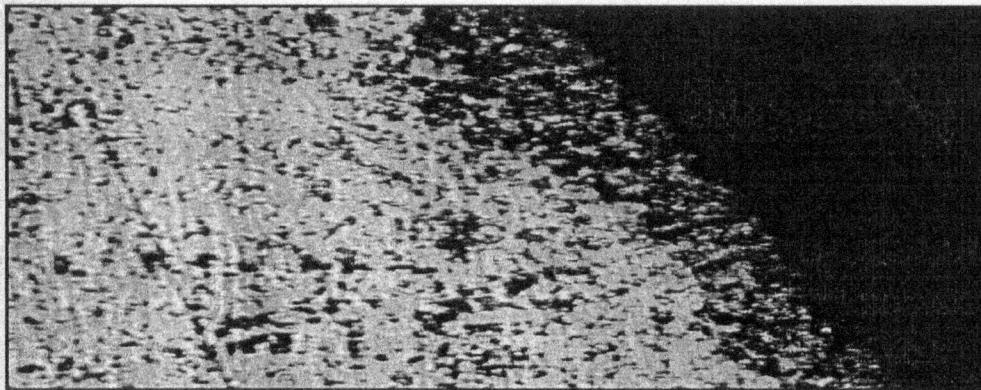

**C**

**Figure 5.** The microstructure of hard alloy inserts whose surfaces were grounded using different methods. A , insert 'dry' ground; B,  insert ground with IGM; C, insert ground with pressurized air.

application of IGM as CF.
3. A coated inserts under "dry turning".
4. A coated inserts used for turning with the application of IGM.

The result indicates first and foremost, that coating can drastically reduce tool wear (flank wear).Secondly, the use of IGM decreases tool wear even more. The obtained

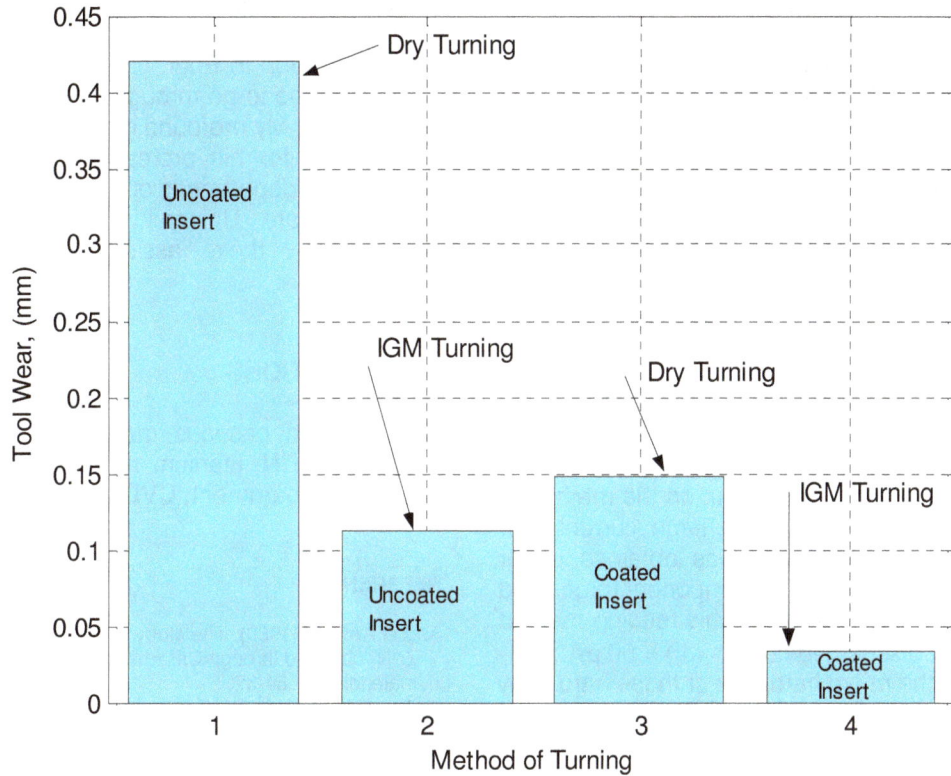

**Figure 6.** The influence of IGM on the tool surface wears.

**Table 1.** The influence of IGM on the component forces, during grinding

| S/N | Corona discharged current ($I_k$), μA | Anticlockwise feeding of GM Component forces, N | | Clockwise feeding of IGM Component forces, N | | Cross feeding of IGM Component forces, N | |
|---|---|---|---|---|---|---|---|
| | | $P_y$ | $P_z$ | $P_y$ | $P_z$ | $P_y$ | $P_z$ |
| 1 | 25 | 38.83 | 19.30 | 52.24 | 29.31 | 46.26 | 23.61 |
| 2 | 50 | 23.71 | 14.62 | 43.67 | 23.96 | 35.05 | 21.61 |
| 3 | 75 | 48.15 | 18.33 | 51.54 | 27.30 | 30.44 | 19.43 |
| 4 | 100 | 26.45 | 16.65 | 35.58 | 20.92 | 67.27 | 27.13 |

result shows that inserts "2" reduction in tool wear is about 4 times compared to insert "1" just with application of IGM without coating. With the application of IGM and coating, insert "4" indicated a decrease in wear to about 10 times lower than insert "1" that is, an increase in tool life of about 10 times greater than inserts "1". Insert "4" has about 4.5 times increase in tool life compared to insert "3" and about 4 times compared to insert "2". Thus, the essence of coating and application of IGM cannot be over emphasized.

## The effect of IGM on the component forces during grinding

The effectiveness of IGM does not only depend on the method of its feeding to the cutting zone, but also the corona discharged current. Therefore, IGM efficiency varies according to the value of corona discharged current ($\tau_k$).

It was noted that IGM was most effective in reducing the component forces (normal and tangential) of grinding when corona discharge current ($\tau_k$) values were 50 to 75 μA for the three methods of its feeding (Table 1).

However, in anticlockwise feeding, the values of the component forces for all values of $\tau_k$ (25, 50, 75, 100) μA, are lower than the values obtained from other methods of feeding for example, when the value of $\tau_k$ = 25 μA, the component forces values for anticlockwise feeding were $P_y$ = 39 N and $P_z$ = 19 N. For clockwise feeding, they were 52 and 29 N and for traverse feeding they were 46 and 24 N for normal ($P_y$) and Tangential ($P_z$) forces, respectively. When the value of $\tau_k = 50$ µA: in the anticlockwise feeding of IGM, $P_y = 24$ N and $P_z = 15$ N; in the clockwise feeding, $P_y = 44$ N and $P_z = 24$ N and in traverse feeding, $P_y = 35$ N and $P_z$ 22 N.

The high reduction in the component forces during anticlockwise feeding of IGM was attributed to its high lubricating and cooling performance due to better penetration. As a result, there was drastic reduction in temperature and friction at the contact area.

## Conclusion

Using IGM to grind the hard alloy materials before coating enhances the quality of coating.

The effectiveness of IGM depends both on the method of its feeding and value of corona discharge current. For instance, the best coating quality was obtained under anticlockwise feeding and both the component forces and the tool wear were very small under this feeding method and with the corona discharged current ($J_k$) = 50 µA.

It was noted that the micro hardness of those hard alloy tools grounded with IGM were higher than those grounded with other methods (compressed air and dry grinding) for example, inserts ground with IGM had micro hardness of 15 and 12% greater than those grounded with compressed air and dry grinding, respectively.

The microphotograph of the structure of the coated inserts indicated that inserts whose surface were grounded with IGM prior to coating have better quality layer, that is, good clinginess to metal base and distinct coated layer from base metal. It was established that the micro hardness of hard alloy tool from grade T14k8 are generally higher than those from grade T5K10.

## RECOMMENDATION

During grinding, IGM should be fed anticlockwise with corona discharged current ($\tau_k$) value = 50 µA and for the other methods (longitudinal and traverse feedings). $\tau_k$ =100 µA for longitudinal and $\tau_k = 50 - 75$ µA for traverse feedings, respectively. Better quality single layer coating is better done using CIB method on HH6.6U1 apparatus under the following conditions: the evaporators' arc current = 100A, the pressure of nitrogen = 0.2 Pa, coating time = 20 min

## ACKNOWLEDGMENTS

I say thank you to Engr. C.O. Izelu and Dr. M.Y. Onimisi for taking time to go through this work and making useful suggestions. My profound gratitude goes to the Prof. A.S. Vereschaka for his professional advice and the entire staff of the department of High Effective Technologies, Moscow State University of Technology, STANKIN, Russia for their assistance in conducting the experiments.

## ABBREVIATIONS

**IGM**, Ionized gaseous medium; **PVD**, physical vapour deposition; **TiN**, titanium nitride; **CIB**, condensation and ionized bombardment; **CVD**, chemical vapour deposition.

### REFERENCES

Andrei AAG IV (1980). The study of some condensers' properties (Ti-N, Zr-N) obtained through CIB method in a book. Phys. Chem. treatment materials, pp. 64-67.

Anikeev AI, Anikin VN, Toropchenov VS (1980). coating as ways of increasing/improving cutting tools efficiency in a book "Modern hard alloy tool and its rational application" Leningrad, pp. 40 – 44.

Elgomayel JI, Radovich JF, Tseung MH (1979). The style of wear mechanism of titanium carbide coated tools. Int. J. machine tools design No. 4:205-219.

Gyrevich DM (1979). Wearing of coated hard alloy tool inserts during turning machine production Bulletin No. 6:45-47.

Lenskaya TG, Toropchenkov VS, Anikeev AS (1982). Tungsten free coated hard alloys in a book "Production and Applications of Hard alloys", Metallurgy Publisher, Moscow, pp. 107 –109.

Samoilov ÉF, Éikhmans VA, Fal K (1988). Hard alloy cutting tool for metal cutting-Reference book, machine production publisher, Moscow. P. 368.

Vereschaka AS, Tretyakov IP (1986). Coated cutting tools. Machine production publisher, Moscow, P.196.

Yakubu SO (2000). Coated hard alloy tools efficiency enhancement by optimizing its surface preparation prior to coating, dissertation thesis, Moscow P. 285.

Yakubu SO, Popov AU (2003). Ecological pure Technology of grinding and Turning machine parts, Science week publisher, Moscow pp. XXII-14.

# 15

# Heavy metal concentration in soil of some mechanic workshops of Zaria-Nigeria

N. N. Garba[1] , Y. A. Yamusa[3], A. Isma'ila[1], S. A. Habiba[1], Z. N. Garba[2], Y. Musa[3] and S. A. Kasim[3]

[1]Department of Physics, Ahmadu Bello University, Zaria, Nigeria.
[2]Department of Chemistry, Ahmadu Bello University, Zaria, Nigeria.
[3]Centre for Energy Research and Training, Ahmadu Bello University, Zaria, Nigeria.

This research paper investigated the elemental composition of soil samples from four selected mechanic workshops in Zaria. A total of eight samples were analyzed at Centre for Energy Research and Training (CERT) Ahmadu Bello University Zaria using standardless X-Ray Fluorescence spectroscopy (XRF). From the result, it was found that Silicon (Si) has the highest mean concentration ranging from 0.0013-0.0024 ppm and Ba, Ni, Cr, Mn, Cu, V, Mo and Zn having very low concentration witha mean of (0.000035 0.000053) ppm, (0.000009 0.000012) ppm, (0.0000054 0.000012) ppm, (0.0000049 0.000012) ppm, (0.0000052 0.000017) ppm, (0.0000052 0.000029) ppm, (0.0000068 0.00007) ppm and (0.00001 0.000055) ppm respectively. Lead was found in only one sampling point (Samaru Dogon Icce) with an abundance of 0.00018 ppm which is less than the maximum permissible limits (MPL) recommended by W.H.O. Hence, the result shows that there were no much toxic elements in some of the mechanic workshops in Zaria. It is advisable that substances containing heavy metals should not be disposed in farm lands or any dumpsites close to residential areas.

Key words: Heavy metals, soil, contamination, mechanic workshop.

## INTRODUCTION

Man's activity in the environment has led to the pollution of soil mainly by chemical contaminants. Presently in developing countries like Nigeria where estimates have been made that; there is large number of illiteracy in the country, lack of knowledge on how to eradicate the problem of soil pollution. The presence of heavy metals in soil can affect the quality of food, groundwater, micro-organisms activity, plant growth etc. (Antoaneta et al., 2009). When contaminated soils are later abandoned and then used for agricultural purposes such as farming, animal breeding, herding etc. plants take in these metals in the process. For the fact that they are not bio-degradable (cannot be broken down into smaller parts by bacteria), can have adverse effect on plants. Also these

heavy metals have toxic effect on living organisms in the soil when permissible concentration levels are exceeded.

In Zaria (Kaduna state), because of the large number of roadsides mechanical workshops where motor oil, body parts, grease, battery electrodes and electrolytes which contained heavy metals are commonly found and used and because most of the activities in the mechanical workshops are carried out on the ground (soil), the soil is mostly contaminated. Generally, the most common of these heavy metals found in the soil include Lead (Pb), Copper (Cu), Zinc (Zn), Cadmium (Cd) etc. Lead and copper are the commonly heavy metals found in the soil. Lead at certain exposure level, is a poisonous substance to animals as well as human beings.

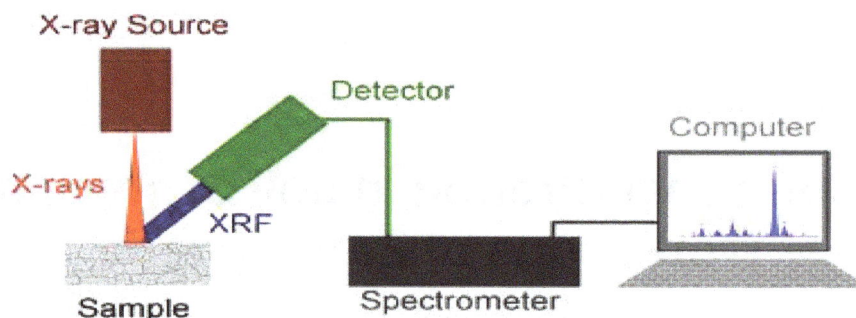

**Figure 1.** Schematic arrangement of Energy Dispersive XRF spectrometer Source: http://www.horiba.com/scientific/products/x-ray-fluorescence-analysis/tutorial/xrf-spectroscopy.

According to USDA (2000), acute (immediate) poisoning from heavy metals is rare through ingestion or dermal contact, but it is possible. Chronic problems associated with long-term heavy metal exposures are mental lapse (lead); toxicological effects on kidney, liver and gastrointestinal tract (cadmium); skin poisoning and harmful effects on kidneys and the central nervous system (Adelekan and Abegunde, 2011). According to an estimate made by the National Institutes of Occupational Safety and Health (NIOSH), more than 3 million workers are potentially exposed to lead in the work place (Binns and Ricks, 2004). In most part of the United States, heavy metal toxicity is an uncommon condition; however, it is a clinically significant condition when it does occur. If unrecognised or inappropriately treated, toxicity can result in significant illness and reduced quality of life (Ferner, 2001). Therefore, it is important for research to be conducted to evaluate and limit exposure of dangerous levels of these heavy metals in the environment.

### EXPERIMENTAL

**Sample collection**

A total of eight samples were collected from four locations (Samaru, Kofar-Doka, Sabon Gari and Tudun Wada). At each sampling point, samples were collected randomly using polythene bags and hand gloves and then transported to the laboratory for analysis.

**Sample preparation**

The samples were homogenised and crushed with an agate mortar grain size less than 125 nm. Three drops of toluene acid (binder) was then added to 0.5 g of the powdered sample and crushing continued until the mixture was returned to fine powder again. The 0.5 g weighed of the crushed sample was placed under a hydraulic press machine and a 10-tone pressure was applied which compressed and converted to fine powder and then into pellet form. The pellets were carefully labelled, covered with Mila and stored in partitioned sample storage plastic containers for analysis.

**Sample analysis**

The analysis was done using Mini pal which is a compact energy dispersive X-ray spectrometer designed for the elemental analysis of a wide range of samples (Figure 1). The system is controlled by a PC running the dedicated Mini pal analytical software. The Mini pal 4 version in use is PW 4030 X-ray spectrometer, which is an energy dispersive microprocessor controlled analytical instrument designed for the detection and measurement of elements in a sample (solids, powders and liquids), from sodium to uranium. The source (X-ray tube in this case) irradiates the sample and the detector measures the irradiation coming from the samples. The detector that is able to measure the different energies of the characteristic radiation coming from the sample directly.

### RESULTS AND DISCUSSION

Two categories of soil samples were collected and analysed for heavy metals from each sampling point. A total of eight samples were obtained, four of which are at the surface and the remaining four are at about 0.5 m beneath the surface of the ground, the results of the analysis are presented in Figures 2 to 5.

The results obtained from Samaru Dogon Icce (Figure 2) workshop shows the presence of lead, this is because Samaru Dogon Icce workshop is one of the busiest workshop in Zaria and its environs, and it is located along several higher institutions and the busiest Zaria - Sokoto express way. This leads to more number of vehicles in the area which constitutes the accumulation of lead. Lead (Pb) is only present in point 1 in Samaru Dogon Icce workshop, because point 1 was collected from the surface of the soil where lead accumulated more while point 2 was collected 0.5 m beneath the soil surface which has lesser content of lead compared to that of point 1.

Lead was only found in Samaru Dogon Icce workshop because all the samples which were analysed was collected during the rainy season which lead to the washing away of top soil accompanied by washing away of some of these metals. Also, the presence of Pb in

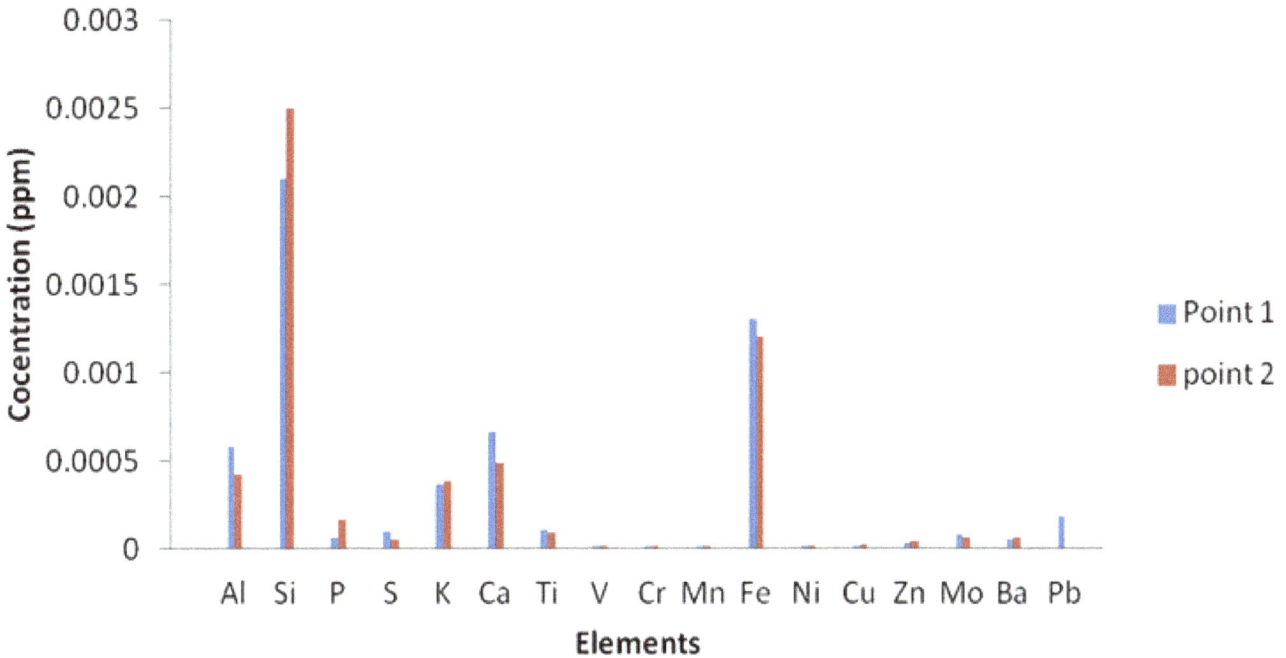

**Figure 2.** Variation of concentration of Elements from Samaru Dogon Icce mechanic workshop.

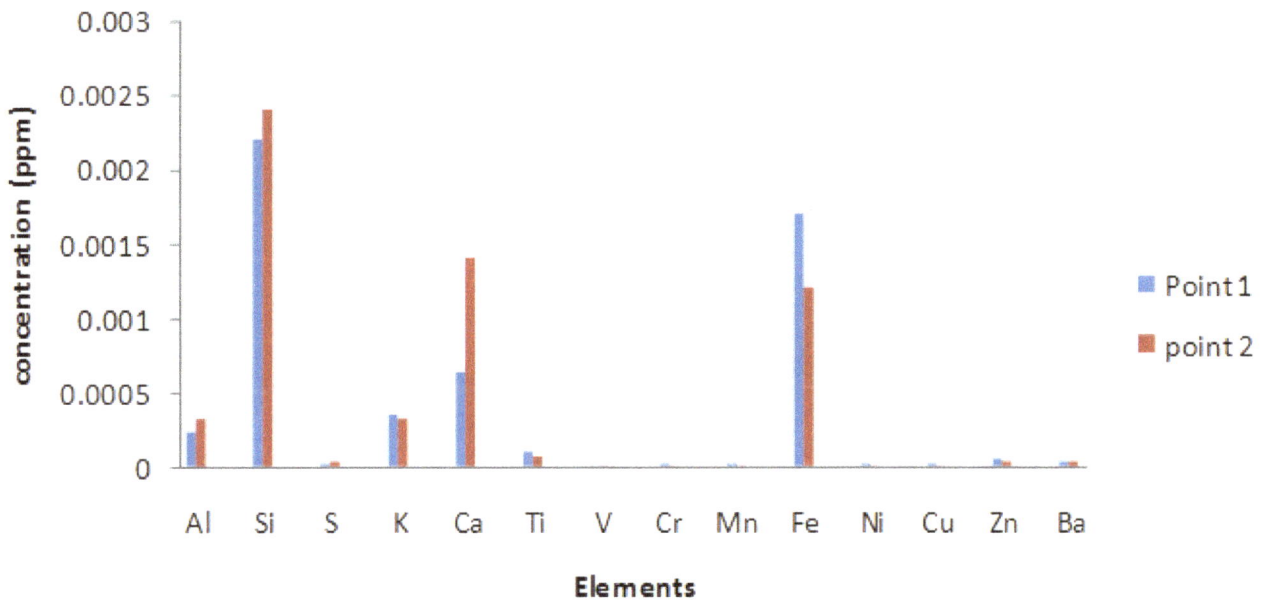

**Figure 3.** Variation of concentration of elements from Sabon Gari mechanic workshop.

auto-repair workshop in Samaru Dogon Icce soils may be due to fall-out of lead from batteries or lead accumulators, which are commonly used and abandoned in the workshops. The presence of iron (Fe) in large concentration from Sabon Gari (Figure 3) and Kofar Doka (Figure 4) mechanic workshops deserves evaluation because of the fact that different types of trees are

present in the site, the dropping and decomposition of their leaves accumulate in the soil.

In all samples from the four mechanic workshops, it can be observed that silicon has the highest concentration with a mean ranging between 0.0013 to 0.0024 ppm. This is due to the fact that silicon is the key component of sand. Some of the soil obtained from these mechanic

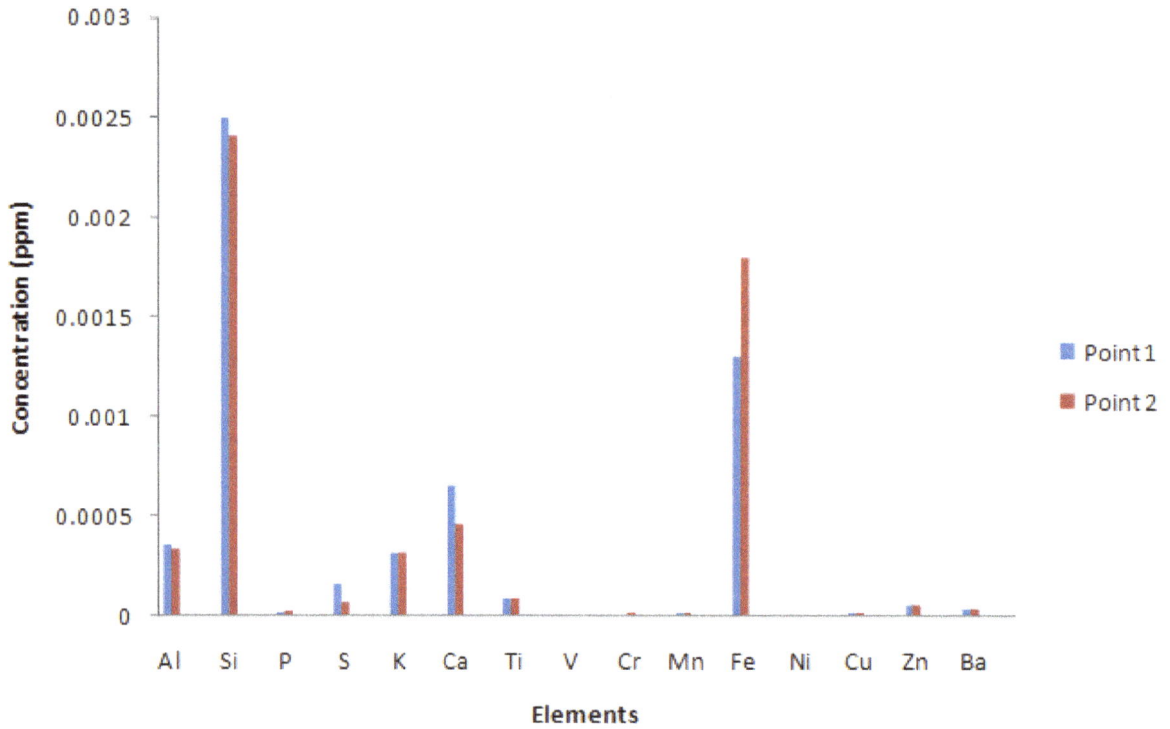

**Figure 4.** Variation of concentration of Elements from Kofar Doka mechanic workshop.

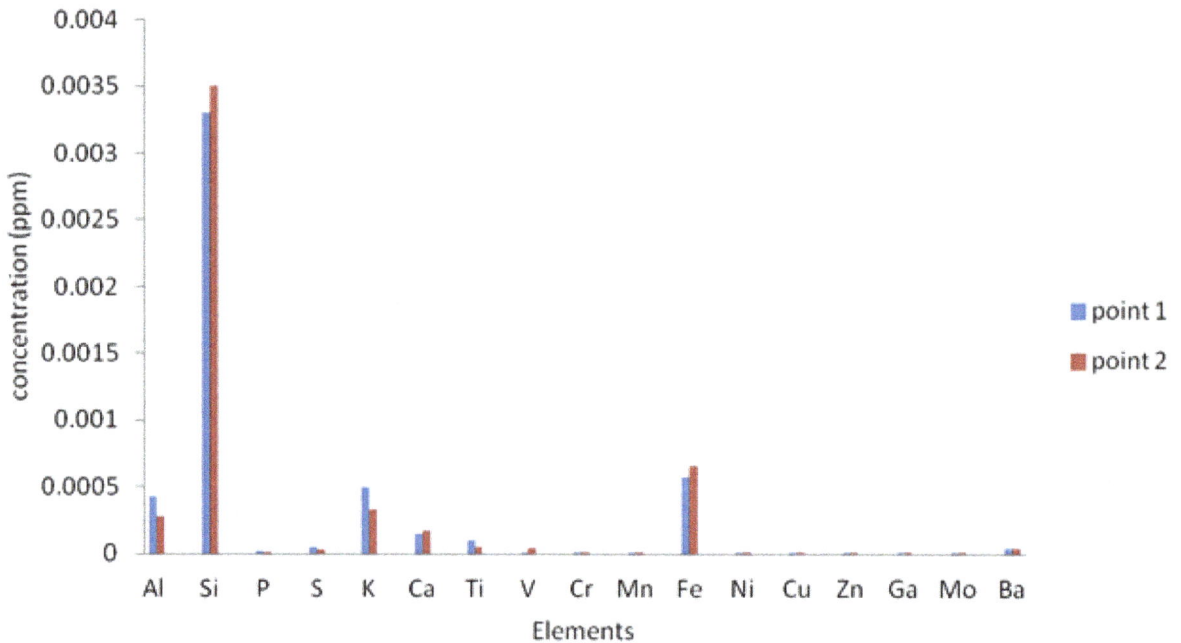

**Figure 5.** Variation of concentration of elements from Samaru Dogon Icce mechanic workshop.

workshops with excess silicon can be used as manure by local farmers to enhance and standardize rice production for the public consumption and industrial purposes. Also any crop that is cultivated using such soil as manure is expected to have high concentration of silicon and when taken by animals it helps in building strong bones and formation of connective tissues, it also assists in healthy growth of hair, skin and finger nails (Buhari, 2011).

By mere observation on the frequency distribution charts, one can see that there is no much difference in the concentrations of both points 1 and 2. Only for the concentration of Point 1 and Point 2 from Kofar Doka and Sabon Gari mechanic workshop; the concentration of Fe of Point 2 is large compared to that of Pont 1. This is because Point 2 was taken from the area very close to trees and iron (Fe) is much more present in the leaves that frequently drop and decomposed. From Figures 2 to 5, concentration of potassium for both points 1 and 2 are low and in very close ranges compared to the remaining elements. This is because in soils, plants absorb potassium in greater amount than any other nutrient. The total K content of soils frequently exceeds 20,000 ppm (parts per million). Nearly all of this is in the structural component of soil minerals and is not available for plant growth. Because of large differences in soil parent materials and the effect of weathering of these materials in the United States, the amount of K supplied by soils varies (George et el; 2002). For children, ingestion contaminated soil is most significant in pathway for land (Chaney et al., 1989; EPA, 1997). Also, the maximum permissible limits (MPL) for lead is 15 ppm (15 part per million) while the abundance recorded in this work is only 0.00018 ppm. This is the indication that the mechanic workshop does not cause much toxicity to the plants and animals in the area even though lead is poisonous no matter the amount of concentration, its toxicity can result in significant illness and reduced quality of life (Ferner, 2001).

It can also be observed from Figures 1 to 4 that Ba, Ni, Cr, Mn, Cu, V, Mo and Zn have very low concentration with a mean of (0.000035-0.000053)ppm, (0.000009-0.000012)ppm, (0.0000054 0.000012)ppm, (0.0000049 0.000012)ppm, (0.0000052 0.000017)ppm, (0.0000052 0.000029)ppm, (0.0000068 0.00007)ppm and (0.00001-0.000055)ppm respectively. This is because soil samples were collected from a depth of 0 to 15 cm and also were collected during the rainy season which may have caused the washing away of top soil leading to washing away of most of these metals from the soil surface and also because heavy metals in auto-repair workshop soils are not significantly derived from the natural geology or the processes of weathering and deposition (Ayodele and Modupe, 2007). From Figures 1 to 5, the order of abundance is Si>Fe>Al>Ca>K>Pb>Ti>P>S>Zn>Ba>Mn>Cu>V>Cr, with an exemption of Pb that is only present in trace amount in the samples collected from Samaru Dogon Icce mechanic workshop. Pb is considered the primary contaminant of most auto-mobile workshops no matter the amount of concentration.

## Conclusion

The heavy metal concentrations in soil samples from some selected mechanic workshops of Zaria and environs were collected and analysed using XRF at Centre for Energy and Training (CERT), Ahmadu Bello University, Zaria. The result obtained from this work shows that the pollution levels within the study area as a result of fall-out of lead from batteries or lead accumulator has not risen to a dangerous level at the moment. But there is also the danger of build-up of small doses either through inhalation or absorption through skin or bio-accumulation. Data obtained from this research work shows that Si has the highest concentration in all the samples analysed with a mean concentration of 0.0013 to 0.0024 ppm. Silicon is also the only element that does not damage plant when accumulated in excess. Therefore some of the soil obtained from these mechanic workshops with excess silicon can be used as manure by local farmers to enhance and standardize rice production for the public consumption and industrial purposes. Also V, Mn, S and Cr have the lowest concentration level in all of the samples collected and analysed from the mechanic workshops. Hence, all these soil samples collected for analysis when used up by humans are less prone to Human Carcinogen (Ayodele et al., 2007) and less exposed to diseases such as brain damage, skin and throat irritation.

Lead derived mostly from exhausts of vehicles is in Nigeria still used as minor additives to gasoline and various auto-lubricants. It is estimated that about 2800 metric tons of vehicular gaseous lead emission is deposited to urban areas in Nigeria annually (Ayodele and Modupe, 2007). Concern for lead concentration in automobile workshop soils may therefore arise principally due to the fact that mechanic workshop could be identified as playground or near residential areas where children play about freely.

However, Pb concentration was only obtained from a sample collected from Samaru Dogon Icce mechanic workshop with a concentration of 0.00018 ppm which is less than the maximum permissible limits (MPL) of Pb recommended by WHO which is 15 ppm (15 part per million). Since lead is a very poisonous element, it is advisable not to use the soil from Samaru Dogon Icce mechanic workshop for crop cultivation since its toxicity can result in significant illness and reduced quality of life (Ferner, 2001).

## RECOMMENDATION

Based on the observations and experience from this work, the following were recommended:

(i) This research work should be carried out from time to time so as to monitor the amount of heavy metals released into the soil to avoid accumulation.

(ii) Also, the research should also be carried out in the dry season or preferably in both dry and rainy season so as to get more accurate results.

(iii) There is need to investigate any water body close to the mechanic workshops so as to assess and monitor the concentration level of heavy metals likely present in the water due to the activities in the mechanic workshops.

## REFERENCES

Adelekan BA, Abegunde KD (2011). Heavy metal contamination of soils and ground water at automobile mechanic villages in Ibadan, Nigeria. Int. J. Phys. Sci. 6(5):1045-1058, 4 March, 2011.

Antoaneta E, Alina B, Georgescu L (2009). Determination of heavy metals in soils using XRF Technique. Rom. J. Phys. 55(7-8):815-820.

Ayodele RI, Modupe D (2007). Heavy metals contamination of topsoil and dispersion in the vicinities of reclaimed auto repair workshops in Iwo, Nigeria. Bull. Chem. Soc. Ethiop. 22(3):339-348.

Binns HJ, Ricks OB (2004). Helping Parents Prevent Lead Poisoning. ERIC Digest.

Buhari U (2011). Determination of heavy elements in dumpsites of Zaria city using X-Ray Fluorescent Spectroscopy. Unpublished B.Sc. project, Department of Physics, A.B.U Zaria.

Chaney RI, Malik M, Li YM, Brown SI, Brewer EP, Angle JS, Baker AJM (1989). Phytoremediation of soil metals. Current opinions biotechnol., 8(3):279.

EPA (Environmental Protection Agency) (1997). Electrokinetic laboratory and field processes applicable to radioactive and hazardous mixed waste in soil and groundwater. EPA 402/R- 97/006. Washington, DC.

Ferner DJ (2001). Toxicity, heavy metals. eMed. J. 2(5):1.

George R, Lowell B, John L, Gyles R, Michael S (2002).The nature phosphorus in soils.Ww-06795-90.

http://www.horiba.com/scientific/products/x-ray-fluorescence-analysis/tutorial/xrf-spectroscopy/ Accessed date 26th November, 2013.

United States Department of Agriculture, USDA (2000). Heavy Metals Contamination, Soil Quality Urban Technical Note 3, Natural Resources Conservation Service.

# 16

# Polyaniline/Fe$_3$O$_4$ coated on MnFe$_2$O$_4$ nanocomposite: Preparation, characterization, and applications in microwave absorption

## Seyed Hossein Hosseini[1] and A. Asadnia[2]

[1]Department of Chemistry, Faculty of Science, Islamshahr Branch, Islamic Azad University, Tehran-Iran.
[2]Young Researchers Club, Center Tehran Branch, Islamic Azad University, Tehran-Iran.

**Conductive polyaniline (PANi)/Fe$_3$O$_4$ is coated on the MnFe$_2$O$_4$ nanocomposite with multi core shell structure was synthesized by in-situ polymerization in the presence of dodecyl benzene sulfonic acid (DBSA) as the surfactant and dopant and ammonium persulfate (APS) as the oxidant. The structure and magnetic properties of Fe$_3$O$_4$ coated on the MnFe$_2$O$_4$ nanoparticles were studied by using powder X-ray diffraction (XRD) and vibrating sample magnetometer (VSM), respectively. The morphology, microstructure and DC conductivity of the nanocomposite were characterized by scanning electron Microscopy (SEM), fourier transform infrared spectroscopy (FTIR) and four-wire-technique, respectively. The microwave absorbing properties of the nanocomposite dispersed in resin acrylic with thickness of 1.4 mm were investigated by a HP 8720B vector network analyzer and standard horn antennas in Anechoic chamber in the frequency range of 8–12GHz. A minimum reflection loss of -18 dB was observed at 8.6 GHz.**

**Key words:** Nano-structures, polymer-matrix composites (PMCs), magnetic properties, microwave absorption (nominated).

## INTRODUCTION

Microwave absorbing material plays a great role in electromagnetic pollution, electromagnetic interference (EMI) shielding and stealth technology, to name but a few. An"ideal"microwave absorbing material owns such advantages as low thickness, low density, wide band width and flexibility simultaneously (Hosseini et al., 2011). In the past decades, the spinel ferrites have been utilized as the most frequent absorbing materials in various forms. Manganeseferrite (MnFe$_2$O$_4$) is a common spinel ferrite material and has been widely used in microwave and magnetic recording applications (Xiao et al., 2006). The absorbing characteristics of the materials depend on the frequency, layer thickness, complex permittivity ($\varepsilon_r$) and complex permeability ($\mu_r$). Fe$_3$O$_4$ is a kind of microwave absorbers with complex permittivity and complex permeability (Li et al., 2008a). The core-shell structure composite nanoparticles often exhibit improved physical and chemical properties over their single-component counterpart and hence are very useful in a broader range of applications (Zhang and Li, 2009). Conducting polymer composites with micro/nanostructures have attracted significant academic and technological attention because of their unique physical properties and potential applications in nanoelectronics, electromagnetics, and biomedical devices. Among these conducting polymers composites decorated with organic nanoparticles are of particular interest because possible interactions between the inorganic nanoparticles and the polymer matrices may generate some unique physical properties upon the formation of various micro/nanocomposites (Yang et al., 2009). Among conducting polymers, polyaniline (PANi) is perhaps the most versatile because of easier and inexpensive preparation methods. Also they have desirable properties, such as thermal and chemical stability, low specific mass, controllable conductivity and high conductivity at microwave frequencies (Phang et al., 2008). PANi is a conducting polymer so it has many potential applications in various fields such as electrical – magnetic shields, microwave absorbing materials, batteries, sensors and corrosion protections. The development of PANi properties has received considerable attention lately. The fabrication of PANi /ferrite nanocomposite has been reported by using different methods such as in situ polymerization of aniline in the presence of $Zn_{0.6}Cu_{0.4}Cr_{0.5}Fe_{1.5}O_4$ nanoparticles, micro emulsion process used to prepare PANi/NiZn ferrite nanocomposite and oxidative electro – polymerization of aniline in an aqueous solution in the presence of MnZn ferrite and NiMnZn ferrite. These studies created organic materials possessing both conducting and ferromagnetic functions. The electromagnetic measurements of the PANi/ferrites were improved and tailored by controlling the addition of the ferrite in the composite. Also, the contribution of ferrite to the PANi led to an increase in its thermal stability, however, it was decreased it electrical conductivity (Farghali et al., 2010). The preceding work, we have investigated microwave absorbing property of PANi– manganese ferrite nanocomposite in the frequency range of 8–12 GHz. We showed the PANi–manganese ferrite nanocomposites are good electromagnetic wave absorbent in the microwave range (Hosseini et al., 2011).

## EXPERIMENTAL

### Materials and instrumentals

Chemicals including metal salts, hexamethylene tetraamine (HMTA), potassium persulfate (KPS), ammonium persulfate (APS) and ethylene glycol (EG), $FeCl_3.6H_2O$, $FeSO_4.H_2O$, $NH_3.H_2O$ (28%), Oleic acid (90%), are analytical grade (Merck) and were used without further purification. Water was deionized, doubly distilled, and deoxygenated prior to use. Styrene and methacrylic acid (analytical grade, Merck) were distilled to remove the inhibitor. Aniline monomer (analytical grade, Merck) was distilled twice under reduced pressure. DBSA and acrylic resin were of industrial grade.

The morphology of coated particles and nanocomposite was observed scanning electron microscopy (SEM) with a JSM-6301F (Japan) instrument operated at an accelerating voltage of 10 kV. X-ray powder diffraction (XRD) patterns of the nanoparticles assembles were collected on a Philips-PW 1800 with Cu-K radiation under Cu Kα radiation (λ=1.5406 Å). Fourier transform infrared

spectroscopy (FTIR) spectra were recorded on a PerkinElmer spectrum FTIR using KBr pellets. The M–H hysteresis loops were measured by vibrating sample magnetometer (VSM) (RIKEN DENSHI Co. Ltd., Japan). Microwave absorbing properties were measured by a HP 8720B vector network analyzer and standard horn antennas in anechoic chamber.

### Synthesis of manganese ferrite ($MnFe_2O_4$) nanoparticles

In a typical experiment, 10 ml styrene, 2 ml methacrylic acid and 0.054 g KPS were added to the flask with 100 ml deionized water. To eliminate oxygen effects the solution was purged with nitrogen before the process was initiated. The mixture was heated to 72 °C and stirred with a magnetic stirrer. The polymerization was continued for 24 h and in the whole procedure the nitrogen was purged. Concentration of PS spheres in solution is 80 mg/ml, which was calculated by drying 5 ml colloid solution and weighing the remained solids (Hosseini et al., 2011).

### Synthesis of coated particles

The coating procedure consisted of controlled hydrolysis of ferrous chloride aqueous solutions and other divalent metal salts in the presence of polystyrene latexes. In a typical preparation process, 2 ml PS colloid solution was diluted with 250 ml deoxygenated distilled water and then mixed with the metal salts solution, which contained 10 mmol $FeCl_2$ and 5 mmol $MnCl_2$. After it dispersed under ultrasonic for several minutes, the mixture was in corporate with 4 g HMTA and 0.5 g potassium nitrate and heated to 85 °C under gentle stirring. After 3 h, the system was cooled to room temperature. The solution was poured in to excess distilled water, then magnetic particles were deposited using magnetic field. The precipitate was washed with distilled water for several times and then dried in oven at 80 °C for 24 h. In addition, to modify the surface chemical properties of the magnetic spheres, 5 ml ethylene glycol (EG) was added in to the reaction solution before the incorporation of HMTA.

### Iron Ferrite ($Fe_3O_4$)

$FeCl_3.6H_2O$ (24.3 g) and $FeSO_4.7H_2O$ (16.7 g) were dissolved in 100 ml de-ionized water under nitrogen gas while stirring vigorously at 80 °C. Then 50 ml of ammonium hydroxide were added rapidly into the solution. The color of the solution turned to black instantly. Oleic acid (3.76 g) was added 30 min later. Then the suspension was kept at 80 °C for 1.5 h. The magnetite nanoparticles were washed with de-ionized water until the pH value of the system reached neutral. The as-synthesized sample was dried in vacuum at room temperature.

### Preparation of $Fe_3O_4$ -coated on the $MnFe_2O_4$ ($MnFe_2O_4/Fe_3O_4$)

0.1 g nano-sized particles of prepared $MnFe_2O_4$ were dispersed in 200 ml water solution of pH 6 under ultra sonification for 2 min. At this pH, the surface charge of $MnFe_2O_4$ in the solution is expected to be positive, and that of $Fe_3O_4$ be negative. The two solutions were mixed at 1/19 volumetric ratio (1/19 weight ratio of $MnFe_2O_4$/ $Fe_3O_4$) and subjected to sonification for 2 min so that $MnFe_2O_4$ particles were coated with $Fe_3O_4$ particles. The solution was then filtered using ultrafiltration membrane and dried at room temperature. We also prepared physically mixed $Fe_3O_4$ and $MnFe_2O_4$ by mixing the two solid $Fe_3O_4$ and $MnFe_2O_4$ samples with 1/19 weight ratio of $MnFe_2O_4$/ $Fe_3O_4$ for 1 h using a rotating machine.

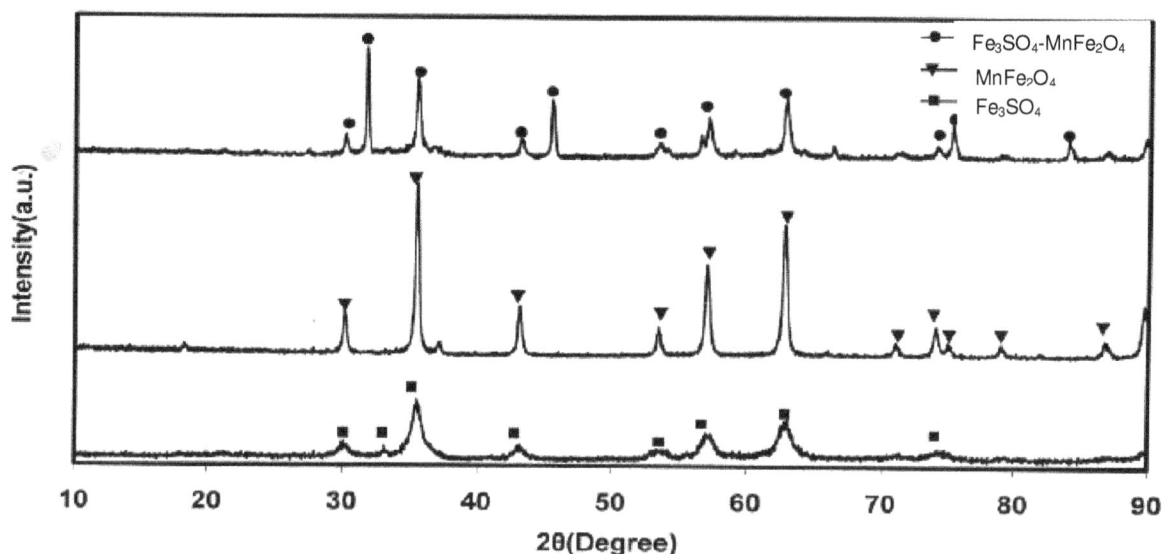

**Figure 1.** X-ray diffraction for $Fe_3O_4$, $MnFe_2O_4$ and $MnFe_2O_4/Fe_3O_4$ nanoparticles.

**Synthesis of $MnFe_2O_4/Fe_3O_4/PANi$ nanocomposite with multi core–shell structure**

$MnFe_2O_4/Fe_3O_4/PANi$ multi core–shell nanocomposites were prepared by in situ polymerization in the presence of DBSA as the surfactant and dopant and APS as the oxidant. The DBSA was dissolved in distilled water with vigorous stirring for about 20 min. The $MnFe_2O_4/Fe_3O_4$ nanoparticles (1.22 g) were added to the DBSA solution under stirring condition for approximately 1 h. Then 8 ml of aniline monomer was added to the suspension and stirred for 30 min. $MnFe_2O_4/Fe_3O_4$ nanoparticles were dispersed well in the mixture of aniline/DBSA under ultra sonicaction for 2 h. 20 g APS in 60 ml deionized water was gradually added drop wise to the stirred reaction mixture. Polymerization was allowed to proceed while stirring in an ice-water bath for 6 h. The nanocomposite was obtained by filtering and washing the suspension with deionized water and ethanol, respectively. The obtained green-black powder containing 15% $MnFe_2O_4/Fe_3O_4$ was dried under vacuum for 24 h.

## RESULTS AND DISCUSSION

### X-ray diffraction analysis

Figure 1 shows the XRD pattern of $Fe_3O_4$, $MnFe_2O_4$ and $MnFe_2O_4/Fe_3O_4$. According to the Figure, cubic ferrite $Fe_3O_4$ and $MnFe_2O_4$ nanoparticles have been obtained. However, it should be noted that there are some peaks of $\alpha$-$Fe_2O_3$ in the XRD pattern for $Fe_3O_4$ ($2\theta=33,54$) and $MnFe_2O_4$ ($2\theta=54$) nanoparticles. All peaks correspond to the characteristic peaks of cubic type lattice for $MnFe_2O_4$ (JCPDS file no. 88-1965) and $Fe_3O_4$ (JCPDS file no. 19-0629). The obtained peak width from XRD patterns addresses to the sizes of nanoparticles. By using Debye–Scherrer equation, the sizes of $MnFe_2O_4$, $Fe_3O_4$ and $MnFe_2O_4/Fe_3O_4$ nanoparticles are calculated as 24.27, 7.38 and 31.65 nm, respectively. The XRD pattern indicates that $MnFe_2O_4/Fe_3O_4$ nanocomposites have

formed. And compared with $MnFe_2O_4$ and $Fe_3O_4$ nanoparticles, the intensity of the characteristic peaks of $\alpha$-$Fe_2O_3$ decreased in the $MnFe_2O_4/Fe_3O_4$ nanocomposites. This may be attributed to the coating of $Fe_3O_4$ nanoparticles on the surface of $MnFe_2O_4$ nanoparticles.

## Magnetic properties

Magnetic properties of the samples were measured at room temperature with a VSM. The hysteresis loops are illustrated in Figure 2a-d. This Figure shows the magnetization (M) versus the applied magnetic field (H) for $Fe_3O_4$, $MnFe_2O_4$, $MnFe_2O_4/Fe_3O_4$ nanoparticles and d) $MnFe_2O_4/Fe_3O_4/PANi$ nanocomposite (15 wt%) respectively. It can be inferred from the hysteresis loops that all the composite magnetic spheres are magnetically soft at room temperature with an applied field -10 kOe≤H≤10 kOe. Figure 2a shows the hysteresis loop of $Fe_3O_4$ (Hosseini et al., 2011). The value of saturation magnetization ($M_s$) is about 66.7 emu/g, the remnant magnetization ($M_r$) and the coercivity field are 17.81 emu/g and 110 Oe respectively. Figure 2b shows clear saturation magnetization ($M_s$) about 60 emu/g and remnant magnetization ($M_r$) and the coercivity field for $MnFe_2O_4$ nanocomposite are about 18 emu/g and 140 Oe respectively. The $Ms$, $Mr$ and $Hc$ are 37 emu/g, 11 emu/g and 155 Oe for $MnFe_2O_4/Fe_3O_4$ nanocomposite that have been shown in Figure 2c, respectively. It is lower than the pure ferrite manganese ferrite (Xiao et al., 2006) nanoparticles.

Although the $MnFe_2O_4/Fe_3O_4$ nanocomposites consist of two magnetic phases, the hysteresis loop shows a single-phase-like behavior, and the magnetization

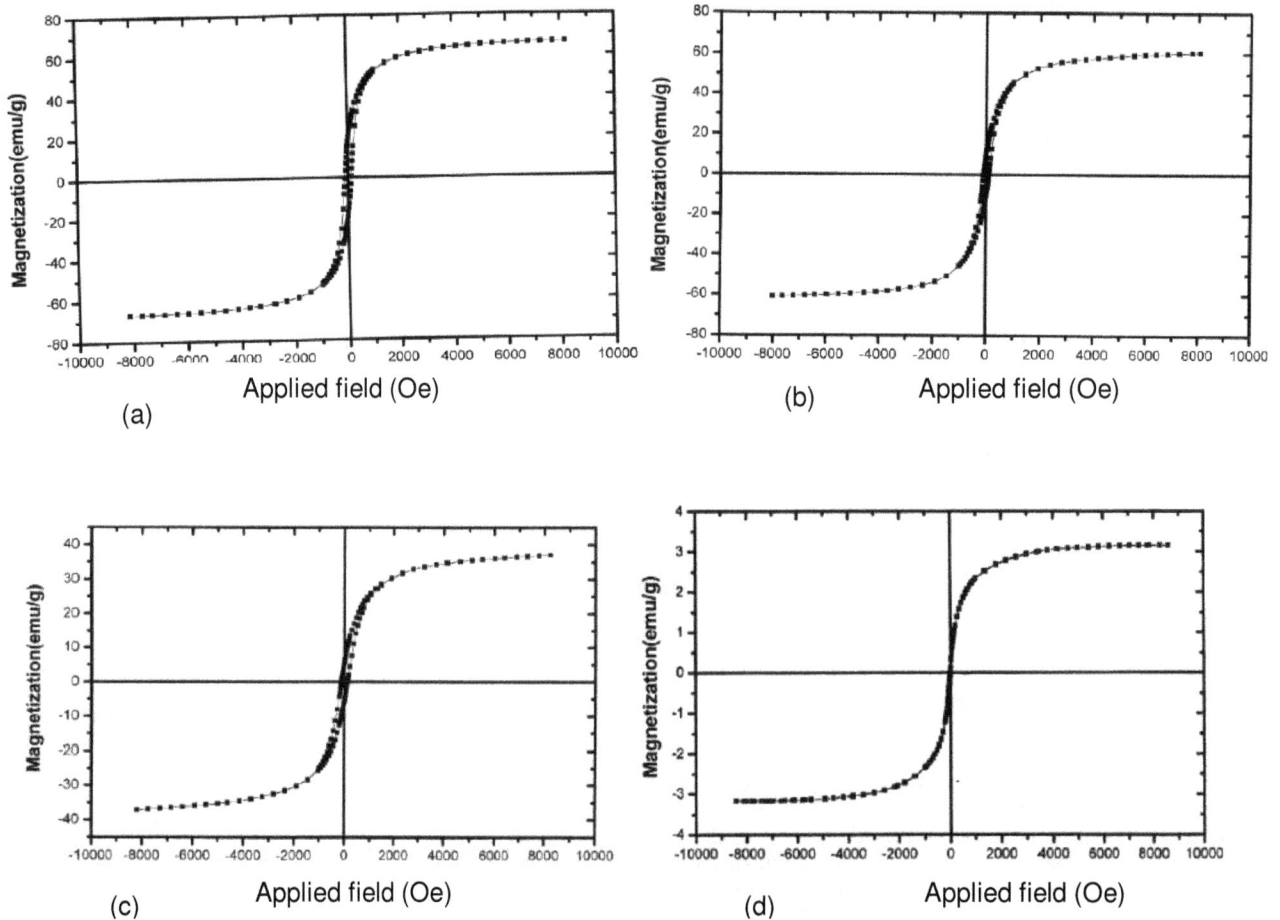

Figure 2. Magnetic hysteresis loop of a) $Fe_3O_4$ nanoparticle, b) $MnFe_2O_4$ nanoparticle, c) $MnFe_2O_4/Fe_3O_4$ and d) $MnFe_2O_4/Fe_3O_4$-PANi nanocomposites.

changes smoothly with the applied field. This indicates that the $MnFe_2O_4$ core and $Fe_3O_4$ first shell contact intimately. They are also clearly seen that the value of $Ms$ decreases from 66.7 emu/g for $Fe_3O_4$ to 37emu/g and 60 emu/g for $MnFe_2O_4$ to 37 emu/g for the core-shell structure nanocomposites. And the $Hc$ of $MnFe_2O_4/Fe_3O_4$ nanocomposites (155 Oe) is near to $Fe_3O_4$ (110 Oe) and higher than $Fe_3O_4$ (110 Oe) and $MnFe_2O_4$ (140 Oe) respectively. The changes in saturation magnetization and the coercivity can be attributed to the existence of $Fe_3O_4$ on the surface of $MnFe_2O_4$ nanoparticles which can result in the interparticle interaction at the interface of two phases. As saturation magnetization, the interphase interaction leads to the non-collinearity of the magnetic moments at the interface of two phases, and then results in there saturation magnetization (Chen et al., 2007). For coercivity, when the particles contact closely the interphase exchange coupling occurs, with which the rotation of the domains on one particle as the field is reversed, induces domains in contiguous particles to rotate, and thereby decreasing the coercivity (Zhang and Li, 2009; Zeng et al., 2004).

Figure 2d shows clear $M_s$ about 3.15 emu/g, $M_r$ about 0.35 emu/g and $Hc$ Oe about 0 for $MnFe_2O_4/Fe_3O_4/PANi$ nanocomposite (15 wt%) which is lower than pure ferrite and manganese ferrite nanoparticles. The magnetization curve of the sample shows weak ferromagnetic behavior, with slender hysteresis. Magnetic properties of nanocomposites containing magnetite or ferrite particles have been believed to be highly dependent on the sample shape, crystallinity, and the value of magnetic particles, so that they can be adjusted to obtain optimum property.

**Morphology investigation**

Figure 3a-c shows the SEM images for a) $MnFe_2O_4$ and b) $Fe_3O_4$ nanoparticles and $MnFe_2O_4/Fe_3O_4/PANi$ nanocomposite. As shown in Figure 3a, the spongy-shaped $MnFe_2O_4$ was seen with a small quantity of amorphous phase. The range of average diameter of spongy-shape is 40 to 50 nm. Figure 3b shows the SEM image for $Fe_3O_4$ nanoparticles. The range of average

(a)

(b)

(c)

Figure 3. SEM microphotographs of a) $Fe_3O_4$ nanoparticle, b) $MnFe_2O_4$ nanoparticle and c) $MnFe_2O_4/Fe_3O_4/PANi$ nanocomposite.

diameter is 30 to 40 nm. In Figure 3c, it is found that the $MnFe_2O_4/Fe_3O_4/PANi$ nanocomposite (15 wt%) still retains the morphology of PANi shape. It is much unknown how to form spongy-shaped composite in the polymerization process. The SEM image clearly shows that the $MnFe_2O_4/Fe_3O_4$ was distributed rather homogeneously, and ultrasonication is effective for dispersing nanoferrite in the polymer matrix.

## FTIR spectra analysis

Figure 4a-d shows the FTIR spectra of $MnFe_2O_4$, $Fe_3O_4$ $MnFe_2O_4/Fe_3O_4$ and $MnFe_2O_4/Fe_3O_4/PANi$ nanocomposite, respectively. In ferrites, the metal ions are usually situated in two different sublattices, designated as tetrahedral and octahedral sites according to the geometrical configuration of the oxygen nearest neighbors (Hosseini et al., 2011). It was observed from Figure 4(a,b) that the peak at 578 $cm^{-1}$ is intrinsic vibrations of Fe-O in $MnFe_2O_4$ and peaks at 586 and 411 $cm^{-1}$ are intrinsic vibrations of Fe-O in $Fe_3O_4$. The peaks at 651 and 562 $cm^{-1}$ are intrinsic vibrations of Fe-O in $Fe_3O_4$ and $MnFe_2O_4$ have been shown in Figure 4c. The characteristic peaks of styrene occur at 1559, 1338-1067 and 851 $cm^{-1}$. The peak at 1559 is attributed to the styrene ring. The peak at 1338 $cm^{-1}$ is attributed to the characteristic C=C stretching ring. The peak at 851 $cm^{-1}$ is related to the C–H outer bending vibrations. As shown in Figure 4d, the characteristic peaks of $MnFe_2O_4/Fe_3O_4/PANi$ nanocomposite occur at 1555, 1483, 1302, 1241, 1122, 1028, 1002, 876, 800, 675 and 580 $cm^{-1}$. The peaks at 1555 and 1483 $cm^{-1}$ are attributed to the characteristic C=C and C-N stretching of the

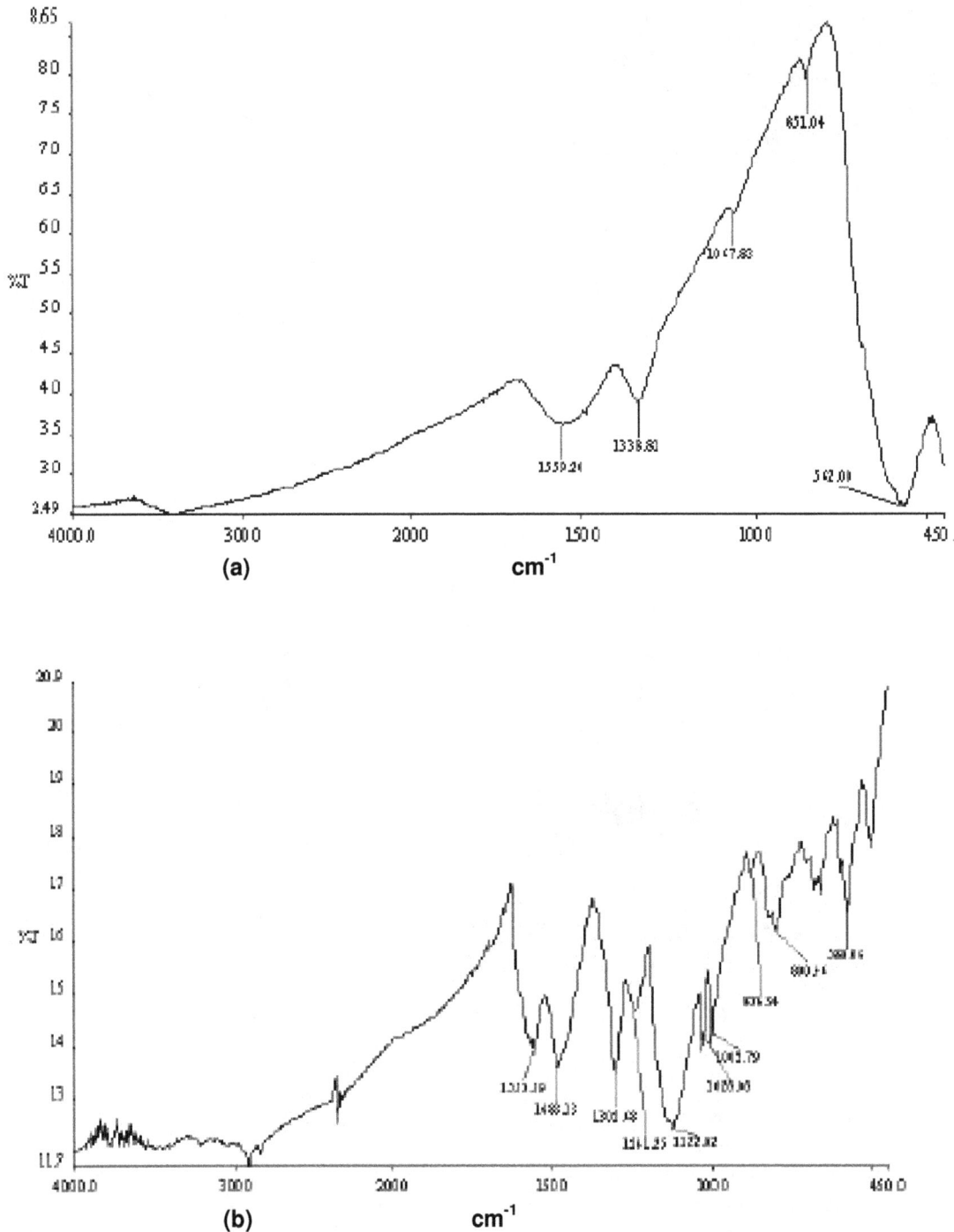

**Figure 4.** FTIR spectra: a) $Fe_3O_4$ nanoparticle, b) $MnFe_2O_4$ nanoparticle c) $MnFe_2O_4/Fe_3O_4$ nanoparticle and d) $MnFe_2O_4/Fe_3O_4/PANi$ nanocomposite.

quinoid and benzenoid rings of polyaniline; the peaks at 1302 and 1241 $cm^{-1}$ correspond to N–H bending and asymmetric C–N stretching modes of the benzenoid ring. The peak around 1122 $cm^{-1}$ is associated with vibrational modes of N=Q=N (Q refers to the quinonic type rings), indicating that PANi is formed in our sample. The peak at 1028 $cm^{-1}$ attributed to the symmetric and anti-symmetric stretching vibration of $SO_3$ group of dopant (DBSA). The peaks at1002, 876 and 800 $cm^{-1}$ are attributed to the p-disubstituted aromatic ring C–H out-of-plane bending. However, the characteristic peaks of $Fe_3O_4$ and $MnFe_2O_4$ can be observed at higher wavenumbers (675 and 580 $cm^{-1}$) indicating that there is an interaction between $MnFe_2O_4/Fe_3O_4$ nanoparticles and PANi chain.

## DC conductivity

DC conductivity of samples at room temperature is shown in Table 1. When the PANi is doped by DBSA, the conductivity was improved to 26 S/cm, which means that doping $H^+$ increase conductivity of PANi. When 15% mass content of $MnFe_2O_4/Fe_3O_4$ nanoparticles was incorporated, the conductivity of $MnFe_2O_4/Fe_3O_4/PANi$ nanocomposite was sharply reduced from 26 to 0.9 S/cm. The decrease in conductivity of $MnFe_2O_4/Fe_3O_4/PANi$ composites may be attributed to the insulting behavior of the ferrite and partial blockage of the conductive path by $MnFe_2O_4/Fe_3O_4$ in the core of the nanoparticles (Li et al., 2008b).

## Reflection loss analysis

According to transmission line theory, the reflection loss (RL) of electromagnetic radiation, under normal wave incidence at the surface of a single-layer material backed by a perfect conductor can be given by:

$$RL = 20 \log \left| \frac{Z_{in} - Z_0}{Z_{in} + Z_0} \right| \tag{1}$$

where $Z_0$ is the characteristic impedance of free space,

$$Z_0 = \sqrt{\frac{\mu_0}{\varepsilon_0}} \tag{2}$$

$Z_{in}$ is the input impedance at free space and materials interface:

$$Z_{in} = \sqrt{\frac{\mu_r}{\varepsilon_r}} \tanh \left[ j \frac{2\pi f t}{c} \sqrt{\mu_r \varepsilon_r} \right] \tag{3}$$

where $\mu_r$ and $\varepsilon_r$ are the complex permeability and permittivity of the composite medium respectively, which can be calculated from the complex scatter parameters, c is the light velocity, f is the frequency of the incidence electromagnetic wave and t is the thickness of composites. The impedance matching condition is given by $Z_{in} = Z_0$ to represent the perfect absorbing properties (Chen et al., 2007). There are two different concepts to satisfy the zero reflection condition. The first concept is the "matched characteristic impedance". The intrinsic impedance characteristic of material is made equal to the impedance characteristic of the free space. The second is the "matched-wave-impedance" concept. The wave impedance at the surface of the metallic substrate layer is made equal to the intrinsic impedance of the free space. In this work, the second concept was applied. The condition of maximal absorption is satisfied at a particular point where thickness and frequency match each other. Ferrites are the only materials that present two matching

frequencies and thicknesses. The first matching at low-frequency is associated with the mechanisms of magnetic resonance and shows a dependence on the chemical composition. The second matching at high-frequency is associated with the thickness of absorbent material. To satisfy the zero-reflection condition where maximum absorption would occur, $Z_{in}$ should be 1 to prevent reflection. This can be ideally achieved when the material presents $|\mu_r| = |\varepsilon_r|$. In this case, the performance of electromagnetic wave-absorbing material increases linearly with the increase in thickness. In practical terms, however, this is rarely achieved because the values of complex permeability and complex permittivity are very different in the frequency range of interest. When $|\mu_r| \neq |\varepsilon_r|$, we should consider two other cases. For materials with intrinsic impedances greater than unity, $|\mu_r| \gg |\varepsilon_r|$, the minimum reflection loss occurs at around a half-wavelength thickness of the material, and for materials with intrinsic impedances lower than unity, $|\mu_r| < |\varepsilon_r|$, the minimum reflection loss occurs at around a quarterwavelength thickness of the material. Within the microwave region, ferrites usually present electromagnetic characteristics of $|\mu_r| < |\varepsilon_r|$, giving rise to the term "quarter-wavelength absorbent". Minimum loss occurs when the thickness is about an odd multiple of one quarter of the wavelength of the incident frequency. It measured inside the absorbing material, and the material has the proper loss factor for this particular thickness .The thickness, d, can be written as Equation (4), where c is the speed of light and f is the frequency of interest (Bueno et al., 2008).

$$d = \frac{c}{4f\sqrt{|\mu_r||\varepsilon_r|}} \tag{4}$$

## Investigation of microwave absorbing properties

Nanocomposite dispersed in acrylic resins then the mixture was pasted on metal plate with the area of 100 ×100 mm as the test plate. The microwave absorbing properties of the nanocomposite with the coating thickness of 1 mm were investigated by using a HP 8720B vector network analyzer and standard horn antennas in anechoic chamber in the frequency range of 8–12 GHz. Figure 5 shows the microwave absorption behavior of the $MnFe_2O_4/Fe_3O_4/PANi$ nanocomposite. For PANi with the coating thickness of 1 mm, the minimum reflection loss is -8 dB at the frequency of 8-12 GHz. For $MnFe_2O_4/Fe_3O_4/PANi$ nanocomposites with the coating thickness of 1 mm, the reflection loss values were obtained less than -10 dB in the frequency of 8–12 GHz and its value of minimum reflection loss are -18 and -17 dB at the frequency of 8.6 and 9.2 GHz, respectively. Compared with the core-shell $MnFe_2O_4/PANi$ (-15.3 dB at 10.4 GHz) (Hosseini et al., 2011) and the new multi core-

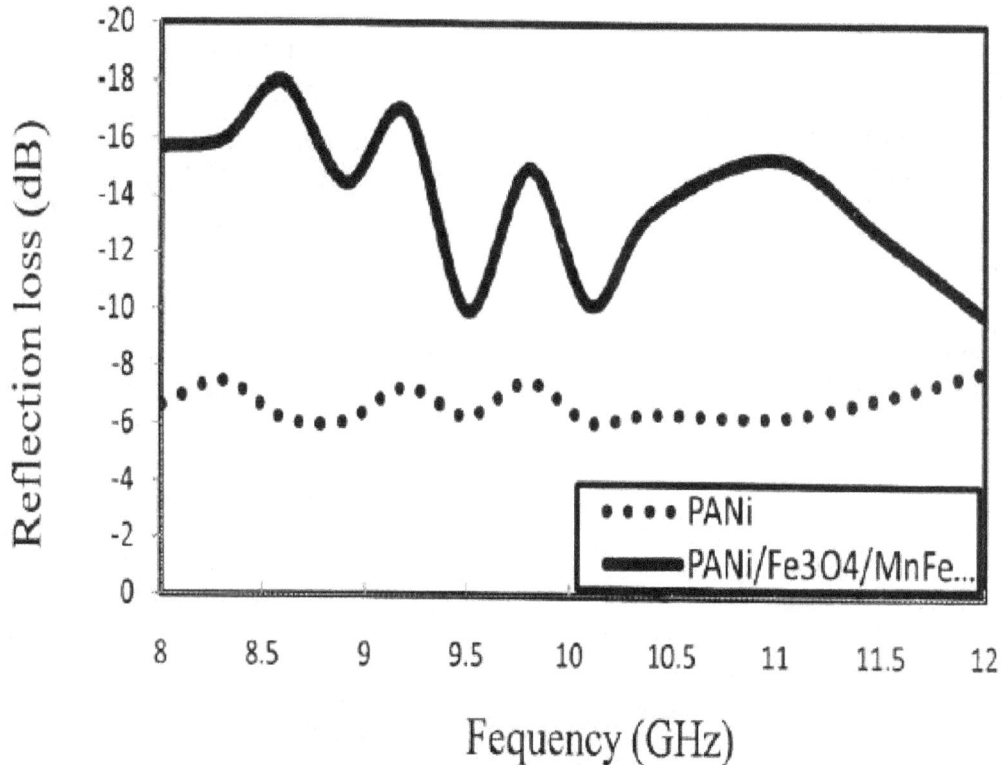

**Figure 5.** Frequency dependence of RL for the $MnFe_2O_4$/PANi nanocomposite.

shell $MnFe_2O_4$/$Fe_3O_4$/PANi nanocomposite (-18 dB at 8.6 GHz and -17 dB at 0.2 GHz), the microwave absorption properties of multi core-shell structure nanocomposite have been improved effectively. The reason for this improvement may be due to the interphase interaction between the cubic $MnFe_2O_4$ and cubic ferrite $Fe_3O_4$. These two kinds of ferrite combine intimately with the two step co-precipitation synthetic method, so they can couple to each other by an exchange through interface of ferrite particles. The interphase interaction which can cause the interphase exchange coupling and the non-collinearity of the magnetic moments at the interface of two phases exists between cubic ferrite materials. The interphase interaction can affect the microwave absorption properties of the two-phase composites had been shown in previous study (Zhang and Li, 2009). As mentioned above, the results of this work are consistent with these previous reports.

## Conclusion

The obtained magnetic nanoparticles are of a diameter of 24.27, 7.38 and 31.65 nm for $MnFe_2O_4$, $Fe_3O_4$ and $MnFe_2O_4$/$Fe_3O_4$, respectively. $MnFe_2O_4$/$Fe_3O_4$/PANi ferrite nanocomposite with the magnetic behavior is successfully synthesized by *in situ* polymerization of aniline in the presence of $MnFe_2O_4$/$Fe_3O_4$ nanoparticles.

The results of spectro-analysis indicate that there is an interaction between PANi chain and ferrite particles. Furthermore; for 1 mm thickness of nanocomposite used, a minimum reflection loss of -18 and -17 dB were observed at 8.6 and 9.2 GHz, respectively. Journal of Nanomaterials, Vol. 2012, 1687-4110 (2012).

## REFERENCES

Bueno AR, Gregori M, No'brega MCS (2008). Microwave-absorbing properties of $Ni_{0.50-x}Zn_{0.50-x}Me_{2x}Fe_2O_4$ (Me=Cu, Mn, Mg) ferrite-wax composite in X-band frequencies, J. Magn. Magn. Mat. 320:864-870.
Chen N, Mu GH, Pan XF, Gan KK, Gu MY (2007). Microwave absorption properties of $SrFe_{12}O_{19}$/$ZnFe_2O_4$ composite powders, Mater. Sci. Eng. B 139:256-260.
Farghali AA, Moussa M, Khedr MH (2010). Synthesis and characterization of novel conductive and magnetic nanocomposites, J. Alloys Comp. 499:98–103.
Hosseini SH, Mohseni SH, Asadnia A, Kerdari K. (2011). Synthesis; characterization and microwave absorbing properties of polyaniline/$MnFe_2O_4$ nanocomposite, J. Alloys Comps. 509:4682-4687.
Li X, Han X, Tan Y, Xu P (2008a). Preparation and microwave absorption properties of Ni–Balloy-coated $Fe_3O_4$ particles, J. Alloys Comp. 464:352–356.
Li Y, Zhang H, Liu Y, Wen Q, Li J (2008b). Rod-shaped polyaniline–barium ferrite nanocomposite: preparation, characterization and properties, Nanotechnology, 19:105605-105610.
Phang SW, Tadokoro M, Watanabe J, Kuramoto N (2008). Microwave absorption behaviors of polyaniline nanocomposites containing $TiO_2$ nanoparticles, Cur. Appl. Phys. 8:391–394.
Xiao HM, Liu XM, Fu SY (2006). Synthesis, magnetic and microwave

absorbing properties of core-shell structured $MnFe_2O_4/TiO_2$ nanocomposites, Comp. Sci. Tech. 66:2003-2008.

Yang C, Li H, Xiong D, Cao Z (2009). Hollow polyaniline/$Fe_3O_4$ microsphere composites: Preparation, characterization, and applications in microwave absorption, Reactive Functional Polym. 69:137–144.

Zeng H, Sun S, Li J, Wang ZL, Liu JP (2004). Tailoring magnetic properties of core/shell nanoparticles, Appl. Phys. Lett. 85:792-794.

Zhang L, Li Z (2009). Synthesis and characterization of $SrFe_{12}O_{19}/CoFe_2O_4$ nanocomposites with core-shell structure, J. Alloys Comp. 469:422–426.

# 17

# Investigation of Kenyan bentonite in adsorption of some heavy metals in aqueous systems using cyclic voltammetric techniques

**Damaris Mbui[1], Duke Omondi Orata[1], Graham Jackson[2] and David Kariuki[1]**

[1]Department of Chemistry, College of Biological and Physical Sciences P. O. Box 00100-30197, University of Nairobi, Kenya.
[2]University of Cape Town P. O. Private Bag Rondebosch 7701, South Africa.

Potential application of Kenyan bentonite for adsorption of iron, cobalt, copper, nickel and zinc and for analysis of electroactive species in water from a polluted water course using a 3-electrode potentiostat and cyclic voltammetry was studied. Polished carbon graphite electrodes were used either bare or modified with Kenyan bentonite using an electrochemically inert adhesive to a thickness of about 0.8 mm. These were used to prepare calibration curves of iron, cobalt, copper, nickel and zinc by plotting cyclic voltammograms of the ions at different concentrations and using 0.1 M sulphuric acid as supporting electrolyte. The slopes of the curves from bentonite-modified electrodes were observed to be higher than those obtained from bare carbon electrodes by a factor of between 1.7 and 24, implying that bentonite enhanced electron transfer kinetics of the metal ions. It was also observed that the magnitude of the ratio depended on the proximity of the element to either filled or half-filled 3d orbitals, which implied that a chemical reaction may have taken place between the bentonite and the ions (chemisorption). Carbon graphite electrodes were modified with bentonite that had been soaked in water samples from a polluted water course at a ratio of 1:1 w/w. The cyclic voltammograms showed clear oxidative and reductive peaks indicating that electroactive species that previously could not be detected on the potentiostat were pre-concentrated on the bentonite and could thus be detected. Thus Kenyan bentonite is observed to chemically adsorb zinc, cobalt, copper, nickel and iron species in aqueous solution and can be used to monitor electroactive pollutants in aqueous systems using electroanalytical techniques.

**Key words:** Bentonite, heavy metals, surface modified electrodes, cyclic voltammetry, preconcentration, adsorption.

## INTRODUCTION

According to the Global Environment Assessment Report (UNEP/GEMS, 1991) in developing countries, untreated water is the most commonly-encountered health threat and causes about 25,000 deaths per day due to water-borne diseases. Some of the pollutants that should be monitored regularly are heavy metals, as they have been known to be toxic especially when present in high levels (Normandin, 2004; Elbetieba and Al-Hammod, 1997;

Tvrda et al., 2013; Lloyd, 1960; Pearce, 2007). Since water resources development and management contribute significantly to socio-economic development in Africa in general and to agricultural development in particular, there is an urgent need to monitor the level of these species in water.

Cyclic voltammetry is a versatile electro-analytical technique for the study of electro-active species, and has been labeled 'electrochemical spectroscopy' (Orata and Segor, 1999). Modified electrodes, that is, electrodes on whose surfaces that chemical species has been deliberately immobilized have been used to facilitate electron transfer from bacteria to electrodes (Heinze and Muller, 1998)., as sensors (Guo et al., 2013; Rezaei et al., 2013), catalysts (Rahul et al., 2007) and to provide new methods of analysis (Nada et al., 2007; Kuralay et al., 2013; Oukil et al., 2007).    Modified electrode surfaces can act as preconcentrating surfaces in which the analyte species is collected and concentrated on the electrode. The collected analyte is subsequently measured by the electrochemical response to a potential step or sweep.

Bentonite is a clay where the principal exchangeable cation is sodium. It has been used together with pure Bottom Ash (BA) as a land fill cover (Puma et al., 2013). It has also been used in modification of electrodes for various purposes, for example catalysis and photocatalysis (Li et al., 2013; Boz et al., 2013).

It has also been used to remove pollutants from waste waters (Jovic-Jovicic et al., 2013; Zhang et al., 2013; Shi et al., 2013; Reitzel et al., 2013). Kenyan bentonite has been used, with good results, to modify electrodes for electrodeposition of polyaniline (Orata and Segor, 1999). However, the same has not been used to monitor other species like cations in aqueous systems. In this project, electrodes modified using Kenyan bentonite were used to monitor adsorption and preconcentration of iron, nickel, copper, cobalt and zinc in aqueous systems. It was used to analyze electroactive species from a polluted water course could through electrochemical means. Cyclic voltammetry was used to monitor heavy metals in aqueous systems using Kenyan bentonite - modified electrodes.

## EXPERIMENTAL

All reagents used in the experiments were of analytical grade and were used without further purification. The water used to prepare the solutions was triply distilled and the experiments were performed in triplicate. The instruments used included a Princeton Applied Research (PAR) model 173 potentiostat/galvanostat, a logarithmic current converter model 396 that controlled the current, a PAR model 175 universal programmer and a PAR RE 0089 X-Y recorder. For the super dry conditions, a PAR model 362 scanning potentiostat/galvanostat and a Fluke and Philips PM 8271XYt recorder were used for electrochemical control and data recording respectively. Bentonite was obtained from Athi River Mining Co. Ltd., (Kenya) with a mesh size of 150 to 200 μm; a Cationic Exchange Capacity, CEC of 1.118 to 1.22 mM/g, a pH of 8.4 to 9.6

**Table 1.** Composition of bentonite.

| Element | Concentration (mg/g) |
|---------|---------------------|
| K | 8.350 |
| Na | 10.855 |
| Ca | 37.000 |
| Mn | 0.581 |
| Fe | 50.500 |
| Al | 171.100 |
| Ni | 0.016 |
| Cu | 0.069 |
| Zn | 0.096 |
| Mg | 16.500 |
| Pb | 0.021 |
| V | 0.124 |
| Cr | 0.068 |
| Rb | 0.069 |
| Sr | 0.313 |
| Zr | 0.142 |
| Ti | 8.100 |
| Ba | 3.776 |
| $SiO_2$ | 469.710 |
| $P_2O_5$ | 0.569 |

and a density of 1.15 $g/cm^3$. The elemental composition is shown in Table 1. The bare carbon graphite electrode of surface area of 0.38 $cm^2$ was polished on a felt polishing cloth containing alumina.

## Determination of type of reaction

To establish whether or not the reaction was diffusion –controlled, a 1,10 - phenanthroline complex of iron (BDH, England) was prepared by mixing 1 mg/L $Fe^{2+}$ solution with 3 mg/L 1,10 – Phenanthroline. The cyclic voltammetric response was recorded by scanning from -0.20V to + 0.85V at scan rates of 5, 10, 20, 50 and 100 mV/s using polished carbon graphite electrodes and 0.1 M $H_2SO_4$ as supporting electrolyte. A plot of peak current versus the square root of the scan rate was plotted.

## Calibration curves

Calibration curves indicating variation of concentration with anodic peak current were plotted for both the non-modified and modified electrodes. 1000 mg/L stock solutions each of copper sulphate, cobaltous sulphate, zinc sulphate, nickel nitrate and ferrous sulphate all from BDH Chemicals were prepared. Lower concentrations ranging from 10 to 50 mg/L of each of the above solutions were obtained by appropriate dilution.

The bentonite-modified electrodes were formed using a slurry solution composed of 1 ml of an electrochemically inert adhesive from Henkel, Kenya with 0.1 g of bentonite. The mixture was spread on the surface of a polished carbon graphite electrode up to a thickness of about 0.8 mm and air-dried for 12 h. The surface area of the modified electrode was approximately 0.64 $mm^2$. For non-modified electrodes, polished carbon graphite electrodes were used. For both working electrodes (modified and bare carbon), cyclic voltammograms of the different concentrations were obtained using the appropriate scan window depending on the oxidative (and

**Table 2.** Scanning potential range for various metal ions.

| Metal ion | Cu | Zn | Ni | Fe | Co |
|-----------|-----|-----|-----|-----|-----|
| Scan range (v) | -0.4- 0.75 | -1.5 - 0.75 | -0.2 - 0.75 | -0.4 - 0.75 | -0.2 - 0.75 |

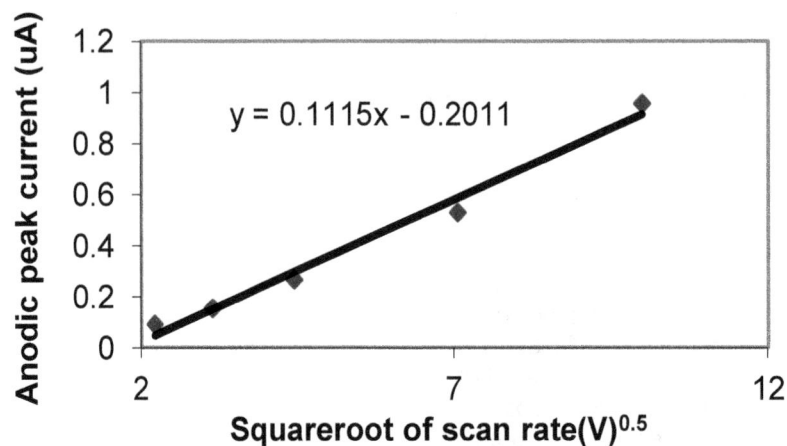

**Figure 1.** Anodic peak current versus square root of scan rate for 1,10-phenanthroline – complex.

reductive) potentials observed for the metal ion (Table 2) at a scan rate of 10 mV/s. All the voltammograms were obtained after 5 min of equilibration.

### Pre-soaking of bentonite on water samples

To investigate the ability of Kenyan bentonite to improve electron-transfer kinetics of the electro-active species in water samples, 50 g of bentonite was added to 50 ml of water from a polluted water course in Nairobi, Kenya. The mixture was continuously agitated for 24 h to ensure homogeneity. The slurry obtained was air-dried for 12 h and crushed to a homogeneous powder. The powder was used to modify electrodes and cyclic voltammograms obtained, using 0.1 M sulphuric acid as supporting electrolyte, from -1.5V to +0.8V.

### RESULTS AND DISCUSSION

### Investigation of the nature of reaction

In the study to determine the nature of reaction of the metal ion at the surface of the bare carbon electrode with 1, 10-phenanthroline – Fe complex, it was observed that the peak current was linear to the square root of scan rate implying that the process was diffusion controlled as per the Cottrell equation (Cottrell, 1903) (Figure 1).

### Calibration curves: Comparison between modified and non-modified electrodes

Cyclic voltammograms obtained for Fe ions using both the non-modified and modified electrodes are shown in Figure 2. The voltammograms for the modified electrodes looked slightly different from those of the modified electrodes and some of them had different oxidation/ reduction peaks (Table 3). This could be attributed to the fact that bentonite is a carrier of electroactive species as part of its elemental composition (Table 1), and also the pH of the bentonite (8.4 – 9.6) may cause a change in the potential of the peaks, since pH generally affects adsorption and catalysis (Jovic-Jovicic et al., 2013; Soetaredjo et al., 2011). pH is also known to affect the stability of various species of elements (Millero, 2001). This may explain the different (predominant) peaks in the ions observed. Copper, for example has the predominant peak at a potential of 0.025 mV vs CE on bare carbon (pH of solution about 6) while at higher pH (pH of bentonite, 8.4 – 9.6) the predominant species is observed to have a potential of about 0.40 mV vs CE.

Another example is zinc, whose predominant species is observed to have a potential of -0.998 mV vs CE at neutral pH, while in the presence of bentonite (higher pH) the predominant species is observed to have a much higher potential of -0.143 mV vs CE. The other metal ions studied also showed some variation, which could likewise be attributed to a pH change. It may also have been due to interaction of the ions with the elements which make up the structure of bentonite (Table 1). Straight line calibration curves were obtained for all the metal ions studied. Some of the curves are indicated in Figure 3. The constants (slopes) for the calibration curves for the various metal ions are shown in Table 4.

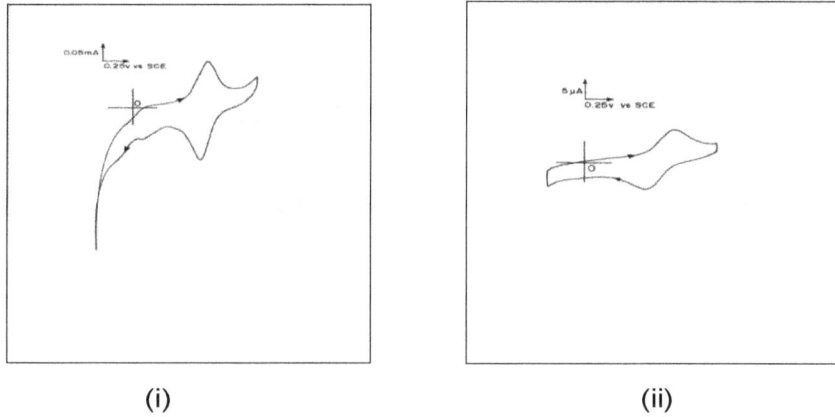
(i)                                                    (ii)

**Figure 2.** Cyclic voltammogram for Fe for unmodified (i) and modified (ii) electrodes.

**Table 3.** Anodic peak potentials of the metal ions $E_p$ = of the non-modified electrode; $E_p$ (CME) = of the modified electrode.

| Metal ion | Cu | Zn | Ni | Fe | Co |
|---|---|---|---|---|---|
| $E_p$ (V) | (0.025±0.001)* | -0.998±0.020 | 0.450±0.025 | 0.420±0.032 | 0.590±0.071 |
| $E_p$(CME)(V) | (0.462±0.010)* | -0.143±0.035 | 0.464±0.040 | 0.414±0.089 | 0.386±0.009 |

* A number of peaks were observed. The largest peak was used to obtain the calibration curve.

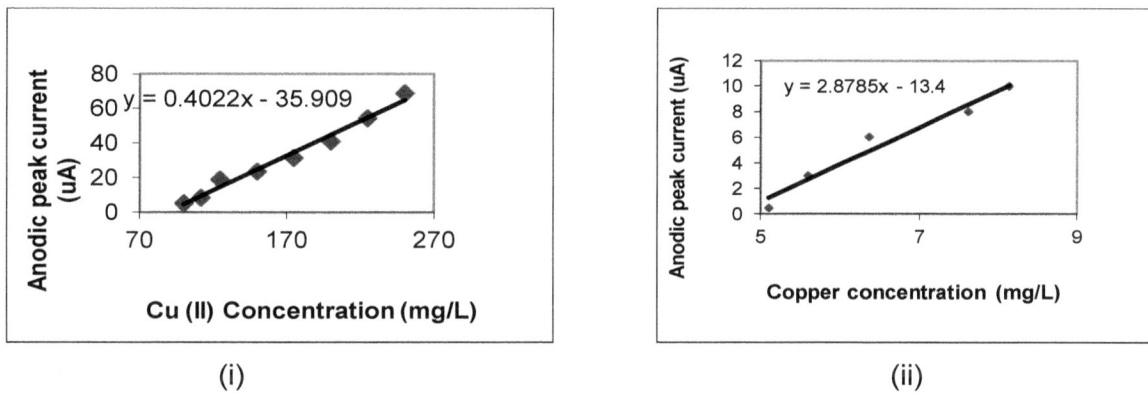
(i)                                                    (ii)

**Figure 3a.** Peak current vs copper concentration for (i) unmodified and (ii) unmodified electrodes.

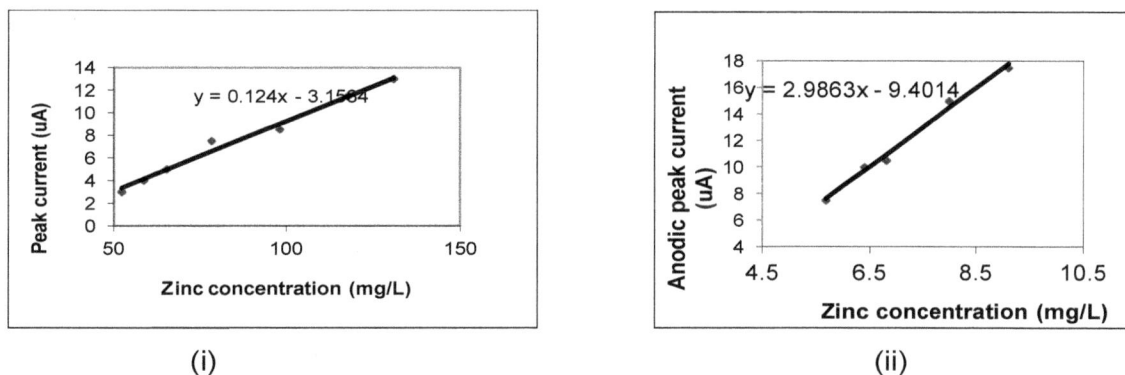
(i)                                                    (ii)

**Figure 3b.** Peak current versus zinc concentration for (i) unmodified and (ii) modified electrodes.

**Table 4.** Values for slopes obtained in calibration curves for bare carbon and for the modified electrodes.

| Slope(μA/mg/L) | Cu | Zn | Ni | Fe | Co |
|---|---|---|---|---|---|
| Slope (bare C) | 0.402 | 0.124 | 0.413 | 0.905 | 0.023 |
| Slope (CME) | 2.878 | 2.986 | 0.736 | 6.67 | 0.196 |
| Ratio: CME/bare C* | 7.159 | 24.081 | 1.782 | 7.370 | 8.521 |

* No units.

**Table 5.** Some peaks observed for the samples from Water samples preconcentrated on bentonite.

| Site number | Anodic peak voltage vs SCE(v) (±0.001) |
|---|---|
| 1 | -0.498, -0.114, 0.057, 0.100, 0.192, 0.541, 0.584, 0.683 |
| 2 | -0.484, -0.107, 0.029, 0.085, 0.256, 0.355,0.427, 0.470 |
| 3 | -0.455, 0.0285, 0.071, 0.114, 0.182, 0.455, 0.541 |
| 4 | -0.498, -0.128, 0.114, 0.142, 0.413, 0.463. |
| 5 | -0.484, -0.228, 0.085, 0.128, 0.185, 0.227, 0.389, 0.427. |
| 6 | -0.455, -0.313, -0.128, 0, 0.043, 0.085, 0.128, 0.399, 0.455. |

It is observed that the CME/bare carbon ratio is greater than one in all the metal ions investigated, indicating that the concentration of the analyte is higher in the bentonite matrix than in solution, a factor that indicates that bentonite enhances the electron transfer kinetics of the metal ions investigated (Orata and Segor, 1999). This may indicate that there is preconcentration of electroactive species due to the presence of bentonite by factors ranging from 1.782 for nickel to 24.081 for zinc.

It is observed that the more stable the element (the higher the proximity to the stable $3d^5$ or $3d^{10}$ configuration), the higher the ratio between the modified electrode and the bare carbon electrode. An observation which implied that the difference in analyte pre-concentration depends on the stability of the metal ion in the periodic table, which may in turn indicate that a chemical reaction takes place between bentonite and the metal ion in question. The mechanism followed by this adsorption may be one proposed by Jovic-Jovicic et al. (2013) and Ma et al. (2011) when they observed adsorption of $Pb^{2+}$ on bentonite. Most probably the adsorption of the cations follow two mechanisms: cation exchange with inter-layer exchangeable cations and bonding to the silanol or alumino groups at the edge smectite. Jovic-Jovicic et al. (2010a) propose that the molecules they were investigating (RB5) could be adsorbed due to electrostatic interaction of $-SO_3^-$ groups from the dye with previously adsorbed $Pb^{2+}$ ions at the edge sites. The reaction they proposed is as follows:

$$S\text{-}OH + Pb^{2+} + \text{-}OSO_2\text{-}dye \longrightarrow S\text{-}O\text{-}Pb\text{-}O\text{-}SO_2\text{-}dye + H^+$$

The same reaction is proposed in the cations involved in the Kenyan bentonite.

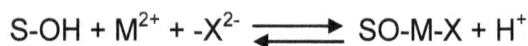

$$S\text{-}OH + M^{2+} + \text{-}X^{2-} \rightleftharpoons SO\text{-}M\text{-}X + H^+$$

Where $M^{2+}$ is the metal ion, $X^{2-}$ is the anion in solution and S is the bentonite surface. It is therefore expected that adsorption would be pH-dependent, given that a hydrogen ion is removed as a by-product. This may explain the shift in anodic peak potential for some ions when adsorbed on bentonite.

### Preconcentration with water samples

Further experiments were performed to investigate pre-concentration of electroactive species from samples of water obtained from sampling sites on a polluted water course in Nairobi, Kenya on bentonite. Some of the peaks observed for each sampling point are given in Table 5. Most aqueous systems contain relatively low concentration of metal ions which may not be detected on a potentiostat, and preconcentration may be necessary in order to elicit a response. It is clear that if bentonite did not adsorb/ preconcentrate [electroactive] species there would be no observable response from the potentiostat. However, for all the bentonite electrodes that had been pre-soaked in the water, clear peaks were obtained (Table 5). These results corroborate those observed by Orata and Segor (1999). Thus bentonite-modified electrodes can be used to monitor pollution levels in, for example, aqueous systems.

### CONCLUSION AND RECOMMENDATIONS

Kenyan bentonite was investigated for adsorption and

preconcentration of metal ions. Straight line calibration curves were obtained with bentonite-modified electrodes with slopes of between 0.196 and 2.986 µA/mg/L. The slopes of the curves of modified electrodes were higher than those of non-modified electrodes by a factor of between 1.7 and 24 which means that bentonite adsorbed and preconcentrated the metal ions. It was also observed that the magnitude of the ratio depended on the proximity of the ion being investigated to the stable 3d configuration (either filled or half-filled 3d orbitals). This may mean that chemical rather than physical adsorption took place. When electrodes were modified with bentonite that had been soaked with water samples from a polluted water course, the reponse indicated that the electroactive species in the water could be detected on a potentiostat. Thus Kenyan bentonite can be used in analysis of species in aqueous solution through preconcentrating the species on the bentonite. Spiking can be used to identify the species observed on preconcentration on water samples on bentonite. It is also recommended that other methods (eg. computational speciation programs) be used to compare the species observed electrochemically with those suggested by such programs.


## ACKNOWLEDGEMENT

Author would like to acknowledge The German Academic Exchange Service (DAAD) for providing funds to carry out the project and also acknowledge the Eric Abraham Academic Visitorship (EAAV) for funds at the University of Cape Town.


## Conflict of Interest

The author(s) have not declared any conflict of interests.

### REFERENCES

Boz N, Degirmenbasi NK, Dihan M (2013). Transesterification reaction of canola oil to biodiesel using calcium bentonite functionalized with K compounds. Appl. Catalysis B-Environ. 138:236–242. http://dx.doi.org/10.1016/j.apcatb.2013.02.043

Cottrell FG (1903). Der Reststrom bein galvanischer Polarisation betrachtet als ein Diffusions problem. Inaugural Dissertation, University of Leipsig. Also published in Zeit. Phys. Chem. 42:385-431.

Elbetieba A, Al-Hammod MH (1997). Long term Exposure of Male and Female Mice to trivalent and hexavalent chromium compounds: Effect of Fertility. Toxicology. 116:19–47.

Guo K, Chen X, Freguia S, Donose B, Dan C (2013). Spontaneous modification of carbon surfaces with neutral red from its diazonium salts for bioelectrochemical systems. Biosensors and bioelectronics. 47:184–189. http://dx.doi.org/10.1016/j.bios.2013.02.051 PMid:23578972

Heinze J, Muller R (1998). Direct Electrochemical Detection of C60 in Solution by steady-state voltammetry at microelectrodes. J.

Electrochem. Soc. 145:1227–1232. http://dx.doi.org/10.1149/1.1838443

Jovic-Jovicic N, Milutinovic-Nikolic A, Bankovic P, Mojovic Z, Zunic M, Grzetic I, Jovanovic D (2010a). Organo-inorganic Bentonite for simultaneous adsorption of acid orange 10 and lead ions. Appl. Clay Sci. 47(3-4):452–456. http://dx.doi.org/10.1016/j.clay.2009.11.005

Jovic-Jovivic NP, Milutinovic-Nikolic AD, Zunic MJ, Mojovic ZD, Bankovic PT, Grzetic IA, Jovanovic DM (2013). Synergic Adsorption of Pb2+ and reactive dye – RB5 on organo-modified bentonites. J. Contaminant Hydrol. 50:1–11. http://dx.doi.org/10.1016/j.jconhyd.2013.03.004 PMid:23624568

Kuralay F, Erkut Y, Lokman U, Adil D (2013). Cibacron Blue F3GA modified disposable pencil graphite electrode for the investigation of affinity binding to bovine serum albumin. Colloids surfaces. B: biointerfaces. 110:270–274. http://dx.doi.org/10.1016/j.colsurfb.2013.04.024 PMid:23732804

Li YM, Zhang Y, Li J, Sheng G, Zheng X (2013). Enhanced reduction of chlorophenols by nanoscale zerovalent iron supported on bentonite. Chemosphere. 92(4). http://dx.doi.org/10.1016/j.chemosphere.2013.01.030

Lloyd R (1960). Great Britain department of scientific and industrial research. Water Pollution Res. P. 83.

Ma J, Cui B, Dai J, Li D (2011). Mechanism of adsorption of anionic dye from aqueous solution onto organo bentonite. J. Hazardous mater. 86(2-3):1758–1765. http://dx.doi.org/10.1016/j.jhazmat.2010.12.073 PMid:21227582

Millero F (2001). Speciation of metals in natural waters. Geochemical Translations. 8. (article) http://dx.doi.org/10.1186/1467-4866-2-57

Nada FA, Soher AD, Sayed EK, Gala A (2007). Effect of Surfactants on the voltammetric response and determination of an antihypersentive drug. J. talanta 72(4):1438–1445.

Normandin L (2004). Manganese distribution in the brain and neurobehavioral changes following inhalation exposure of rats to three chemical forms of manganese. Neurotoxicology. 25(3):433-441. http://dx.doi.org/10.1016/j.neuro.2003.10.001 PMid:15019306

Orata D, Segor F (1999) Bentonite (Clay montmorillonite) as a template for electrosynthesis of thyroxine, Catalysis Lett. 58:157-162. http://dx.doi.org/10.1023/A:1019038219556

Oukil D, Makhloufi L, Saidani B (2007). Preparation of polypyrrole films containing ferrocyanide ions deposited onto thermally pretreated and untreated iron substrate. Application in the electroanalytical determination of ascorbic acid. J. Sensors Actuators B: Chemical. 123(2)1083–1089. http://dx.doi.org/10.1016/j.snb.2006.11.014

Pearce JMS (2007). "Burton's line in lead poisoning". European neurol. 57(2):118–119. http://dx.doi.org/10.1159/000098100 PMid:17179719

Puma S, Marchese F, Dominijanni A, Manassero M (2013). Reuse of MSWI bottom Ash mixed with natural sodium bentonite as landfill cover material. Waste Manage. Res. 31(6):577–584. http://dx.doi.org/10.1177/0734242X13477722 PMid:23478909

Rahul MK, Purvi BD, Ashwini KS (2007). Behaviour of riboflavin on plain carbon paste and aza macrocycles based chemically modified electrodes. J. Sensors Actuators B: Chemical. 124(1):90–98. http://dx.doi.org/10.1016/j.snb.2006.12.004

Reitzel K, Andersen FO, Egemose S, Jensen HS (2013). Phosphate adsorption by Lanthanum modified bentonite clay in Fresh and brakish water. Water Res. 47(8):2787–2796. http://dx.doi.org/10.1016/j.watres.2013.02.051 PMid:23521977

Rezaei B, Elaheh H, Ensafi AA (2013). Stainless Steel modified with an aminosilane layer and gold nanoparticles as a novel disposable substrate for impedimetric immunosensors. Biosensors and Electronics. 58:61–66.

Shi LN, Zhou Y, Chen Z, Megharey M, Naidu R (2013). Simultaneous adsorption and degradation of Zn2+ and Cu2+ from Waste waters using nanoscale zero—valent iron impregnated with clays. Environ. Sci. Pollution Res. 20(6):3639–3648. http://dx.doi.org/10.1007/s11356-012-1272-7 PMid:23114838

Soetaredjo FE, Ayucitra A, Ismadji S, Mankar AL (2011). KOH/bentonite catalysts for transesterification of palm oil to biodiesel. 53:341–346.

Tvrda E, Zuzana K, Jana L, Monika S, Zofia G, Agniezka G, Csaba S, Peter M, Norbert L (2013). The impact of lead and cadmium on selected motility, proxidant and antioxidant parameters of bovine seminal plasma and spermatozoa. J. Environ. Sci. Health part A –

Toxic/Hazardous Substances Environ. Eng. 48(10):1292–1300.

UNEP/GEMS (1991). Fresh water Pollution. UNEP/GEMS Environmental Library No. 6. UNEP.

Zhang YJ, Liu LC, Chen DP (2013). Synthesis of Cd/S bentonite nanocomposite powders for H − 2 production by photocatalytic decomposition of water. Powder Technol. 241:7–11. http://dx.doi.org/10.1016/j.powtec.2013.02.031

# 18

# Research in physical properties of $Al_xGa_{1-x}As$ III-V Arsenide ternary semiconductor alloys

**Alla Srivani, Vedam Ram Murthy and G. Veera Raghavaiah**

Department of Physics, T. J. P. S College and Sri Mittapalli college of Engineering, Guntur, Andhra Pradesh, India.

**General description of III-Arsenide semiconductors is presented and significance of the present work is stressed. The electrical and optical properties of III-Arsenide from binary semiconductors are evaluated using the principle of additivity involving quadratic expressions. The electrical and optical properties studied in this group include refractive index, optical polarizability, absorption coefficient and energy gap. A comparison of these data is made with reported data wherever available. The significance of the present method developed from refractive indices with out need for sophisticated experimental methods is stressed. The advantage of this group alloys is also outlined.**

**Key words:** Physical properties, III-V group, ternary semiconductors, alliminium, gallium, arsenic.

## INTRODUCTION

III-Arsenide has important position in Science and Technology of Compound semiconductors in modern electronic and optical devices. Semiconductor alloys, which are solid solutions of two or more semiconducting elements, have important technological applications, especially in the manufacture of electronic and electro-optical devices. One of the easiest ways to change artificially the electronic and optical properties of semiconductors is by forming their alloys; it is then interesting to combine two different compounds with different optical band gaps and different rigidities in order to obtain a new material with intermediate properties. Hence the major goal in materials engineering is the ability to tune the band gap independently in order to obtain the desired properties. The zinc-blende compounds AlAs, GaAs, InAs and BAs have been arousing increasing interest, both theoretically and experimentally, because of their potentially inherent

advantages. There is a considerable interest in the study of ternary alloy semiconductors such as $Al_xGa_{1-x}As$, $In_xGa_{1-x}As$ with the ultimate object of providing device materials with a specific band gap and band structure. The maximum direct energy gap requirement for solar cells, light emitting diodes and semiconductor lasers has led to the use of $Al_xGa_{1-x}As$ alloys, where the band gap can be increased by addition of Al. The purpose of using Group-III Arsenide alloys is to obtain a material which consumes the minimum of power with maximum brightness. Liquid phase epitaxy compared with molecular beam epitaxy or metal organic chemical vapour deposition still continues being a useful technique for obtaining optoelectronic devices based on III–V ternary compounds. The main interest of the work is the development of the III-Arsenides for applications on current quantum well technology. In this work we report the optical and electrical characterization of $Al_xGa_{1-x}As$

epitaxial layers (Puron et al., 1999).

## Arsenides

We present a study of the optical and electrical properties of the AlAs, GaAs, InAs semiconductors and their alloys $Al_xGa_{1-x}As$. Because of the technological importance of $Al_xGa_{1-x}As$, its various properties have been extensively studied. In particular, parameters of the $Al_xGa_{1-x}As$ band structure have been determined from a variety of measurements, including photo response, optical transmission and photoluminescence and variation of Hall electron concentration with temperature. We present a comprehensive up-to-date compilation of band parameters for the technologically important III-V zinc blende compound semiconductors: GaAs, AlAs and InAs along with their ternary alloys. The III–V Arsenide semiconductors are important materials in the fields of fabrication of microwave, optoelectronic, and electronic devices. The film materials of devices are usually obtained by several techniques, such as metal organic vapour phase epitaxy (MOVPE), molecular beam epitaxy (MBE) and liquid phase epitaxy (LPE). Semiconductor material selection plays a vital role in developing semiconductor devices. Extensive research in materials has produced a number of compound semiconductors (Adachi, 1992).

## $Al_xGa_{1-x}As$

In metal organic chemical vapour deposition (MOCVD), $Al_xGa_{1-x}As$ thin films were characterized using Raman and Hall measurements. The $Al_xGa_{1-x}As$ thin films were grown by (MOCVD using metallic arsenic instead of arsine as the arsenic precursor. Some difficulties in the growth of $Al_xGa_{1-x}As$ by MOCVD are the composition homogeneity of the layers and the oxygen and carbon incorporation during the growth process. The composition homogeneity of the films was demonstrated by the Raman measurements. Hall measurements on the samples showed highly compensated material. Samples grown at temperatures lower than 750°C were highly resistive. Independently of the V/III ratio; the samples grown at higher temperatures were n-type. As the growth temperature is increased the layers compensation decreases but the Raman spectra show that the layers become more defective (Díaz-Reyes, 2002).

## METHODOLOGY

The refractive index, optical polarizability, absorption coefficient and energy gap of Arsenide semiconductor alloys are evaluated by using principle of additivity and quadratic expressions. The principle of additivity is used to study physical properties even at very small compositions. The calculated properties of refractive index, optical polarizability, absorption coefficient and energy gap versus

**Table 1.** Values of refractive index of binary semiconductors.

| Compound | Refractive index $n$ |
|----------|------------------------|
| AlAs | 3.00 |
| GaAs | 3.03 |
| InAs | 3.50 |

concentrations was fitted by equations.

**Method 1**

$A_{12}=A_1*x+A_2*(1-x) + 1/1000*SQRT (A_1*A_2)*x*(1-x)$

**Method 2**

$A_{12}=A_1*x+A_2*(1-x) +1/1000* SQRT (A_1*A_2*x*(1-x))$

**Method 3**

$A_{12}=A_1*x+A_2*(1-x) - 1/1000*SQRT (A_1*A_2)*x*(1-x)$

**Method 4**

$A_{12}=A_1*x+A_2*(1-x) - 1/1000*SQRT (A_1*A_2*x*(1-x))$

## Additivity

$A_{12}=A_1*x+A_2*(1-x)$

Where $A_{12}$ denotes refractive index ($n_{12}$), optical polarizability ($\alpha_{m12}$), absorption coefficient ($\alpha_{12}$) and energy gap ($Eg_{12}$). $A_1$ and $A_2$ denotes refractive index (n), optical polarizability ($\alpha_m$), absorption coefficient ($\alpha$), energy gap (Eg) of two binary compounds forming ternary compound.

## Refractive index

The refractive index of semiconductors represents a fundamental physical parameter that characterizes their optical and electronic properties. It is a measure of the transparency of the semiconductor to incident spectral radiation. In addition, knowledge of the refractive index is essential for devices such as photonic crystals, wave guides, solar cells and detectors (Yadav et al., 2012). The refractive index values of Arsenide semiconductor alloys are evaluated by using additivity and quadratic expressions of the equations by replacing A by n the refractive index from the reported values of refractive index of binary semiconductors (Naser et al., 2009)(Table 1).

## Optical polarizability

Optical polarizability ($\alpha_m$) is used to study optical behaviour of binary and ternary semiconductors belonging to III-arsenide ternary semiconductor alloys.

## Optical polarizability of binary compounds

### Lorentz-Lorenz relation

The mean optical polarizability $\alpha_M$ for binary semiconductors is obtained by using Lorentz-Lorenz relation (Sathyalatha, 2012) given below:

$$\alpha_m = \left(\frac{n^2-1}{n^2+1}\right) \frac{M}{\rho} \quad \frac{3}{4\pi N}$$

Where M, N, n, $\rho$ refer to Molecular weight, Avogadro number, Refractive index and density

### New dispersion principle

The equation of motion of the electron may be written as:

$$mZ + mbZ + \omega_0^2 mZ = eE_0 e^{i\omega t}$$

Here

$Ee^{i\omega t}$ refers to the electric force, $\omega = 2\pi v$, m is the electron mass, $\omega_0$ is Natural frequency of the electron and mbZ represents the damping term.

By solving the above equation, value of z will be obtained in the form as:

$$Z = \frac{\left(\frac{e}{m}\right) E_0 e^{i\omega t}}{\omega_0^2 - \omega^2 + i\omega b}$$

Thus the moment induced (P) per unit volume will be

P= ΣZe

$$P = \frac{v\left(\frac{e^2}{m}\right) E_0 e^{i\omega t}}{\omega_0^2 - \omega^2 + i\omega b}$$

Here v is Loschmidt number. Displacement vector is obtained as:

D=E+4πp

$$D = E + 4\pi \left[ \frac{v\left(\frac{e^2}{m}\right) E_0 e^{i\omega t}}{\omega_0^2 - \omega^2 + i\omega b} \right]$$

$$D = E \left[ 1 + \frac{4\pi v\left(\frac{e^2}{m}\right)}{\omega_0^2 - \omega^2 + i\omega b} \right]$$

$$\frac{D}{E} = (n - ik)^2$$

$$\frac{D}{E} = \left[ 1 + \frac{4\pi v\left(\frac{e^2}{m}\right)}{\omega_0^2 - \omega^2 + i\omega b} \right]$$

The expression of n can be obtained by separating real and imaginary parts in the above equations that is,

$$n = 1 + 2\pi \frac{e^2 v}{m} \frac{\omega_0^2 - \omega^2}{\left(\omega_0^2 - \omega^2\right)^2 + \omega^2 b^2}$$

If incident frequency $\omega < \omega_0$ then $\omega^2 b^2$ can be neglected. Thus the above equation can be written as:

$$n = 1 + \frac{2\pi e^2 v}{m(\omega_0^2 - \omega^2)}$$

Rearranging the terms in the above equation, we get:

$$\left[ \frac{1}{\lambda_0^2} - \frac{1}{\lambda^2} \right] = \frac{e^2 v}{2\pi c^2 m(n-1)}$$

$$\frac{1}{\lambda^2} = \alpha + \frac{\beta}{(n-1)}$$

Optical polarizability of binary compounds can be calculated by using new dispersion relation (Murthy et al., 1986) by knowing α and β values

$$\frac{1}{\lambda^2} = \alpha + \frac{\beta}{n-1}$$

Where $\alpha = \frac{1}{\lambda 0^2}$ and $\beta = -\frac{e^2 v}{2\pi m c^2}$

Dividing through out by $\beta$ and rearranging the terms $n_\alpha$

$$\frac{1}{n-1} = \frac{1}{\beta \lambda^2} - \frac{\alpha}{\beta}$$

This equation is of the form Y=mx+c

$$Lt(\lambda \to \infty, \frac{1}{\beta \lambda^2} \to 0)$$

Hence $\frac{1}{n\infty - 1} = -\frac{\alpha}{\beta} = \gamma$

$$n_\infty - 1 = \frac{1}{\gamma}$$

$$n_\infty = \frac{1+\gamma}{\gamma}$$

**Table 2.** Optical Polarizability of compounds.

| Compound | Optical polarizability $\alpha_m$ $(cms)^3$ [] |
|----------|-----------------------------------------------|
| AlAs | 71.14 |
| GaAs | 82.75 |
| InAs | 104.90 |

$$\frac{n_{\infty}^2-1}{n_{\infty}^2+2}=\frac{\frac{(1+\gamma)^2}{\gamma^2}-1}{\frac{(1+\gamma)^2}{\gamma^2}+2}$$

Substitute the value of $\frac{n_{\infty}^2-1}{n_{\infty}^2+2}$ in Lorentz-Lorenz formula, we get

$$\alpha_m = \left(\frac{(\gamma+1)^2-\gamma^2}{(\gamma+1)^2+2\gamma^2}\right)\frac{M}{\rho}\quad\frac{3}{4\pi N}$$

Where M, N and ρ are molecular weight, Avogadro number and density of binary semiconductors and $v=-\frac{\alpha}{\beta}$

Here α is Y-Intercept and β is the slope (Table 2).

**Absorption coefficient**

Lorentz-Lorenz relation for solids is represented as follows:

$$\left(\frac{n^2-1}{n^2+1}\right)=\frac{4\pi v\alpha_m}{3}$$

$$(n^2-1)=\frac{4\pi v\alpha_m.3}{3-4\pi v\alpha_m}=\left(\frac{1}{3}-\frac{1}{4\pi v\alpha_m}\right)^{-1}$$

$$\frac{1}{3}-\frac{1}{4\pi v\alpha_m}=\left(\frac{1}{n^2-1}\right)$$

$$(n^2-1)=\frac{12\pi v\alpha_m.}{3-4\pi v\alpha_m}$$

The absorption coefficient $\alpha=2k = \frac{32\pi^3.}{3v\lambda^4}(n-1)^2$

From the above two equations we get

$$\frac{(n^2-1)}{(n-1)^2}=\left(\frac{12\pi v\alpha_m}{3-4\pi v\alpha_m}\right)^1\frac{32\pi^3.}{3v\lambda^4\alpha}$$

$$\frac{(n^2-1)}{(n-1)^2}=f(\text{consider})$$

$$\frac{n+1}{n-1}=f \text{ Or } n=\frac{f+1}{f-1}$$

$$n=\left(\frac{\frac{128\pi^4\alpha_m}{\alpha\lambda^4(3-4\pi v\alpha_m)}+1}{\frac{128\pi^4\alpha_m}{\alpha\lambda^4(3-4\pi v\alpha_m)}-1}\right)$$

$$\frac{n+1}{n-1}=\frac{128\pi^4\alpha_m}{\alpha\lambda^4(3-4\pi v\alpha_m)}$$

or

$$\alpha=\frac{128\pi^4\alpha_m}{\alpha\lambda^4(3-4\pi v\alpha_m)}\left(\frac{n-1.}{n+1}\right)$$

$$\alpha=\frac{128\pi^4\alpha_m}{\lambda^4}\left(\frac{n-1.}{n+1}\right)\left(\frac{M}{3M-4\pi N\rho\alpha_m}\right)$$

Here $v=\frac{N\rho}{M}$

Where N is Avogadro number ρ is the density and M is molecular weight of the semiconductor.

Thus the expression for absorption coefficient of binary semiconductor is given as (Sathyalatha, 2012):

$$\alpha = \left(\frac{128\pi^4\alpha_m}{\lambda^4}\right)\left(\frac{n-1}{n+1}\right)\left(\frac{M}{3M-4\pi N\rho\alpha_m}\right)$$

Where $\alpha_m$, n, M, ρ and λ refer to the optical polarizability, refractive index, molecular weight, density and wavelength of binary semiconductors. N is Avogadro number.

Similarly for ternary semiconductors, the expression for absorption coefficient can be given as:

$$\alpha = \left(\frac{128\pi^4\alpha_{m12}}{\lambda^4}\right)\left(\frac{n_{12}-1}{n_{12}+1}\right)\left(\frac{M_{12}}{3M_{12}-4\pi N\rho_{12}\alpha_{m12}}\right)$$

Where $\alpha_{m12}$, $n_{12}$, $M_{12}$ and $\rho_{12}$ are optical polarizability, refractive index, molecular weight and density of ternary semiconductor alloys and N is Avogadro number. They are calculated by using different additivity relations and quadratic expressions (Sathyalatha, 2012).

**Energy gap**

The electrical conductivity of semiconductors depends on width of energy gap and it is affected by Dopant composition, temperature, pressure, magnetic and electrical fields. Indirect band gap semiconductors is inefficient for emitting light. Semiconductors that have direct band gap are good light emitters. A wide band gap (WBG) semiconductor is a semiconductor with an energy band gap wider than about 2 eV, suitable for microwave devices. A narrow band semiconductor has energy band gap narrower than about 2 eV suitable for tunnel devices and infrared technology. Band gap is measured by both spectroscopic and conductivity methods.

**Table 3.** Energy gap of compounds.

| Compound | Energy gap Eg  e.v[] |
|----------|----------------------|
| AlAs | 2.95 |
| GaAs | 1.42 |
| InAs | 0.36 |

### Energy gap of ternary semiconductors

The formula used for calculation of Energy gap of ternary semiconductors are given below:

$$E_g = \left\{ \frac{28.8}{((2^{x_m} - x_n)^2)^{\frac{1}{4}}} \left[ \frac{1 - \Phi_{12}}{1 + 2\Phi_{12}} \right] \right\} \left[ \frac{x_M}{x_N} \right]^2$$

Where $x_M$ and $x_N$ are the electro negativities of the constituent atoms of ternary semiconductor

$$\Phi_{12} = \left[ \frac{4\pi N}{3} \right] \left[ \frac{\alpha_{M_{12}} \rho_{12}}{M_{12}} \right]$$

Where $\alpha_{M_{12}}$, $\rho_{12}$, $M_{12}$ and N are Optical polarizability, density, molecular weight and Avogadro number of ternary semiconductor Alloys (Table 3).

## RESULTS AND DISCUSSION

The refractive index values of binary Arsenide compound semiconductors are taken from reference (Naser et al., 2009) and are given in tables. The refractive index values of ternary semiconductor alloys are calculated by using different expressions of for whole composition range (0<x<1) and are presented in tables. These values are compared with literature reported data (http://www.cleanroom.byu.edu/EW_ternary.phtml-BRIGHAM; Sathyalatha, 2012). It is found that calculated values are in good agreement with reported values. Graphs are drawn for all these alloys by taking their composition values on x axis and refractive index values on y axis.

The refractive indices at various wavelengths for the binary semiconductors are taken from hand book of optical constants of solids (Edward, 1991) are presented in table along with $\frac{1}{n-1}$ and $\frac{1}{\lambda^2}$ values. The graphs drawn between $\frac{1}{n-1}$ and $\frac{1}{\lambda^2}$ for these semiconductors are shown in figures. From these graphs intercept α values and the slope β of the straight line are calculated. All these values are given from the tables.

The evaluated optical polarizabilities of binary semiconductors by using equation are also from the tables. The computed optical polarizabilities by new dispersion relations are compared with reported values.

The values of Molecular weight (M), density (ρ) and refractive index (n) of the semiconductors which are required for evaluation of $\alpha_m$ are taken from CRC Hand book (William and David, 2010). The energy gap values of $Al_xGa_{1-x}As$ are calculated by using different additivity expressions and presented in tables. These values are compared with Reported data (http://www.cleanroom.byu.edu/EW_ternary.phtml-BRIGHAM; Sathyalatha, 2012).

Graphs are drawn for the above $Al_xGa_{1-x}As$, alloys with variation of Dopant compositions and are given in figure. Calculated values of energy gap is taken on x axis and their composition values are taken on y axis. The refractive indices at various wavelengths for the binary semiconductors are taken from hand book of optical constants of solids (Edward, 1991) are presented in table along with $\frac{1}{n-1}$ and $\frac{1}{\lambda^2}$ values.

The graphs drawn between $\frac{1}{n-1}$ and $\frac{1}{\lambda^2}$ for these semiconductors are shown in figures. From these graphs intercept α values and the slope β of the straight line are determined and γ values are calculated. All these values are given from the tables. The evaluated optical polarizabilities of binary semiconductors by using equation are also from the table. The computed optical polarizabilities by new dispersion relations are compared with reported values. The values of molecular weight (M), density (ρ) and refractive index (n) of the semiconductors which are required for evaluation of $\alpha_m$ are taken from CRC Hand book (William and David, 2010) (Tables 4 to 11).

The applications of III-V Arsenide ternary semiconductor alloys of $Al_xGa_{1-x}As$, $In_xGa_{1-x}As$, $Al_xIn_{1-x}As$, $InP_{1-x}As_x$, $GaAs_xP_{1-x}$, $AlAs_xP_{1-x}$ as electronic, optical and optoelectronic devices are determined by elementary material properties of refractive index, optical polarizability, absorption coefficient, energy gap and mobility. Photonic crystals, wave guides and solar cells require knowledge of refractive index and energy gap of all above arsenide group alloys. The energy gap of semiconductor alloys determines threshold for absorption of photons in semiconductors. Refractive index is measure of transparency of semiconductor alloys to incident radiation. Refractive index and energy gap of ternary semiconductor alloys has significant impact on band structure. High absorption coefficient semiconductor alloys can be used for fabricating in thin film hetero junction photovoltaic (PV) devices.

Narrow band gap semiconductor alloys of $InP_{1-x}As_x$, $In_xGa_{1-x}As$ are used for photo catalytic applications. Wide band gap semiconductor alloys of $GaAs_xP_{1-x}$, $Al_xIn_{1-x}As$, $Al_xGa_{1-x}As$ are investigated for devices that allow one to

**Table 4.** Optical polarizability, absorption coefficient and energy Gap of $Al_xGa_{1-x}As$ X=0.09.

| Wave length $\lambda$ (A$^0$) | $\frac{1}{\lambda^2}$ In $(10)^8$ $(cms)^2$ | R.I value n | $\frac{1}{n-1}$ | Optical polarizability $\alpha_m$ $(10)^{-25}(cms)^3$ | | Absorption coefficient ($\alpha$) $(10)^{-1}$ cms$^-$ | Energy gap (e.v) | |
|---|---|---|---|---|---|---|---|---|
| 4133 | 5.854 | 4.963 | 0.252 | Calculated | Reported Sathyalatha, 2012) | 4.317 | Calculated | Reported |
| 4275 | 5.472 | 4.838 | 0.261 | 82.0 | 82.03 | 3.747 | 1.50 | 1.42 |
| 4428 | 5.100 | 4.725 | 0.268 | | | 3.236 | | |
| 4592 | 4.742 | 4.518 | 0.284 | | | 2.764 | | |
| 4769 | 4.397 | 4.353 | 0.298 | | | 2.351 | | |
| 4959 | 4.066 | 4.220 | 0.311 | | | 1.991 | | |
| 5166 | 3.747 | 4.111 | 0.321 | | | 1.676 | | |
| 5391 | 3.441 | 4.018 | 0.331 | | | 1.403 | | |
| 5636 | 3.148 | 3.940 | 0.340 | | | 1.166 | | |
| 5904 | 2.869 | 3.876 | 0.348 | | | 0.962 | | |
| 7293 | 1.880 | 3.678 | 0.373 | | | 0.123 | | |
| 8266 | 1.463 | 3.572 | 0.389 | | | 0.055 | | |

**Table 5.** X=0.198.

| Wave length $\lambda$ | $\frac{1}{\lambda^2}$ In $(10)^8$ $(cms)^2$ | R.I value n | $\frac{1}{n-1}$ | Optical polarizability $\alpha_m$ $(10)^{-25}(cms)^3$ | | Absorption coefficient ($\alpha$) $(10)^{-1}$ cms$^{-1}$ | Energy gap (e.v) | |
|---|---|---|---|---|---|---|---|---|
| 4133 | 5.854 | 4.943 | 0.254 | Calculated | Reported (Sathyalatha, 2012) | 2.657 | Calculated | Reported |
| 4275 | 5.472 | 4.757 | 0.266 | 80.0 | 81.17 | 1.943 | 1.35 | 1.75 |
| 4428 | 5.100 | 4.547 | 0.282 | | | 1.645 | | |
| 4592 | 4.742 | 4.375 | 0.296 | | | 1.389 | | |
| 4769 | 4.397 | 4.235 | 0.309 | | | 1.170 | | |
| 4959 | 4.066 | 4.118 | 0.321 | | | 0.984 | | |
| 5166 | 3.747 | 4.022 | 0.331 | | | 0.822 | | |
| 5391 | 3.441 | 3.940 | 0.340 | | | 0.684 | | |
| 5636 | 3.148 | 3.871 | 0.348 | | | 0.566 | | |
| 5904 | 2.869 | 3.815 | 0.355 | | | 0.465 | | |
| 7293 | 1.880 | 3.635 | 0.379 | | | 0.591 | | |
| 8266 | 1.463 | 3.457 | 0.407 | | | 0.220 | | |
| | | | | | | 0.193 | | |
| | | | | | | 0.112 | | |

attain frequencies that span over a wide range and attain Terahertz. Applications on these ternary semiconductor alloy span from communications to biomedical engineering. Narrow band gap semiconductor alloys allow hetero junction bipolar transistors to present terahertz (THz) operation capability. Sensors of this type exploit the unique piezoelectric, polarization characteristics, as well as the high temperature stability of wide-band gap semiconductors in order to allow stable operation with high sensitivity. Using this material system one can also explore the possibility of developing fundamental sources operating in the terahertz regime

**Table 6.** X=0.315.

| Wave length $\lambda$ (A$^0$) | $\frac{1}{\lambda^2}$ In $(10)^8$ (cms)$^2$ | R.I value n | $\frac{1}{n-1}$ | Optical polarizability $\alpha_m$ $(10)^{-25}$ (cms)$^3$ | | Absorption coefficient ($\alpha$) $(10)^{-1}$ cms$^{-1}$ | Energy Gap e.v | |
|---|---|---|---|---|---|---|---|---|
| | | | | Calculated | Reported (Sathyalatha, 2012) | | Calculated | Reported |
| 4133 | 5.854 | 4.781 | 0.264 | | | 2.604 | | |
| 4275 | 5.472 | 4.582 | 0.279 | 78.2 | 80.23 | 2.539 | 1.42 | 1.85 |
| 4428 | 5.100 | 4.404 | 0.294 | | | 2.478 | | |
| 4592 | 4.742 | 4.258 | 0.307 | | | 2.424 | | |
| 4769 | 4.397 | 4.135 | 0.319 | | | 2.378 | | |
| 4959 | 4.066 | 4.032 | 0.330 | | | 2.336 | | |
| 5166 | 3.747 | 3.945 | 0.339 | | | 2.300 | | |
| 5391 | 3.441 | 3.872 | 0.348 | | | 2.269 | | |
| 5636 | 3.148 | 3.815 | 0.355 | | | 2.244 | | |
| 5904 | 2.869 | 3.758 | 0.362 | | | 2.218 | | |
| 7293 | 1.880 | 3.509 | 0.398 | | | 2.159 | | |
| 8266 | 1.463 | 3.404 | 0.416 | | | 2.334 | | |

**Table 7.** X=0.419.

| Wave length $\lambda$ (A$^0$) | $\frac{1}{\lambda^2}$ In $(10)^8$ (cms)$^2$ | R.I value n | $\frac{1}{n-1}$ | Optical polarizability $\alpha_m$ $(10)^{-25}$ (cms)$^3$ | | Absorption coefficient ($\alpha$) $(10)^{-1}$ cms$^{-1}$ | Energy Gap e.v | |
|---|---|---|---|---|---|---|---|---|
| | | | | Calculated | Reported (Sathyalatha, 2012) | | Calculate | Reported |
| 4133 | 5.854 | 4.605 | 0.277 | | | 2.239 | | |
| 4275 | 5.472 | 4.430 | 0.291 | 79.44 | 79.40 | 1.911 | 1.30 | 1.05 |
| 4428 | 5.100 | 4.280 | 0.305 | | | 1.626 | | |
| 4592 | 4.742 | 4.159 | 0.317 | | | 1.379 | | |
| 4769 | 4.397 | 4.047 | 0.328 | | | 1.666 | | |
| 4959 | 4.066 | 4.957 | 0.338 | | | 0.982 | | |
| 5166 | 3.747 | 3.881 | 0.347 | | | 0.823 | | |
| 5391 | 3.441 | 3.820 | 0.355 | | | 0.687 | | |
| 5636 | 3.148 | 3.747 | 0.364 | | | 0.670 | | |
| 5904 | 2.869 | 3.686 | 0.372 | | | 0.654 | | |
| 7293 | 1.880 | 3.422 | 0.413 | | | 0.099 | | |
| 8266 | 1.463 | 3.341 | 0.427 | | | 0.090 | | |

and employing micro-electro mechanical systems (MEMS) approaches.

Recent progress and new concepts using narrow and wide-band gap ternary semiconductor alloys of $Al_xGa_{1-x}As$, $In_xGa_{1-x}As$, $Al_xIn_{1-x}As$, $InP_{1-x}As_x$, $GaAs_xP_{1-x}$, $AlAs_xP_{1-x}$ and device concepts such quantum wells with very high mobility and plasma waves will lead in Terahertz detectors and emitters. Semiconductors of this type may also be used for other novel applications such as spintronics and field emission. Terahertz signal sources based on super lattices have explored applications cover a wide range of devices, circuits and components for communications, sensors and biomedical engineering.

Research on physical properties of III-Arsenide semiconductor alloys is due to operating characteristics of semiconductor devices depend critically on the

**Table 8.** X=0.491.

| Wave length λ (A⁰) | $\frac{1}{\lambda^2}$ In $(10)^8$ $(cms)^2$ | R.I value n | $\frac{1}{n-1}$ | Optical polarizability $\alpha_m$ $(10)^{-25}(cms)^3$ | | Absorption coefficient (α) $(10)^{-1}$ cms$^{-1}$ | Energy Gap e.v | |
|---|---|---|---|---|---|---|---|---|
| | | | | Calculated | Reported (Sathyalatha, 2012) | | Calculated | Reported |
| 4133 | 5.854 | 4.483 | 0.287 | | | 2.032 | | |
| 4275 | 5.472 | 4.328 | 0.300 | 81.49 | 78.83 | 1.738 | 1.25 | 1.65 |
| 4428 | 5.100 | 4.195 | 0.313 | | | 1.481 | | |
| 4592 | 4.742 | 4.081 | 0.324 | | | 1.258 | | |
| 4769 | 4.397 | 3.985 | 0.335 | | | 1.064 | | |
| 4959 | 4.066 | 3.903 | 0.344 | | | 0.898 | | |
| 5166 | 3.747 | 3.838 | 0.352 | | | 0.753 | | |
| 5391 | 3.441 | 3.761 | 0.362 | | | 0.626 | | |
| 5636 | 3.148 | 3.696 | 0.371 | | | 0.178 | | |
| 5904 | 2.869 | 3.665 | 0.375 | | | 0.274 | | |
| 7293 | 1.880 | 3.368 | 0.422 | | | 0.172 | | |
| 8266 | 1.463 | 3.283 | 0.438 | | | 0.102 | | |

**Table 9.** x=0.59.

| Wave length λ (A⁰) | $\frac{1}{\lambda^2}$ In $(10)^8$ $(cms)^2$ | R.I value n | $\frac{1}{n-1}$ | Optical polarizability $\alpha_m$ $(10)^{-25}(cms)^3$ | | Absorption coefficient (α) $(10)^{-1}$ cms$^{-1}$ | Energy Gap e.v | |
|---|---|---|---|---|---|---|---|---|
| | | | | Calculated | Reported (Sathyalatha, 2012) | | Calculated | Reporte |
| 4133 | 5.854 | 4.343 | 0.299 | | | 1.801 | | |
| 4275 | 5.472 | 4.208 | 0.312 | 79.20 | 78.04 | 1.543 | 2.30 | 2.05 |
| 4428 | 5.100 | 4.092 | 0.323 | | | 1.317 | | |
| 4592 | 4.742 | 3.992 | 0.334 | | | 1.120 | | |
| 4769 | 4.397 | 3.909 | 0.344 | | | 0.949 | | |
| 4959 | 4.066 | 3.837 | 0.352 | | | 0.801 | | |
| 5166 | 3.747 | 3.758 | 0.362 | | | 0.671 | | |
| 5391 | 3.441 | 3.690 | 0.372 | | | 0.582 | | |
| 5636 | 3.148 | 3.658 | 0.376 | | | 0.464 | | |
| 5904 | 2.86 | 3.54 | 0.393 | | | 0.377 | | |
| 7293 | 1.880 | 3.313 | 0.432 | | | 0.153 | | |
| 8266 | 1.463 | 3.287 | 0.447 | | | 0.091 | | |

physical properties of the constituent materials. The high electron mobility of InAs, is due to its narrow band gap, makes this compound useful for very high-speed and low-power electronic and infrared optoelectronic devices.

The energy band gap of Group III-V Arsenide narrow band gap semiconductor alloys $InP_{1-x}As_x$ and $In_xGa_{1-x}As$ reduces significantly by adding a small amount of Arsenic to InP and Indium to GaAs, the band gaps of these alloys are expected to vary from 1.974 eV (InP) to 1.833 eV (InAs) in $InP_{1-x}As_x$ and 1.42 eV (GaAs) to 0.36 eV (InAs) by increasing As and In Concentrations. These ternary alloys are used for manufacturing infrared detectors, gas sensors. The energy band gaps of above two alloys decrease rapidly leading to a strong disorder when a small amount of Phosphorus atoms in InP is replaced by Arsenic and when small amount of Ga atoms are

**Table 10.** X=0.7.

| Wave length $\lambda$ (A$^0$) | $\frac{1}{\lambda^2}$ In $(10)^8$ (cms)$^2$ | R.I value n | $\frac{1}{n-1}$ | Optical polarizability $\alpha_m$ $(10)^{-25}$(cms)$^3$ | | Absorption coefficient ($\alpha$) $(10)^{-1}$ cms$^{-1}$ | Energy Gap e.v | |
|---|---|---|---|---|---|---|---|---|
| | | | | Calculated | Reported (Sathyalatha, 2012) | | Calculated | Reported |
| 4133 | 5.854 | 4.196 | 0.313 | | | 5.873 | | |
| 4275 | 5.472 | 4.084 | 0.324 | 77.96 | 77.16 | 5.042 | 2.50 | 2.36 |
| 4428 | 5.100 | 3.987 | 0.338 | | | 4.290 | | |
| 4592 | 4.742 | 3.906 | 0.344 | | | 3.677 | | |
| 4769 | 4.397 | 3.823 | 0.354 | | | 3.114 | | |
| 4959 | 4.066 | 3.746 | 0.364 | | | 2.625 | | |
| 5166 | 3.747 | 3.696 | 0.371 | | | 2.208 | | |
| 5391 | 3.441 | 3.595 | 0.388 | | | 1.817 | | |
| 5636 | 3.148 | 3.500 | 0.400 | | | 1.497 | | |
| 5904 | 2.869 | 3.425 | 0.412 | | | 1.222 | | |
| 7293 | 1.880 | 3.225 | 0.449 | | | 0.500 | | |
| 8266 | 1.463 | 3.153 | 0.464 | | | 0.297 | | |

**Table 11.** X=0.804.

| Wave length $\lambda$ (A$^0$) | $\frac{1}{\lambda^2}$ In $(10)^8$ (cms)$^2$ | R.I value n | $\frac{1}{n-1}$ | Optical polarizability $\alpha_m$ $(10)^{-25}$(cms)$^3$ | | Absorption coefficient ($\alpha$) $(10)^{-1}$ cms$^{-1}$ | Energy gap e.v | |
|---|---|---|---|---|---|---|---|---|
| | | | | Calculated | Reported (Sathyalath a, 2012) | | Calculated | Reported |
| 4133 | 5.854 | 4.050 | 0.328 | | | 5.678 | | |
| 4275 | 5.472 | 3.961 | 0.338 | 75.66 | 76.33 | 4.889 | 2.45 | 2.67 |
| 4428 | 5.100 | 3.872 | 0.348 | | | 4.181 | | |
| 4592 | 4.742 | 3.787 | 0.359 | | | 3.559 | | |
| 4769 | 4.397 | 3.783 | 0.365 | | | 3.032 | | |
| 4959 | 4.066 | 3.635 | 0.379 | | | 2.541 | | |
| 5166 | 3.747 | 3.519 | 0.397 | | | 2.106 | | |
| 5391 | 3.441 | 3.440 | 0.410 | | | 1.745 | | |
| 5636 | 3.148 | 3.378 | 0.420 | | | 1.440 | | |
| 5904 | 2.869 | 3.322 | 0.431 | | | 0.180 | | |

replaced by In. This occurs due to the large disparity in the electro negativity and the atomic size between P and As in $InP_{1-x}As_x$ and between In and Ga in $In_xGa_{1-x}As$. The Arsenic atom and Indium atom induces several perturbations in the host crystal (Abbasi et al., 2010).

The energy band gap of Group III-V Arsenide wide band gap semiconductor alloys $Al_xGa_{1-x}As$, $Al_xIn_{1-x}As$ and $GaAs_{1-x}P_x$ increases significantly by adding small amount of Al to GaAs, InAs and by adding As to GaP. The band gaps of these alloys are expected to vary from 1.42 eV (GaAs) to 2.67 ev (AlAs), 0.36 eV (InAs) to 2.95 eV (AlAs) and 1.42 eV (GaAs) to 2.78 eV (GaP) by increasing Al and P Concentrations. The energy band gaps of above alloys increases rapidly leading to a strong disorder when a small amount of Gallium atoms in GaAs is replaced by Al and when small amount of In atoms are replaced by Al and when small amount of As atoms are replaced by p. This occurs due to the large disparity in the electro negativity and the atomic size between Al and Ga in $Al_xGa_{1-x}As$, between Al and In in $Al_xIn_{1-x}As$ and between As and P in $GaAs_{1-x}P_x$. The Al atom and P atom induces several perturbations in the host crystal of above alloys (Djurišić, 2002).

The binding which was totally covalent for the elemental semiconductors, has an ionic component in III-V Arsenide ternary semiconductor alloys. The percentage of the ionic binding energy varies for various semiconductor alloys. The percentage of ionic binding energy is closely related to electro negativity of the elements and varies for various compounds. The electro negativity describes affinity of electrons of the element. In a binding situation the more electro negative atoms will be more strongly bind to the electrons than its partner and therefore carry net negative charge. The difference in electro negativity of the atoms in a compound semiconductor gives first measure for energy gap. A more electro negative element replacing a certain lattice atom will attract the electrons from the partner more strongly, become more negatively charged and thus increase the ionic part of the binding. This has nothing to do with its ability to donate electrons to conduction band or accept electrons from the valence band.

Mobility at high doping concentration is always decreased by scattering at the ionized dopants. Band gap increases with Electro negativity difference between the elements. Bond strength decreases with decrease of orbital overlapping. Large band gap in $Al_xGa_{1-x}As$, $Al_xIn_{1-x}As$ and $GaAs_{1-x}P_x$ is due to high degree of orbital overlapping. Electro negativity affects the width of the band gap. Electrons are more stabilized by more electro negativity atom. Pure semiconductors are located in Group 3 and 4 of the periodic table. The band gaps of these materials are less influenced by electro negativity. They are influenced by configuration of crystal lattice, valence shell electrons and hybridization of orbitals.

Semiconductor materials with higher absorption coefficients more readily absorb photons, which excite electrons into the conduction band. Knowing absorption coefficients of III-Arsenide ternary semiconductor alloys of $InP_{1-x}As_x$, $In_xGa_{1-x}As$, $Al_xGa_{1-x}As$, $Al_xIn_{1-x}As$ and $GaAs_{1-x}P_x$ aids engineers in determining which material to use in their solar cell designs. The absorption coefficient determines how far into a material light of a particular wavelength can penetrate before it is absorbed. In a material with a low absorption coefficient, light is only poorly absorbed, and if the material is thin enough, it will appear transparent to that wavelength. The absorption coefficient depends on the material and also on the wavelength of light which is being absorbed. III-V Arsenide ternary semiconductor alloys have a sharp edge in their absorption coefficient, since light which has energy below the band gap does not have sufficient energy to excite an electron into the conduction band from the valence band. Consequently this light is not absorbed.

The plot of hv versus $(\alpha hv)^2$ of III-Arsenide ternary semiconductor alloys of $InP_{1-x}As_x$, $In_xGa_{1-x}As$, $Al_xGa_{1-x}As$, $Al_xIn_{1-x}As$ and $GaAs_{1-x}P_x$ at various concentrations of As, In, Al and P forms a straight line, it can normally be inferred that there is a direct band gap, measurable by extrapolating the straight line to the $\alpha=0$ axis. On the other hand, if a plot of hv versus $\alpha hv^{1/2}$ forms a straight line, it can normally be inferred that there is an indirect band gap, measurable by extrapolating the straight line to $\alpha=0$ axis. Measuring the absorption coefficient for ternary semiconductor alloys gives information about the band gaps of the material. Knowledge of these band gaps is extremely important for understanding the electrical properties of a semiconductor. Measuring low values of Absorption coefficient ($\alpha$) with high accuracy is photo thermal deflection spectroscopy which measures the heating of the environment which occurs when a semiconductor sample absorbs light (Priester and Grenet, 2000, 2001).

The energy levels adjust with alloy concentration, resulting in varying amount of absorption at different wavelengths in III-Arsenide ternary semiconductor alloys of $InP_{1-x}As_x$, $In_xGa_{1-x}As$, $Al_xGa_{1-x}As$, $Al_xIn_{1-x}As$ and $GaAs_{1-x}P_x$. This variation in optical properties is described by the material optical constants, commonly known as refractive index (n). The optical constants shape corresponds to the material's electronic transitions. Thus, the optical constants become a "fingerprint" for the semiconductor alloys. In $Al_xGa_{1-x}As$, the direct band gap shifts toward shorter wavelengths with increasing Al concentration. The low band gap semiconductors used in infrared detectors will absorb over most conventional ellipsometer wavelengths in $In_xGa_{1-x}As$ (Das et al., 2007).

The refractive index of Group III-V Arsenide ternary semiconductor alloys $InP_{1-x}As_x$ and $In_xGa_{1-x}As$, $Al_xGa_{1-x}As$ and $GaAs_{1-x}P_x$ reduces significantly by adding a small amount of Arsenic to InP, Indium to GaAs, Al to GaAs and P to GaAs. The refractive index of these alloys are expected to vary from 4.433(InP) to 3.157 (InAs) in $InP_{1-x}As_x$ and 4.484 (GaAs) to 4.229 (InAs) by increasing As and In concentrations. These ternary alloys are used for manufacturing infrared detectors, gas sensors. This occurs due to the large disparity in the electro negativity and the atomic size between P and As in $InP_{1-x}As_x$ and between In and Ga in $In_xGa_{1-x}As$. The Arsenic atom, Indium atom, Aluminium atom and Phosphorus atoms induces several perturbations in the host crystal.

## Conflict of Interests

The author(s) have not declared any conflict of interests.

## REFERENCES

Abbasi FM, Ahmad H, Perveen F, Inamullah, Sajid M, Brar DS (2010). Assesment of genomic relationship between *Oryza sativa* and *Oryza australiensis*. Afr. J. Biotechnol. 9(12):1312-1316.

Adachi S (1992). "Physical Properties of III-V Semiconductor Compounds, "John Wiley & Sons, New York.

Djurišić AB (2002). Progress in the room-temperature optical functions of semiconductors. Materials Science and Engineering R Reports 38:237.

Priester C, Grenet G (2000). "Surface roughness and alloy stability interdependence in lattice-matched and heteroepitaxy". Phys. Rev. B. 61(23)15.

Priester C, Grenet G (2001). "Surface roughness and alloy stability interdependence in lattice-matched and heteroepitaxy". Phys. Rev. B. 61(23)15.

Das TD, Mondal A, Dhar S (2007). IEEE International Workshop on the Physics of Semiconductor Devices (IWPSD), P. 511.

Díaz-Reyes J (2002). "Raman and Hall characterization of AlGaAs epilayers grown by MOCVD using elemental arsenic", Superficies y Vacío15, 22-25, diciembre de äSociedad Mexicana de Ciencia de Superficies y de Vacío.

Edward DP (1991). Handbook of Optical Constants of Solids, Volume 2, Academic press, 21 march 1991. http://www.cleanroom.byu.edu/EW_ternary.phtml-BRIGHAM YOUNG UNIVERSITY, Department of Electrical and Computer Engineering, "Direct Energy Band Gap in Ternary Semiconductors".

Murthy VR, Jeevan kumar R, Subbaiah DV (1986). "New dispersion relation: Relation to ORD, MORD and Molecular Polarization' Proc viii Annual conference on IEEE/EMBS, Texas, XIV, P. 1636-1639.

Naser MA, Zaliman S, Uda H, Yarub A (2009). Investigation of the absorption coefficient, refractive index, energy band gap, and film thickness for Al0.11Ga0.89N, Al0.03Ga0.97N, and GaN by optical transmission method, Int. J. Nanoelect. Mater. 2:189-195.

Puron E, Martinez-Criado G, Riech I (1999). "Growth and Optical characterization of indirect-gap AlxGa1-xAs alloys J. Appl. Phys. 86(1):1.

Sathyalatha KC (2012). PhD Thesis"Optical and related properties of few II-VI and III-V Semiconductors" SKU University, Anantapuram.

William MH, David RL (2010). CRC Hand book of Physics and chemistry, Taylor & Francis Group, 91st Edition, 26-May-2010.

Yadav DS, kumar C, Singh J Parashuram, Kumar G (2012). "Optoelectronic properties of zinc blende and wurtzite structured binary solids". J. Eng. Comput. Innov. 3(2):26-35.

# 19

# Kinetics study of copolymerization of 2-anilinoethanol onto chitosan by ammonium peroxydisulfate as a initiator

**Seyed Hossein Hosseini**

Department of Chemistry, Faculty of Science, Islamic Azad University, Islamshahr Branch, Tehran, Iran.

**Graft copolymerization of 2-anilinoethanol (2AE) onto chitosan (Chit-g-P2AE) was carried out by using ammonium peroxydisulfate (APS) as a long initiator under nitrogen atmosphere. Evidence of grafting was confirmed by comparison of FTIR spectrum of chitosan and the grafted copolymer. The grafting kinetics in the different conditions was studied as well. The effects of concentration of APS, 2AE, reaction time and temperature on graft copolymerization were studied by determining the grafting percentage, grafting efficiency and percentage add-on. With other conditions kept constant, the optimum grafting conditions were obtained as follows: Chitosan = 1 g, APS = 0.1 M, and 2AE = 0.213 mol/L, reaction temperature = 25°C, and reaction time = 5 h. Electrical conductivities of samples were measured by four probe method.**

**Key words:** Graft copolymerization, ammonium peroxydisulfate (APS), 2-anilinoethanol (2AE), onto chitosan (Chit-g-P2AE).

## INTRODUCTION

Many graft copolymers of chitosan and vinyl monomers have been synthesized and evaluated as flocculants, paper strengthener and drug-releaser (Singh et al., 2006). It has potential applications ranged from biomedicine and pharmacy to water treatment (Yu et al., 2007). Chitosan has both reactive amino and hydroxyl groups that can be used to chemically alter its properties under mild reaction conditions (Hosseini et al., 2010). Kinetics of graft copolymerization investigated from different methods (Li et al., 2002; Mahdavinia et al., 2004).

In addition, among various methods, graft copolymerization is most attractive because it is a useful technique for modifying the chemical and physical properties of natural polymers (Hosseini and Entezami, 2005; 2003; Hosseini and Gohari, 2013; Hosseini, 2006; 2013; Armes and Miller, 1988). Polyaniline are commonly synthesized by chemical or electrochemical oxidation of aniline in acidic aqueous solution (Abdolahi et al., 2012; Hosseini et al., 2010, 2013a). However, other polymerization techniques have now been developed such as enzymatic (Silva et al., 2005), ultrasonic irradiation (Liu et al., 2002), polymerization using electron acceptors (Su and Kuramoto, 2001) and under electric and magnetic fields (Hosseini et al., 2009, 2013b).

Although, the method of preparation is easy, the process ability is found to be very difficult. In order to

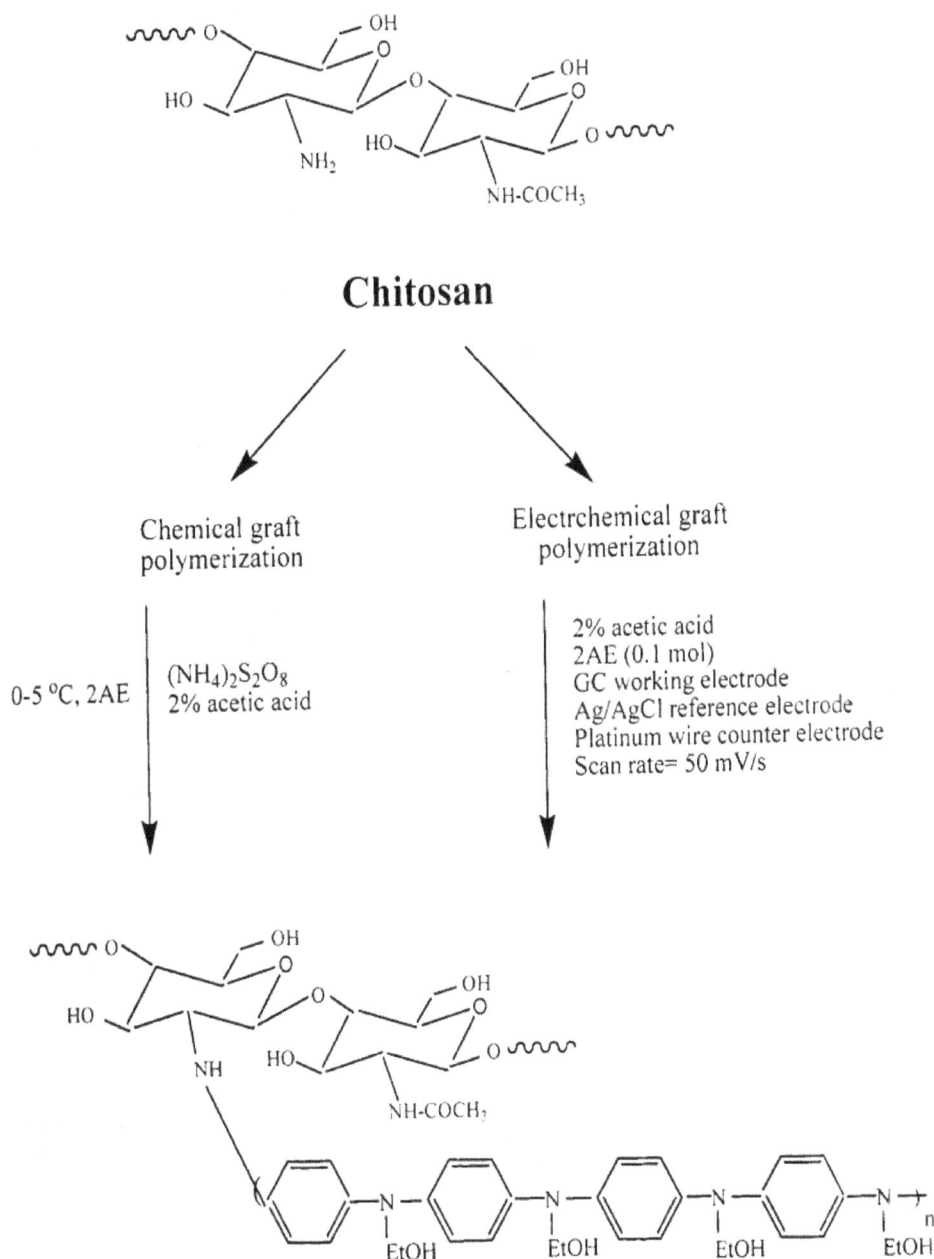

**Figure 1.** Scheme of formation all steps of Chit-*g*-P2AE.

diversify these difficulties, the electrically conducting polymers are made blend with conventional polymers.

In the preceding works, the authors have reported chemical and electrochemical synthesis of conducting graft copolymer of vinyl acetate with pyrrole and studied of its gas and vapor sensing (Hosseini et al., 2013a). In continuation, we have synthesized PANI grafted onto polyvinylpropionate (Hosseini and Entezami, 2005). Then investigation of sensing effects of polyaniline grafted on polystyrene for cyanide compounds (Hosseini, 2006) and graft copolymer of polypyrrole grafted on polystyrene for some of toxic gases (Hosseini and Entezami, 2005) were

reported too. In this work, conducting polymer was employed to initiate the graft copolymerization of 2AE onto chitosan. So, the grafting P2AE onto chitosan was carried out by chemical polymerization. Therefore, effects of concentration of ammonium peroxydisulfate (APS), 2AE, reaction time and temperature on graft copolymerization were studied by determining the grafting percentage, grafting efficiency and percentage add-on. In continuation, efficiency and grafting percentages and electrical conductivities of graft copolymer was measured. Figure 1 showed all steps of formation of Chit-g-P2AE.

## EXPERIMENTAL

### Grafting procedure

A typical graft copolymerization study was carried out as follow. Chitosan ($W_1$ g) was immersed in definite concentration of acetic acid (2%) in a glass tube and thermo stated at 45°C for 30 min. The solution was deaerated by passing pure nitrogen gas for 30 min. Required amount of monomer, 2AE was added and de-aerated for another 15 min. Graft copolymerization was initiated by the addition of calculated volumes of APS (using standard solutions). The time of adding the oxidizing agent, APS was taken as the starting time for the reaction. The polymerization conditions were selected in such a way that no polymerization occurred in the absence of added oxidant. This was ascertained by a separate experiment. At the end of the reaction time, the reaction was arrested by blowing air into the glass tube to freeze further reactions. The grafted chitosan fiber along with the homopolymer poly 2-anilinoethanol (P2AE) were filtered from the reaction mixture using a G4 sintered crucible and washed well with 1M HCl for several times, dried (at 80°C for 4 h) and weighed till to get constant weight. The weight of chitosan to be ($W_1$, g) and the total weight of the grafted polymer along with the homopolymer was called $W_2$, g.

The mixture of the grafted chitosan/P2AE and the homopolymer, P2AE was soxhlet extracted with N-methyl pyrrolidone (NMP) for several hours to separate the homopolymer. The extraction process was repeated till the separation of the homopolymer from the grafted sample was completed. This was ascertained by drying the polymer grafted in vacuum till to get constant weight ($W_3$, g). The difference in $W_3$-$W_1$ gives the weight of the grafted P2AE. The difference in $W_2$-$W_3$ gives the weight of the homopolymer, P2AE, formed and $W_4$ is the weight of monomer used.

### Electrochemical synthesis of Chit-g-P2AE

Electrochemical polymerization carried out by coating chitosan on surface GC disk working electrode, then growth P2AE onto chitosan in acidic solution. Chit-g-P2AE was prepared by applying intended potential to the electrode using potentiostate. In this electrolysis, a standard three-electrode cell, without any cell partition using a GC working electrode and Ag/AgCl as a reference electrode were employed. The electrolyte solution consisted of 0.1 M 2-anilinoethanol in 20 mL of 2% acetic acid in water. The potential range for electrochemical polymerization and the scan rate were -0.5 to 2 V (versus Ag/AgCl) and 50 mVs$^{-1}$, respectively.

## RESULTS AND DISCUSSION

Graft copolymerization of synthetic polymers onto chitosan can introduce desired properties and enlarge the field of the potential applications of them by choosing various types of side chains.

## IR Spectroscopy

Structural changes of Chit-g-P2AE were confirmed by FTIR spectroscopy (Figure 2a and b). The spectrums of chitosan (Figure 2a) observe the characteristic absorption bands around 3413 and 1624 cm$^{-1}$. The strong peak around 3413 cm$^{-1}$ due to the stretching vibration of O–H, the extension vibration of N–H, and inter hydrogen bonds

of the chitosan. In graft copolymer, the peak at 3100-3420 cm$^{-1}$ is of quite reduced intensity and broad, (due to overlapping of O–H and -NH$_2$ stretching groups of chitosan) (Figure 2b). There is a sharp peaks in 1510 and 1389 cm$^{-1}$, which is related to aniline ring and 1101 cm$^{-1}$ related to C-N stretching vibration bond. Reduced intensity of this peak with respect to chitosan shows that appreciable N–H at chitosan has been grafted with P2AE chain. From the IR data, it is clear that the grafted copolymer Chit-g- P2AE had characteristic peaks of P2AE of chitosan, which could be strong evidence of grafting.

## Scanning electron microscopy

Figure 3(a-c) shows the SEM micrographs of chitosan and Chit-g-P2AE after purification. Figure 3a shows chitosan surface is monotonous. Figure 3b provides direct evidence that polymer films are monotonous and unruffled. This sample has distinct one-phase morphology; so, it can confirm grafting of polymer without homopolymer. The photograph in Figure 3c shows that the cast film of Chit-g-P2AE is homogeneous and continuous. It is well known that chitosan has the good ability of degradation. Therefore, P2AE grafted with chitosan can improve its biodegradability. Chit-g-P2AE can play an important role in enhancing the compatibility physical properties of chitosan and conducting polymers.

## Conductivity measurements

The conductivities of both grafted and ungrafted chitosan were measured. The Chit-g-P2AE showed a good conductivity value better than pure chitosan. It was found that the conductivity values increased with increase in percentage grafting (Table 1). This confirms the chemical grafting of P2AE onto chitosan matrix.

## Study of cyclic voltammetry

The cyclic voltammograms for Chit-g-P2AE films grown and blank at different scan rates in acidic solution, GC disk as working electrode and Ag/AgCl reference electrode, with a reduced anodic potential limit are presented of Figure 4(a-c). Two pairs of well resolved oxidation peaks A and B which progressively developed at 0.9 and 1.3 V can be seen in Figure 4(a). First curve shows oxidation of chitosan moieties in the precursor, start to be oxidized for polymerization and red-ox behavior as well and it is electroactive completely.

This strongly resembles the behavior of P2AE. Only a peak, the so called "middle peaks" B, which belong to the degradation products (Hosseini, 2006) is poorly recognized. The potentials of the first peaks A in Chi-g-

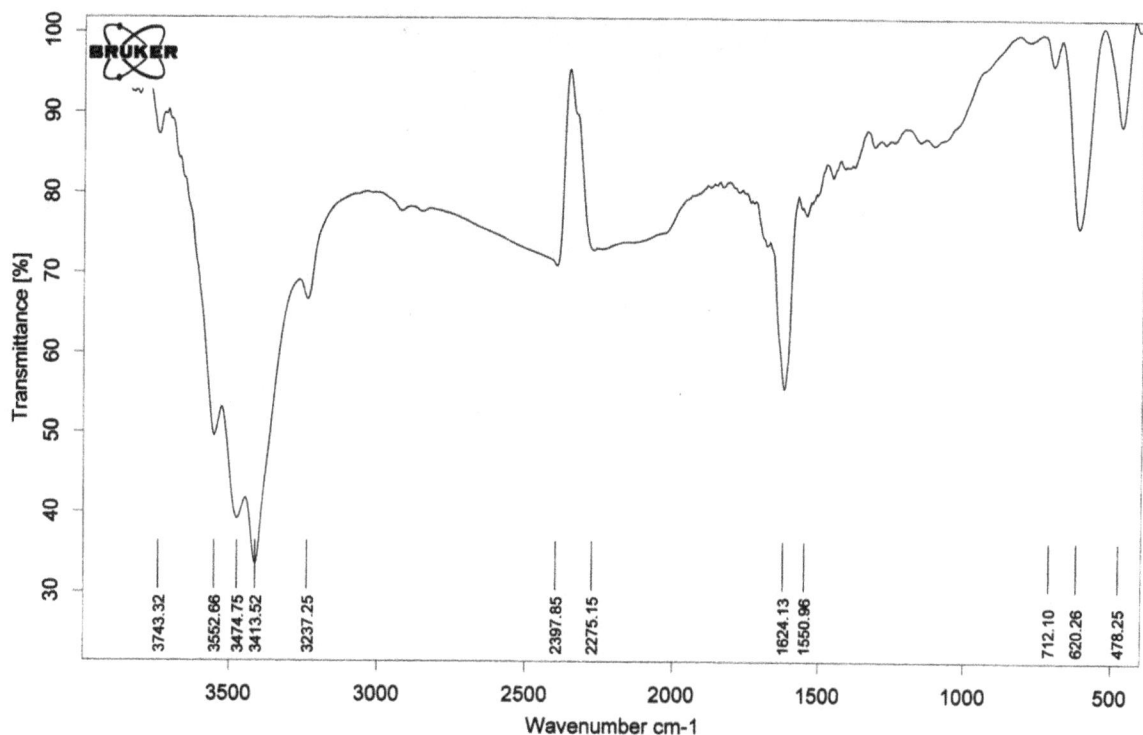

a

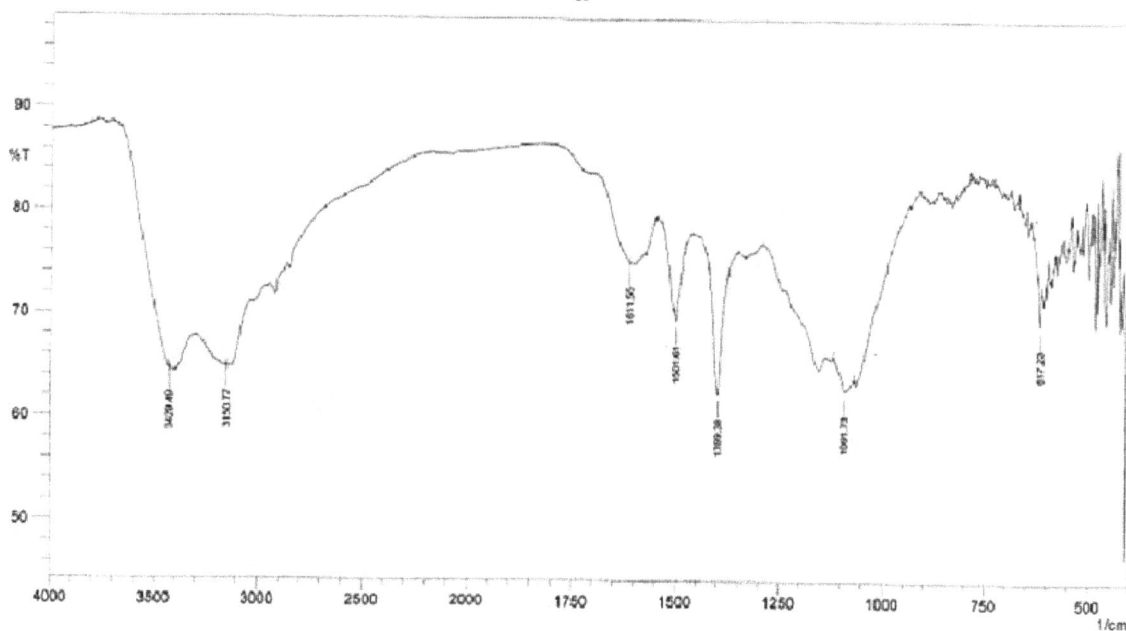

b

**Figure 2.** FTIR of a) Chitosan and b) Chit-*g*-P2AE film.

P2AE are slightly less positive than the corresponding values in P2AE. With the same electrochemical conditions, the curve shapes of Chit-g-P2AE almost replicate the electrochemical behaviour of P2AE, confirming once more that the mechanism of the oxidation processes in Chit-g-P2AE are the same as in the case of P2AE. Figure 4(b, c) showed that cyclic voltammograms of Chit-g-P2AE in blank and different scan rates, respectively. Herein, leucoemeraldine is converted into emeraldine in form of oxidation doping in peak and in continuation of the fall in peak, emeraldine is oxidized to pernigraniline and then, in the emeraldine is

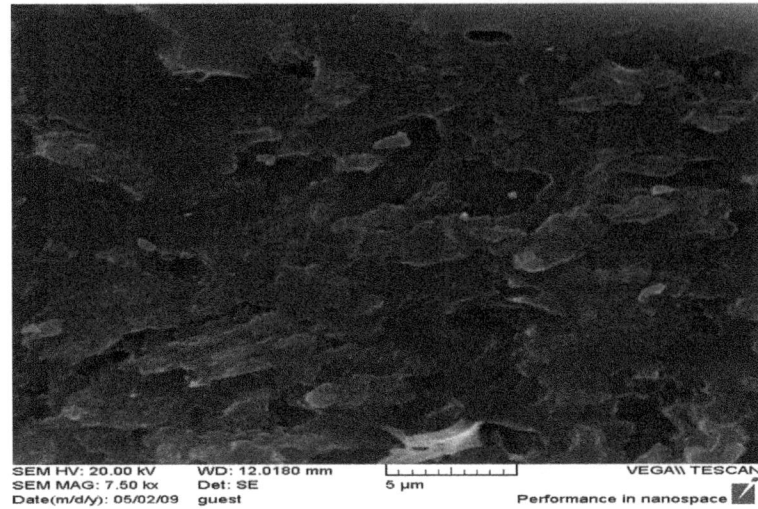

a

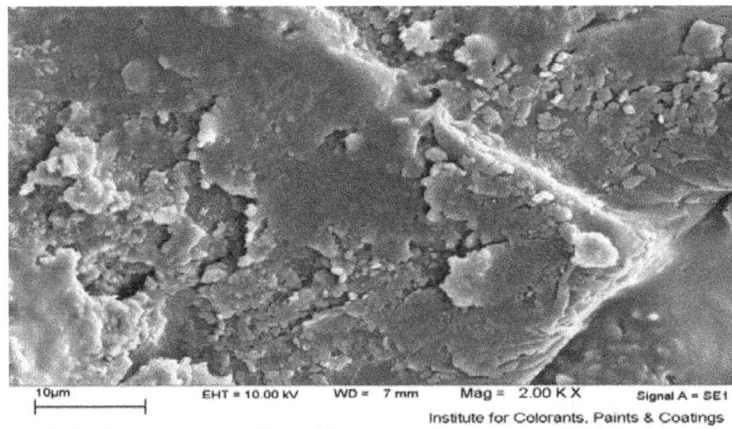

b

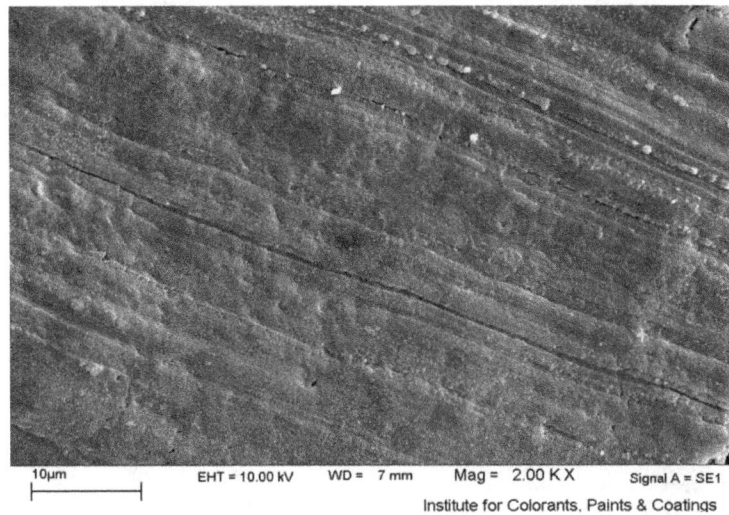

c

**Figure 3.** SEMs of a) Chitosan, b and c) Chit-*g*-P2AE.

reduced to leucoemeraldine.

In both reduction systems, during transferring electron from polymer chain, to neutralize the load, the anions in the electrolyte exits to the polymer structure. In Figure 4a,

**Table 1.** Electrical conductivities of samples.

| Polymer | % Grafting | Electrical conductivity (S/cm) |
| --- | --- | --- |
| Chitosan | - | $1.9 \times 10^{-8}$ |
| P2AE | - | $3.6 \times 10^{-3}$ |
| Chit-$g$-P2AE | 9.3 | $4.3 \times 10^{-5}$ |
| Chit-$g$-P2AE | 11.7 | $8.3 \times 10^{-5}$ |
| Chit-$g$-P2AE | 15.3 | $3.2 \times 10^{-4}$ |
| Chit-$g$-P2AE | 23.7 | $6.4 \times 10^{-4}$ |

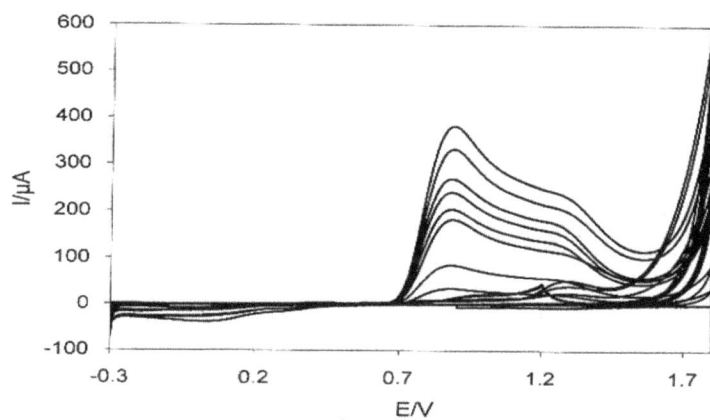

a

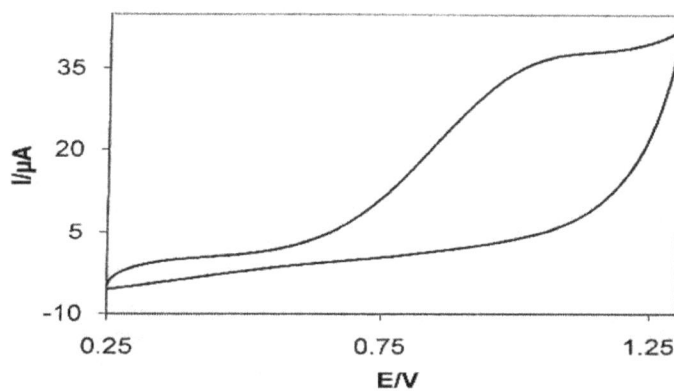

b

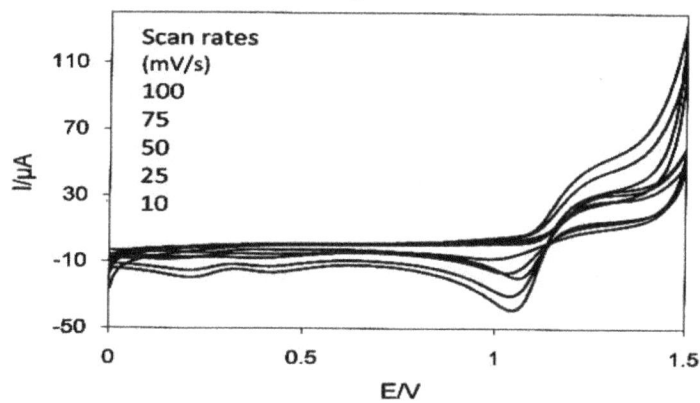

c

**Figure 4.** Cyclic votammograms of a and b) formation and blank of Chit-$g$-P2AE film in scan rate of 50 mV/s and c) different scan rates using a GC disk electrode versus Ag/AgCl and 2% acetic acid.

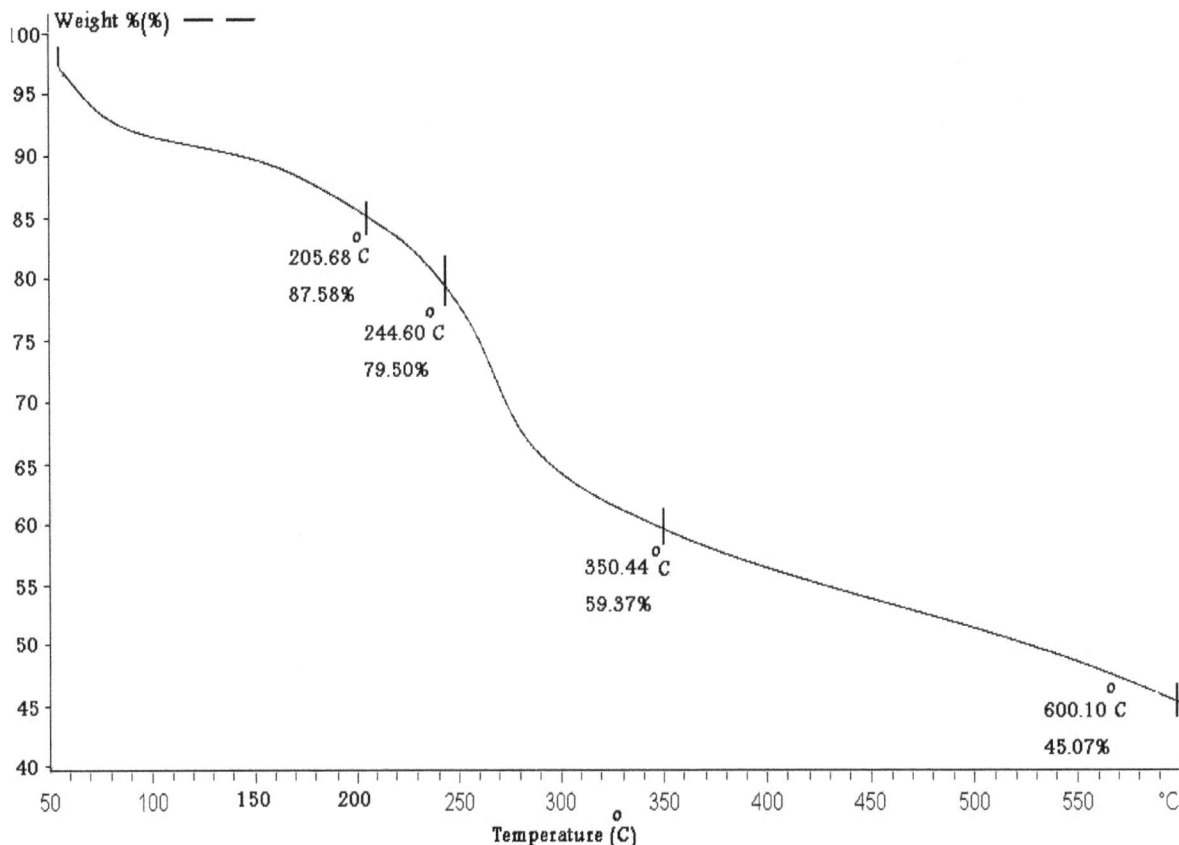

**Figure 5.** TGA of Chit-*g*-P2AE film.

reduction peaks in 50 mV/s scan rate was not seen, but reduction peak increased by decreasing scan rates. So, we can see redox peak in 10 mV/s, as well.

**Differential scanning calorimetry (DSC) and thermogravimetric analysis (TGA) investigation**

Figures 5 and 6 showed TGA and DSC of Chit-g-P2AE. Figure 5 shows that, polymer started to become soften from 30°C up to 205°C, first. It approximately loses 23% of its weight, which is due to humidity and existing solvent in polymer chains. Secondly, at temperature about 205 to 244°C approximately loses 8% of its weight, which is due to degradation of chitosan and side chain of P2AE. Third, at temperature of about 244 to 350°C, it starts to experience structural degradation and approximately loses 20% of its weight, which is due to degradation of polymer backbone. Chit-g-P2AE is thermally stable at temperatures below 205°C and in temperatures above 205°C; the polymer starts degradation and completely decomposed at above 350°C.

As shown in Figure 6 for Chit-g-P2AE, DSC curve, the endothermic peaks appear in 56 to 123°C (80°C), 162 to 230°C (207°C) and 245 to 300°C (270°C) regions. The endothermic peaks are related to TGA degradation steps.

The TGA and DSC curves show that Chit-g-P2AE has more thermal resistance than chitosan and P2AE. This curve has a nearly softening temperature in 200°C and destruction initiation in above than 200°C and in both polymers chitosan and P2AE is less resistance. Figure 7 showed STA (DSC and TGA) thermmograms of chitosan for comparison. The STA curves show that Chit-g-2AE has a little more thermal resistance than chitosan and lower than P2AE. The TGA curve for chitosan has a softening temperature in 194°C and destruction initiation in less than 200°C and for Chit-g-P2AE degradation initiation over than 205°C.

**Rate measurements**

The rate of grafting ($R_g$), rate of homopolymerization ($R_h$), grafting and efficiency percentages were calculated as follows:

$$R_g = \frac{W_3 - W_1}{V.T.M_{2AE}} * 1000$$

$$R_h = \frac{W_2 - W_3}{V.T.M_{2AE}} * 1000$$

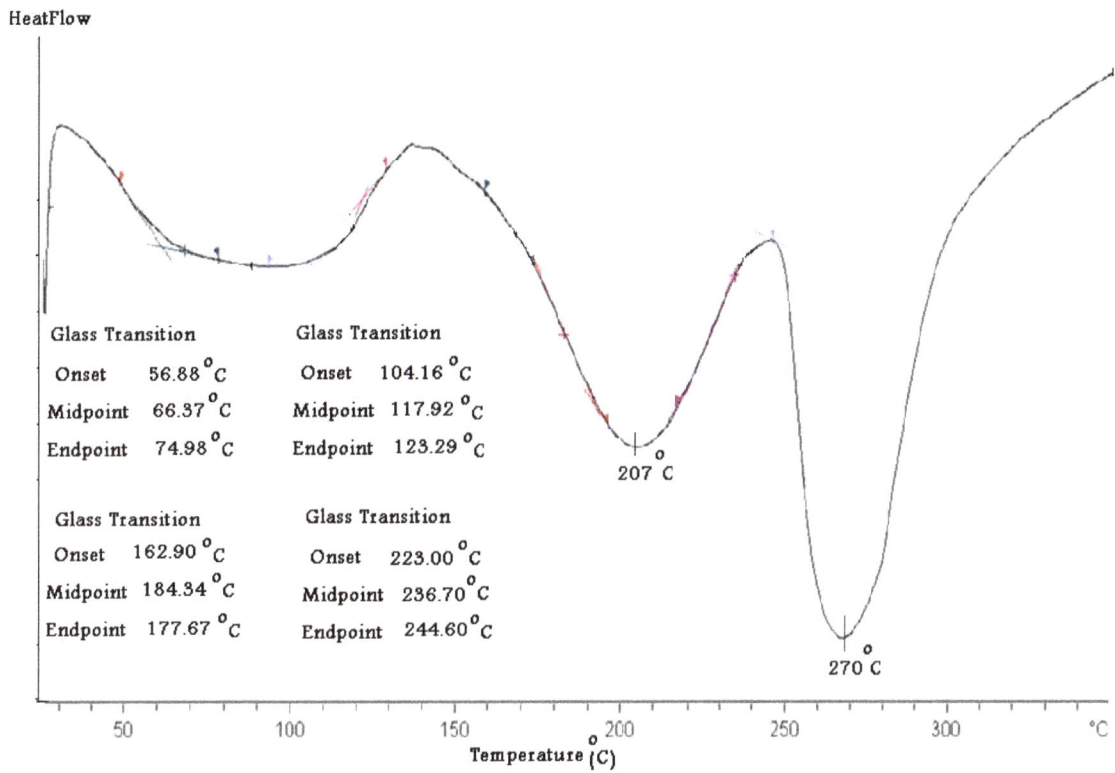

**Figure 6.** DSC of Chit-*g*-P2AE film.

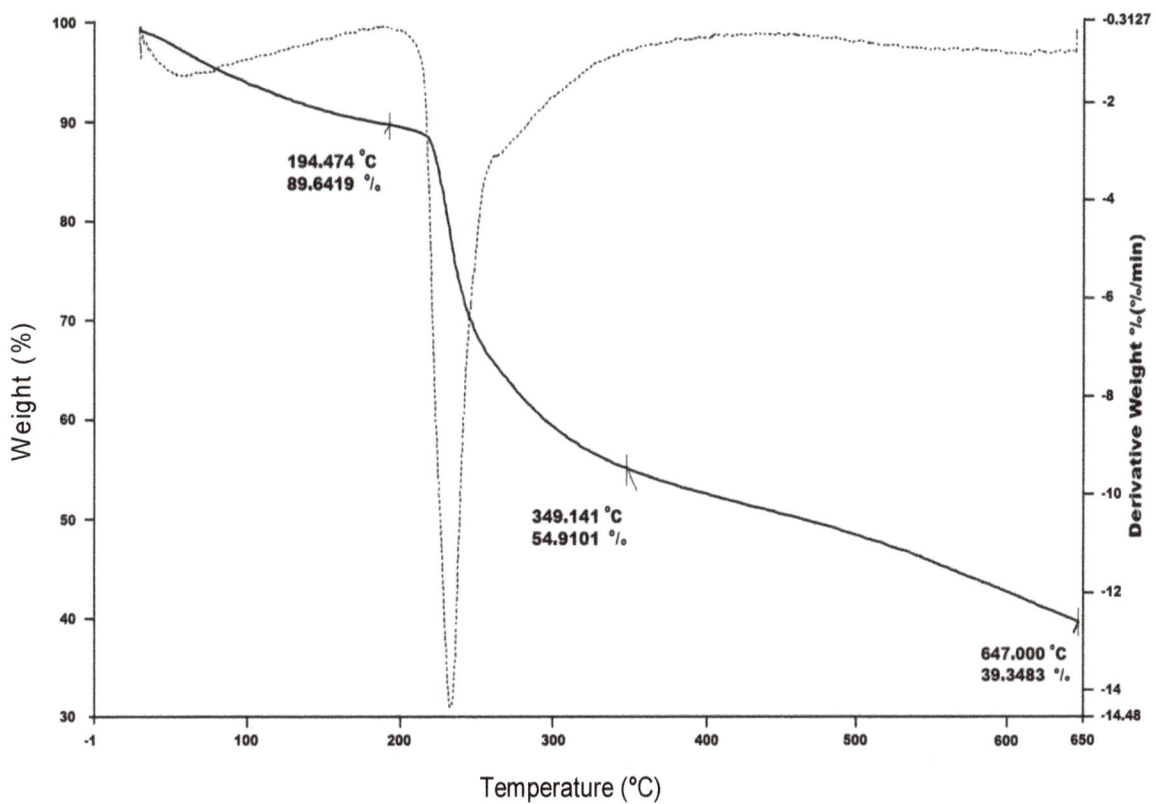

**Figure 7.** STA (TGA and DSC) of Chitosan film.

**Table 2.** Effect of [2AE] on $R_h$ and graft parameters.

| [2AE] (mol L$^{-1}$) | $R_h \times 10^7$ (mol L$^{-1}$ S$^{-1}$) | $R_g \times 10^7$ (mol L$^{-1}$ S$^{-1}$) | % Grafting | % Efficiency |
|---|---|---|---|---|
| 0.1 | 3.5 | 1.5 | 9.3 | 7.8 |
| 0.2 | 7.0 | 3.5 | 14.5 | 11.1 |
| 0.3 | 9.1 | 5.3 | 19.2 | 14.5 |
| 0.4 | 12.5 | 7.2 | 23.7 | 17.8 |

[APS] = 1 mmolL$^{-1}$, HCl = 0.1 M, weight of chitosan = 0.1 g.

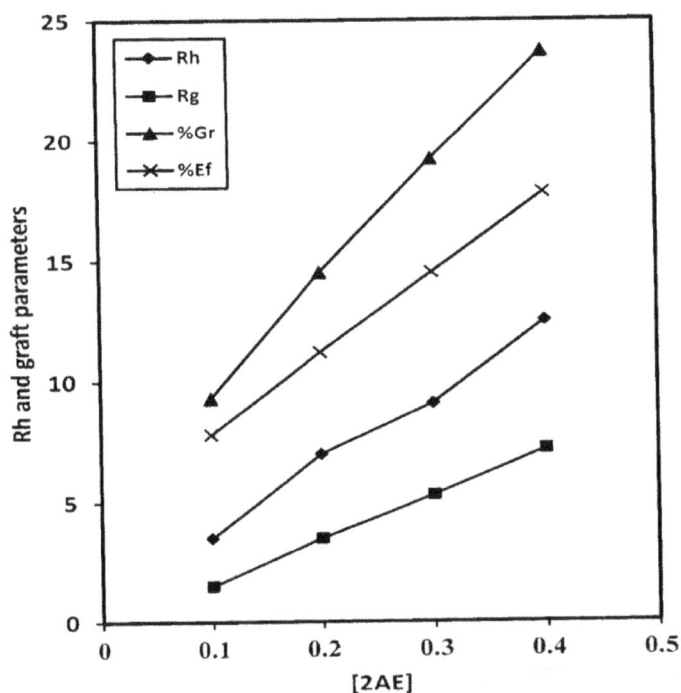

**Figure 8.** The plots of variation of $R_h$ and graft parameters ($R_g$, %Gr, %Ef) at different [2AE] in presence of fixed of [APS] and [chitosan].

$$\% \text{ grafting} = \frac{W_3 - W_1}{W_1} * 100$$

$$\% \text{ efficiency} = \frac{W_2 - W_3}{W_4} * 100$$

where t = reaction time, M = molecular weight of the monomer and V = total volume of the reaction mixture.

## Effect of monomer concentration on $R_h$ and graft parameters

Experimental results obtained by changing the [2AE] in the range from 0.1 to 0.4 mol L$^{-1}$ using APS as an initiator is given in Table 2 while keeping other experimental conditions as constant. The effect of varying the [2AE] on

$R_h$ and graft parameters such as $R_g$, % grafting (%Gr) and % efficiency (%Ef) are represented in Table 2 as well. The $R_h$ and graft parameters value increased with increase in [2AE]. In an attempt to have further confirmation about the dependence of $R_h$ and graft parameters ($R_g$, %Gr, %Ef) on [2AE] a different set of experimental conditions were made as represented in Figure 8. The plots of $R_h$ and graft parameters vs. [2AE] were drawn. The plot indicates the first order dependence of $R_h$ and $R_g$ on [2AE] and intercept of the plots $R_h$ and $R_g$ vs. [2AE] were noted.

## Effect of initiator concentration on $R_h$ and graft parameters

The effects of varying the [APS] on $R_h$ and graft

**Table 3.** Effect of [APS] on $R_h$ and graft parameters.

| [APS] (mmol L$^{-1}$) | $R_h \times 10^7$ (mol L$^{-1}$ S$^{-1}$) | $R_g \times 10^7$ (mol L$^{-1}$ S$^{-1}$) | % Grafting | % Efficiency |
|---|---|---|---|---|
| 0.5 | 4.1 | 2.3 | 7.7 | 11.5 |
| 1 | 5.8 | 2.9 | 11.2 | 15.0 |
| 2 | 8.0 | 4.0 | 17.0 | 22.0 |
| 3 | 11.5 | 6.3 | 20.7 | 31.0 |

[2AE] = 0.2 molL$^{-1}$, HCl = 0.1 M, weight of chitosan = 0.1 g.

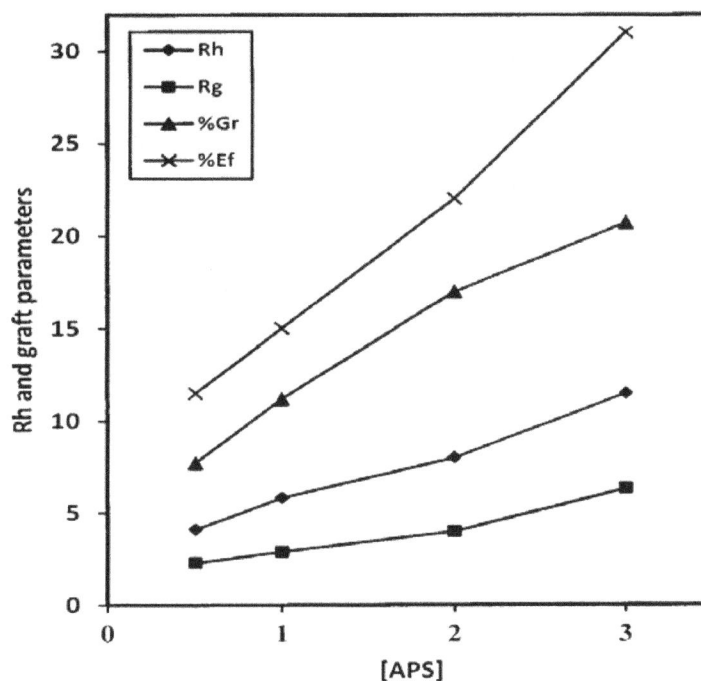

**Figure 9.** The plots of variation of $R_h$ and graft parameters ($R_g$, %Gr, %Ef) at different [APS] in presence of fixed of [2AE] and [chitosan].

parameters are presented in Table 3. The [APS] was varied from 0.5 to 3 mmol L$^{-1}$ while keeping other experimental conditions as constant. Here again, the $R_h$ and $R_g$ value showed increasing trend with [APS]. In a separate set of experimental conditions different from the above mentioned conditions, the effect of [APS] on $R_h$ and graft parameters were studied (Figure 9). The plots of $R_h$, $R_g$, %Gr and %Ef vs. [APS] were drawn. Then direct plots were found to be linear and passing through the origin. Figure 8 indicated the first order dependence of $R_h$ and $R_g$ on [APS].

### Effect of amount of chitosan on $R_h$ and graft parameters

The effect of amount of chitosan on $R_h$ and graft parameters were studied under the conditions mentioned in Table 4. The chitosan weight was varied between

0.075 to 0.30 g while keeping other experimental conditions as constant. $R_h$ and $R_g$ increased with increase in amount of chitosan. In an attempt to quantify the order dependences, the effect of the amount of chitosan on $R_h$ and graft parameters were studied under a set of different experimental conditions as specified in Figure 10. The plots of $R_h$ and graft parameters were drawn. The slope values of the plots were found to be close to one indicating first order dependence of $R_h$ and $R_g$ on weight of chitosan. These plots were found to be linear and passing through the origin. The linear plots support the clear first order dependence of $R_h$ and $R_g$ on backbone amount.

### Mechanism of graft copolymerization

A probable mechanism is proposed here to explain the experimental results obtained. The mechanism

**Table 4.** Effect of [Chitosan] on $R_h$ and graft parameters.

| [Chitosan] (g) | $R_h \times 10^7$ (mol L$^{-1}$ S$^{-1}$) | $R_g \times 10^7$ (mol L$^{-1}$ S$^{-1}$) | % Grafting | % Efficiency |
|---|---|---|---|---|
| 0.075 | 0.20 | 0.70 | 2.4 | 1.0 |
| 0.15 | 1.9 | 1.7 | 5.5 | 7.0 |
| 0.23 | 4.2 | 2.7 | 8.8 | 12.0 |
| 0.30 | 5.8 | 3.5 | 11.5 | 17.5 |

[2AE] = 0.2 molL$^{-1}$, [APS] = 1 mmolL$^{-1}$, HCl = 0.1 M.

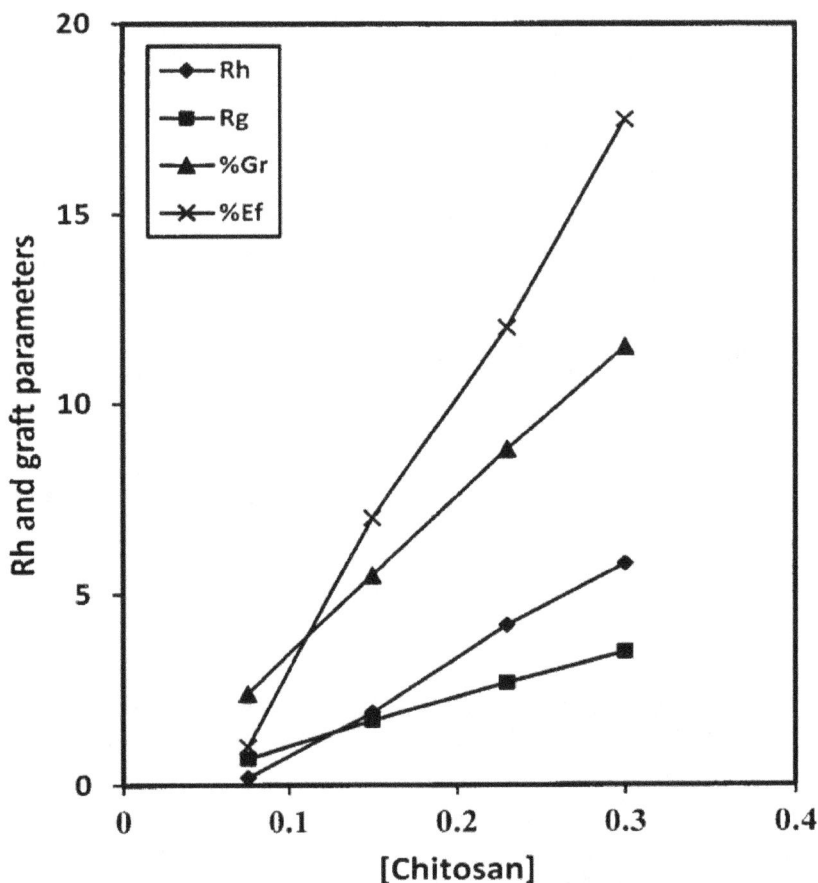

**Figure 10.** The plots of variation of $R_h$ and graft parameters ($R_g$, %Gr, %Ef) at different [chitosan] in presence of fixed of [2AE] and [APS].

suggested for graft copolymerization of P2AE onto chitosan in this paper is based on the mechanism proposed by few research groups. Shim et al. (1990), Bhadani et al. (1996) and Anbarasan et al. (2000) explained the formation of homopolymer via radical cation and mechanism for the graft copolymerization of PANI onto various natural polymers by chemical and electrochemical methods. Taking the above mechanisms as basis, a probable mechanism is suggested here to explain the modification of chitosan through chemical grafting. Probable mechanism for APS initiated graft copolymerization of 2AE onto chitosan.

**Initiation reactions**

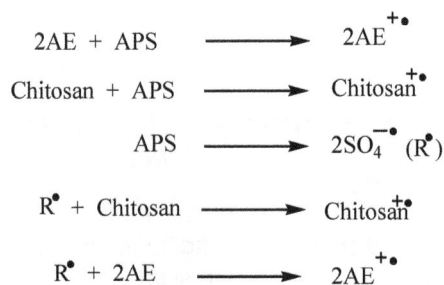

$$2AE + APS \longrightarrow 2AE^{+\bullet}$$
$$Chitosan + APS \longrightarrow Chitosan^{+\bullet}$$
$$APS \longrightarrow 2SO_4^{-\bullet} (R^\bullet)$$
$$R^\bullet + Chitosan \longrightarrow Chitosan^{+\bullet}$$
$$R^\bullet + 2AE \longrightarrow 2AE^{+\bullet}$$

## Homopolymerization

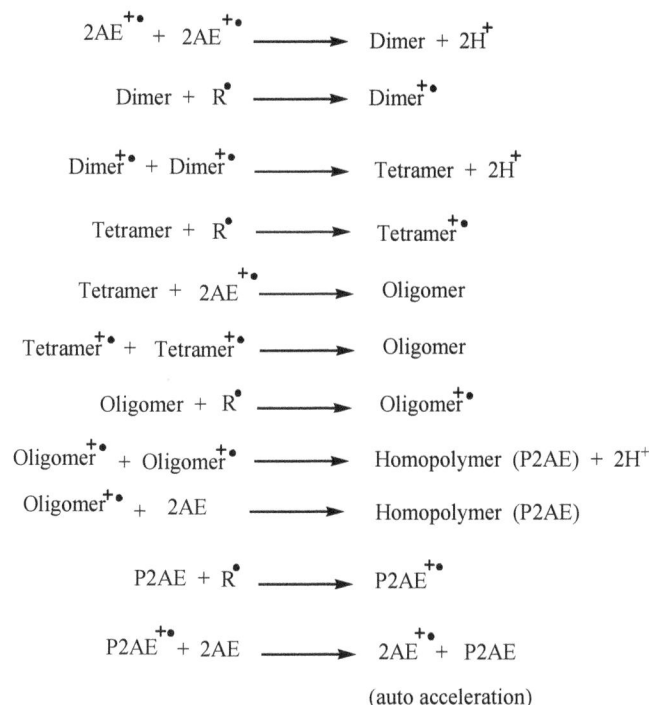

$$2AE^{+\bullet} + 2AE^{+\bullet} \longrightarrow Dimer + 2H^+$$

$$Dimer + R^\bullet \longrightarrow Dimer^{+\bullet}$$

$$Dimer^{+\bullet} + Dimer^{+\bullet} \longrightarrow Tetramer + 2H^+$$

$$Tetramer + R^\bullet \longrightarrow Tetramer^{+\bullet}$$

$$Tetramer + 2AE^{+\bullet} \longrightarrow Oligomer$$

$$Tetramer^{+\bullet} + Tetramer^{+\bullet} \longrightarrow Oligomer$$

$$Oligomer + R^\bullet \longrightarrow Oligomer^{+\bullet}$$

$$Oligomer^{+\bullet} + Oligomer^{+\bullet} \longrightarrow Homopolymer\ (P2AE) + 2H^+$$

$$Oligomer^{+\bullet} + 2AE \longrightarrow Homopolymer\ (P2AE)$$

$$P2AE + R^\bullet \longrightarrow P2AE^{+\bullet}$$

$$P2AE^{+\bullet} + 2AE \longrightarrow 2AE^{+\bullet} + P2AE$$

(auto acceleration)

## Graft copolymerization

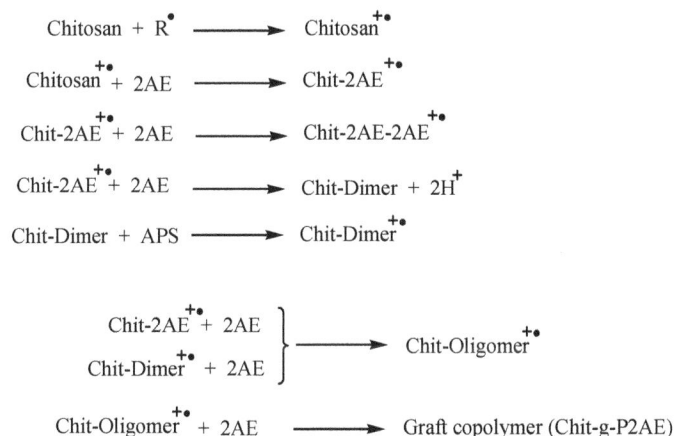

$$Chitosan + R^\bullet \longrightarrow Chitosan^{+\bullet}$$

$$Chitosan^{+\bullet} + 2AE \longrightarrow Chit\text{-}2AE^{+\bullet}$$

$$Chit\text{-}2AE^{+\bullet} + 2AE \longrightarrow Chit\text{-}2AE\text{-}2AE^{+\bullet}$$

$$Chit\text{-}2AE^{+\bullet} + 2AE \longrightarrow Chit\text{-}Dimer + 2H^+$$

$$Chit\text{-}Dimer + APS \longrightarrow Chit\text{-}Dimer^{+\bullet}$$

$$\left.\begin{array}{l} Chit\text{-}2AE^{+\bullet} + 2AE \\ Chit\text{-}Dimer^{+\bullet} + 2AE \end{array}\right\} \longrightarrow Chit\text{-}Oligomer^{+\bullet}$$

$$Chit\text{-}Oligomer^{+\bullet} + 2AE \longrightarrow Graft\ copolymer\ (Chit\text{-}g\text{-}P2AE)$$

$$R_g = \frac{W_3 - W_1}{V.T.M_{2AE}} * 1000 = 10.20\ molL^{-1}s^{-1}$$

$$R_h = \frac{W_2 - W_3}{V.T.M_{2AE}} * 1000 = 16.66\ molL^{-1}s^{-1}$$

In a grafting system of 2AE, APS and chitosan, the relation of the rate of grafting ($R_g$) with the monomer, chitosan and initiator concentrations after calculations of slop of lines from Tables 2 to 4 can be written as:

$$R_g\ \alpha\ [APS]^{1.750},\ R_g\ \alpha\ [2AE]^{0.969},\ R_g\ \alpha\ [Chit]^{1.188}\ so;$$

$$R_g = K[APS]^{1.750}[2AE]^{0.969}[Chit]^{1.188}$$

## Conclusion

Graft copolymerization was employed as an important technique to obtain a chemically and electrochemically modified chitosan. The 2AE was successfully grafted onto the chitosan backbone in an aqueous acidic medium condition. There are higher graft percentage and lower homopolymer formation. The grafting process was confirmed by IR analysis. Based on the TGA and DSC results, it was found that the grafted chitosan was more thermally stable than ungrafted one due to the incorporation of 2AE, which may broaden the range of chitosan application. In addition, the SEM micrographs indicate that the graft copolymer is efficient to improve the compatibility of binary blend of chitosan and P2AE.

## Conflict of Interest

The authors have not declared any conflict of interest.

### REFERENCES

Abdolahi A, Hamzah E, Ibrahim Z, Hashim S (2012). Synthesis of uniform polyaniline nanofibers through interfacial polymerization. Materials 5:1487-1494.

Anbarasan R, Vasudevan T, Gopalan A (2000). Peroxosalts initiated graft copolymerization of aniline onto wool fibre a comparative kinetic study. J. Mat. Sci. 35:617-625.

Armes SP, Miller JF (1988). Optimum reaction conditions for the polymerization of aniline in aqueous solution by ammonium persulphate. Synth. Met. 22:385-393.

Hosseini SH, Entezami AA (2003). Chemical and electrochemical synthesis of conducting graft copolymer of vinyl acetate with pyrrole and studies of Its gas and vapor sensing. J. Appl. Polym. Sci. 90:40-48.

Hosseini SH, Malekdar M, Naghdi Sh (2010). Chemical and electrochemical synthesis of cross-linked aniline sulfide resin. Polymer J. 56:1-8.

Hosseini SH, Ansari R, Noor P (2013a). Application of polyaniline film as a sensor for stimulant nerve agents. Phosphorus, Sulfur, Silicon Related Elements. 188(10):1394-1401.

Hosseini SH, Entezami AA (2005). Graft copolymers of polystyrene and polypyrrole and studies of its gas and vapor sensing. Iran Polym. J. 14(2):101-110.

Hosseini SH (2006). Investigation of sensing effects of polystyrene-graft-polyaniline for cyanide compounds. J. Appl. Polym. Sci. 101(6):3920-3926.

Hosseini SH, Simiari J, Farhadpour B (2009). Chemical and electrochemical graft copolymerization of aniline onto chitosan. Iran Polym. J. 18:3-13.

Hosseini SH (2013). Studies of conductivity and sensing behavior of polyaniline grafted on polyvinylpropionate for pesticide poisons. Synthesis and Reactivity in Inorganic, Metal-Organic, and Nano-Metal Chemistry 43(7):852-860.

Hosseini SH, Gohari SJ (2013). Electrical field influence on molecular mass and electrical conductivity of polyaniline. Polym. Sci. Ser. B. 55(7–8):467-471.

Hosseini SH, Asadi G, Gohari SJ (2013b). Electrical characterization of conducting poly(2-ethanolaniline) under electric field. Int. J. Phys. Sci. 8(22):1218-1227.

Li P, Zhu J, Sunintaboon P, Harris FW (2002). New route to amphiphilic core-shell polymer nanospheres: Graft copolymerization of methyl methacrylate from water-soluble polymer chains containing amino groups. Langmuir. 18:8641-8646.

Liu H, Hu XB, Wang JY, Boughton RI (2002). Structure, conductivity, and thermopower of crystalline polyaniline synthesized by the ultrasonic irradiation polymerization method. Macromolecules 35:9414-9419.

Mahdavinia GR, Pourjavadi A, Hosseinzadeh H, Zohuriaan MJ (2004). Modified chitosan 4 Superabsorbent hydrogels from poly(acrylic acid-co-acrylamide) grafted chitosan with salt and pH-responsiveness properties. Europ. Polym. J. 40:1399-1407.

Shim YB, Won MS, Park SM (1990). Electrochemistry of conductive polymers VIII: In situ spectroelectrochemical studies of polyaniline growth mechanisms. J. Electrochem. Soc. 137:538-544.

Silva RC, Garcia JR, Sanchez JA (2005). Template-free enzymatic synthesis of electrically conducting polyaniline using soybean peroxidase. Eur. Polym. J. 41:1129-1135.

Singh V, Tiwari A, Tripathi DN, Sanghi R (2006). Microwave enhanced synthesis of chitosan-graft-polyacrylamide. Polymer 47:254-260.

Su SJ, Kuramoto N (2001). Optically Active Polyaniline Derivatives Prepared by Electron Acceptor in Organic System: Chiroptical Properties. Macromolecules 34:7249-7256.

Yu H, Chen X, Lu T, Sun J, Tian H, Hu J, Wang Y, ZhangP, Jing X (2007). Poly(L-lysine)-Graft-Chitosan Copolymers: Synthesis, characterization, and gene transfection effect. Biomacromolecules 8:1425-1435.

# 20

# Studies on zinc oxide thin films by chemical spray pyrolysis technique

**Prabakaran Kandasamy and Amalraj Lourdusamy**

Research Center in Physics, VHNSN College, Virudhunagar – 626001, Tamilnadu, India.

**Zinc oxide (ZnO) thin films were deposited by chemical spray pyrolysis (CSP) technique using zinc acetate dihydrate solutions on microscopic glass substrates by varying the precursor concentration. The prepared films were characterized structurally and optically, using the powder X-ray diffraction (XRD) and UV analysis and Photoluminescence analysis. Crystallographic properties were analyzed through powder XRD. The XRD patterns shows a hexagonal structure with c-axis orientation (0 0 2) on self texturing phenomenon. Optical transmittance properties of the optimized ZnO thin films were investigated by using UV-Vis spectroscopy. The optical studies predicated a maximum transmittance in the range of above 70% with direct band gap values in the range of 2.9 to 3.2eV for the zinc oxide thin films. Under excitation of 300 nm radiations, sharp deep level emission peak at 2.506 eV dominates the photoluminescence spectra with weak deep level and near band edge emission peak at 3.026 and 3.427 eV respectively.**

**Key words:** Transparent conducting oxide (TCO), Zinc Oxide thin film, CSP technique, X-ray diffraction (XRD), UV-Vis, Photoluminescence.

## INTRODUCTION

Metal oxide semiconductor thin films have been widely researched and have received considerable attention in recent years due to their optical and electrical properties. Because they are good candidates for transparent conducting oxide (TCO) films (Hongxia et al., 2005). Zinc oxide is (ZnO) a wide direct band gap (~3.37eV at T=300 K) semiconductor (II-VI) which has been widely investigated in the past years for more literature. ZnO also has a high exciton binding energy of 60 meV which is higher than the values of other widely used wide band gap materials, such as ZnSe (20 meV) and GaN (21 meV). The large exciton binding energy can ensure efficient excitonic emission at room temperature (Zahedi and Dariani, 2012). Exciton provides a sensitive indicator of material quality. Photoluminescence (PL) is very sensitive to the quality of crystal structure and to the presence of defects (Sagar et al., 2007). ZnO has a large transparency in the visible region, high natural abundance, absence of toxicity, low cost compared to other oxide materials such as $SnO_2$, $In_2O_3$, $TiO_2$ (Mahalingam et al., 2003; Gaikwad et al., 2012). In generally, thin films are depends not only on the morphology of the sample, but also on the deposition parameters, thickness of the sample and grain sizes

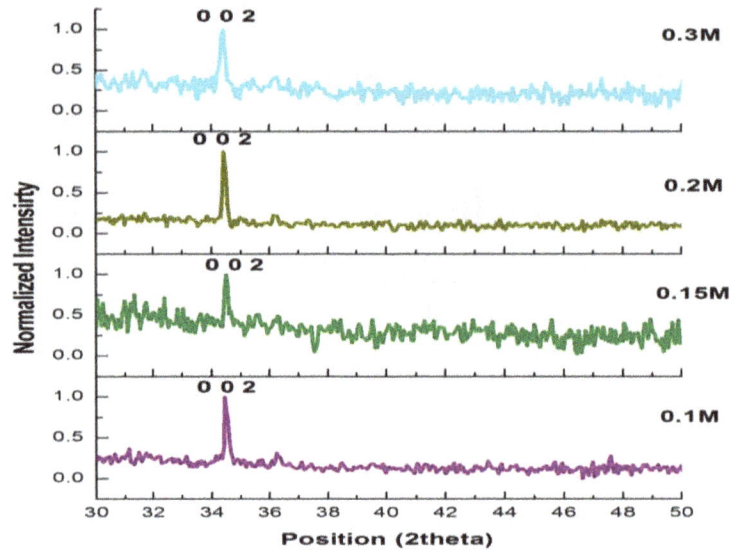

**Figure 1.** X-Ray pattern of ZnO thin film at (a) 0.1 M, (b) 0.15 M, (c) 0.2 M, (d) 0.3 M.

(Godbole et al., 2011). In recent years, the ZnO based films has potential applications in the field of electronic devices such as gas sensors, solar cells, optoelectronics, thin film transistors (Kuo et al., 2006; Chu et al., 2009; Van Heerden et al., 1997; Joseph et al., 1999a; Wei et al., 2007; Mani et al., 2006). Many deposition techniques have been used to prepared zinc oxide thin films in order to improve their properties such as, chemical bath deposition (CBD) (Kathirvel et al., 2009) Sol-gel spin coating (Natsume and Sakata, 2000), metal organic chemical vapour deposition (Wang et al., 2004), RF-magnetron sputtering (Shiyi et al., 2009), spray pyrolysis (Achour et al., 2007; Joseph et al., 1999a; Yoon and Cho, 2000; Ayouchi et al., 2003). Compared to the others, the spray technique has many advantages: it is easy, inexpensive, safety and the low cost of the apparatus and the raw materials, no sophisticated instrument such as vacuum systems etc., large area coating of thin film and well adopted for mass fabrication (Manouni et al., 2006; Gencyilmaz et al., 2013; Saleem et al., 2012). Furthermore, the optical, structural properties of thin films can be easily controlled by the quantity of sprayed solution in this technique. In general, the films produced with this technique are polycrystalline, stable, adherent, and hard. Spray pyrolysis has been developed as a powerful tool for preparation on various kinds of technological materials such as metals, metal oxides, superconducting materials, and nanophase materials (Saleem et al., 2012). In the present work, we report the thickness on mainly structural and optical properties involved during the effect of precursor concentration on the X-ray diffraction (XRD), UV-Visble and photoluminescence behavior of ZnO thin films deposited by chemical spray pyrolysis technique.

**EXPERIMENTAL PROCEDURES**

The zinc oxide films were deposited on microscopic glass substrates (Thickness 1.35 mm) at a constant temperature of 380°C with an accuracy of ±5°C by a chemical spray pyrolysis technique at various precursor concentrations. A solution of 0.1, 0.15, 0.2 and 0.3 M zinc acetate dihydrate was used as a precursor prepared by dissolving in mixture of deposited water and isopropyl alcohol (1:3) volume ratio. The nozzle was at a distance of 24 cm from the substrate during deposition. The solution flow rate was held 4±0.5 ml/min. Compressed air was used as the carrier gas. The pressure of the carrier gas should be 0.7 Kg/cm$^2$. The deposited films were allowed to cool down to room temperature.

The structure of the films were examined by using XPERT – PRO' X-Ray diffractometer with CuKα$_1$ radiation (λ=1.54056 A°). The thickness of the films was determined by stylus profilometer. The optical transmission spectroscopic measurements of the zinc oxide thin films were carried out at room temperature using SHIMADZU - 1800 Double beam Spectrophotometer in the wavelength range between 300 – 700 nm. VARAIN CARRY ECLIPSE Spectrophotometer excited with 300 nm wavelength from a xenon lamp was used to record photoluminescence spectra of the films.

**RESULTS AND DISCUSSION**

**XRD analysis**

X-Ray diffractograms of films prepared at different precursor concentration (0.1 to 0.3 M) was shown on Figure 1. All the diffractograms of the prepared films clean indicate the polycrystalline nature of zinc oxide films with prominent diffraction peak from crystal plane (0 0 2) on self-texturing phenomenon (Chougule et al., 2010) had also obtained a single peak in their XRD pattern by Sol-gel spin coating method with (0 0 2)

**Table 1.** XRD data of ZnO thin film.

| Precursor concentration (M) | Observed values | | Thickness (nm) | Grain size (nm) (k=0.9) |
|---|---|---|---|---|
| | 2θ (deg) | lattice constant (c) A° | | |
| 0.1 | 34.4582 | 5.2013 | 103 | 70.12 |
| 0.15 | 34.4779 | 5.1986 | 110 | 46.23 |
| 0.2 | 34.4005 | 5.2098 | 158 | 84.38 |
| 0.3 | 34.3203 | 5.2216 | 145 | 11.55 |

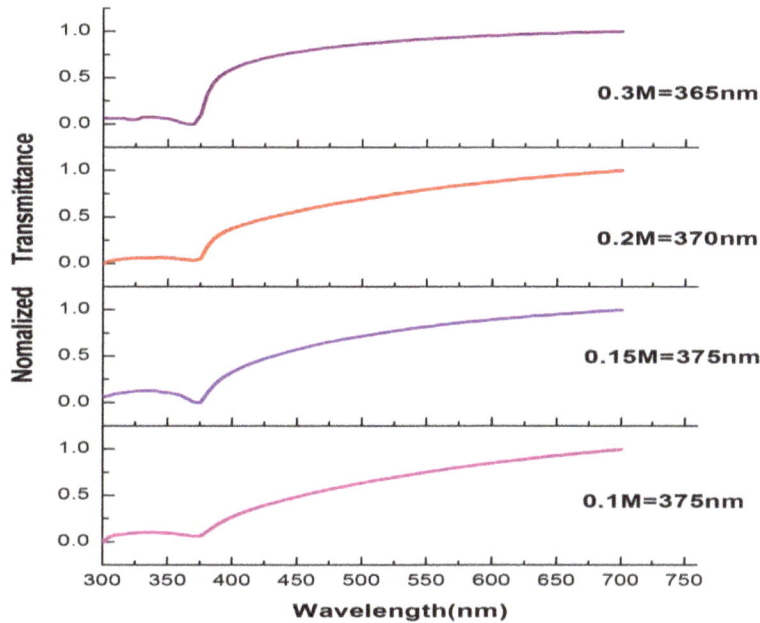

**Figure 2.** Transmittance spectrum of ZnO thin film.

plane reflection. The deposition of precursor concentration increased without the appearance of any new reflections. Thus no other phases were formed. The phase identification revealed that only hexagonal crystal system based zinc oxide (JCPDS File No.75-1526, 80-0075, 80-0074) was conformed. The lattice constant "C" was calculated by using the following equation:

$$\frac{1}{d^2} = \frac{4}{3}\left[\frac{h^2 + hk + k^2}{a^2}\right] + \frac{L^2}{C^2} \tag{1}$$

The calculated values of the lattice constant (c) are found to be close to those of the Joint Committee on Powder Diffraction Standard (JCPDS) data reported for zinc oxide sample. The average grain size was measured using Debye-Scherer's formula (Cullity, 1978):

$$Grainsize(D) = \frac{0.9 * \lambda}{\beta \cos\theta} \; (nm) \tag{2}$$

Where $\lambda$ is the wavelength of Cu Kα1 radiation (1.54056 A°), $\beta$ is the FWHM value. The variation of film thickness, grain size, lattice constant (c), two theta with precursor concentration are shown in Table 1.

**Optical properties**

The transmittance spectrum of ZnO thin films in the wavelength range 300 - 800 nm are shown in the Figure 2. Optical properties of the zinc oxide thin films were studied with the help of transmission spectrum in the UV-visible region. Figure 2 shows the transmittance spectrum of zinc oxide films deposited at different precursor concentration recorded in the range 300 to 700 nm. The spectrum shows a maximum transparency of >70% in the visible region. Sharp ultraviolet absorption edges at $\lambda$=375 nm are observed with the absorption edge being shifted to shorter wavelength at higher concentration. It can be clearly seen the blue shift in band edge in Figure 3. The optical absorption coefficient can be calculated

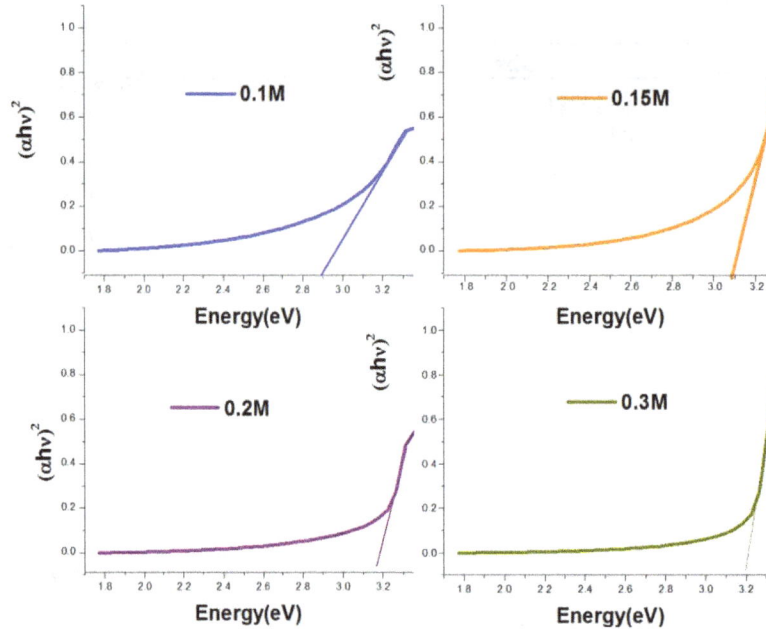

**Figure 3.** Variation of $(ahv)^2$ verus hv of the zinc oxide thin film.

**Table 2.** Transmittance at maximum, 550 nm and direct band gap for different precursor concentration.

| Precursor concentration (M) | Direct band gap (eV) | Transmittance (%) $T_{max}$ | Transmittance (%) $T_{550nm}$ |
|:---:|:---:|:---:|:---:|
| 0.1 | 2.9 | 73.345 | 61.88 |
| 0.15 | 3.1 | 88.603 | 79.93 |
| 0.2 | 3.18 | 74.602 | 61.51 |
| 0.3 | 3.20 | 87.372 | 82.78 |

using the Lambert law relation

$$\alpha = t * \ln\left(\frac{1}{T}\right) \tag{3}$$

where t is the thickness of the film and T the transmittance. The relation between absorption coefficient and incident photon energy can be written as:

$$(\alpha hv) = A(hv - E_g)^{\frac{1}{2}} \tag{4}$$

Where A is a constant, E is the band gap of the material and h is the planks constant. In the present case, the plot of $(ahv)^2$ verus hv, indicates the direct band gap nature of the films. By extrapolating the linear portion of the curve onto the X-axis the energy band gap of the films is determined. The transmittance at maximum 700 and 550

nm, and direct band gap for different precursor concentration was shown in Table 2. Thus on increase in band gap was observed, when the concentration was changed from 0.1 to 0.3 M. This may be due to the hexagonal phase. The effect of bandgap widening is attributed primarily to the Moss-Burstein shift in semiconductors. The bandgap changes found is in good agreement with reported values 3.14 to 3.26 eV by the same spray pyrolysis technique (Joseph et al., 1999b).

The photoluminescence property of film has a close relation with the crystalline, because the density of defects in the film reduces with an improvement of the crystallinity. Room temperature PL emission spectrum for all the samples was measured in the wavelength range of 310 to 550 nm at the excitation wavelength of 300 nm. PL spectra of the zinc oxide thin films deposited on glass substrates at various precursor concentrations are shown in Figure 4. The zinc oxide emission is generally classified into two categories one is the UV emission of all zinc oxide thin films at 3.427 eV and the other is the

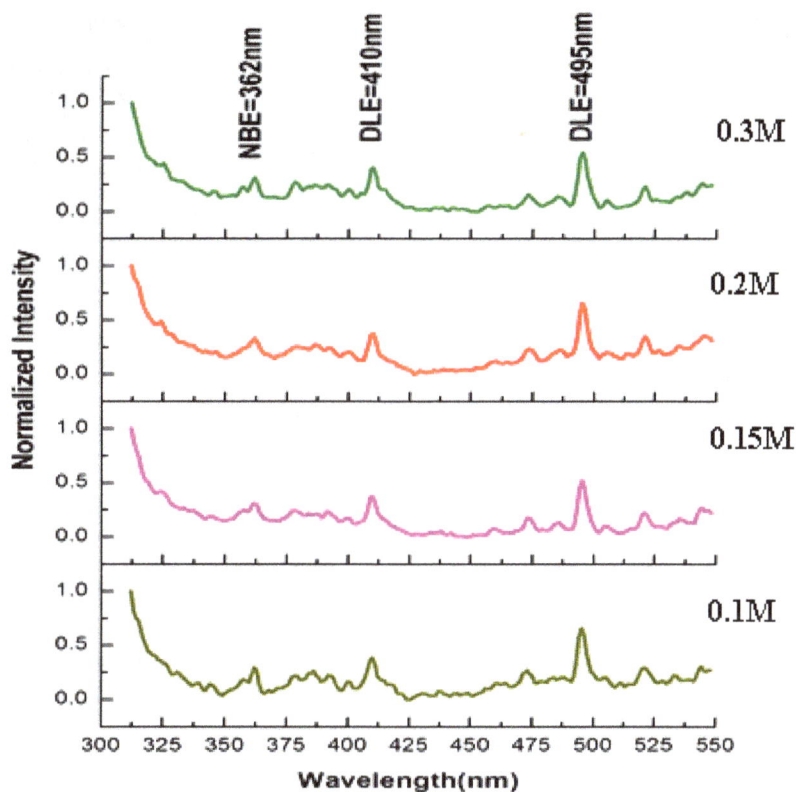

**Figure 4.** Comparison of the variation of peak intensity with different concentrations.

deep level (DL) emission related to the defect emission in the visible range. In the deep level emission, two emission peaks at 3.026 and 2.506 eV appear in the PL spectra shows a strong blue emission band (Gao et al., 2004) around 2.506 eV in all samples. By comparing the results from absorption spectra and PL spectra, pronounced exciton absorption peak in the UV spectra located at around 3.4 eV is assigned to the exciton effect in ZnO. It is clear that the above results exists a stokes shift between the PL spectra and the absorption spectra. The stokes shift is related to many effects such as electron–phonon coupling, lattice distortions, interface defects and point defects that may cause the blue shift of emission line from absorption edge (Sagar et al., 2007). In the case of blue shift, the filling of the conduction band by electrons will generally result in the NBE emission. The stokes shift were calculating the following equation:

$$\Delta E = E_{Abs} - E_{PL} \qquad (5)$$

The stokes shifts were calculated to be much larger value respectively for increase in the molar concentration for ZnO thin films. The crystalline quality of ZnO film grain size has a larger distribution leading to PL band (Shan et al., 2006).

## Conclusion

Zinc oxide thin film was deposited on the glass substrate by chemical spray pyrolysis technique. The XRD studies show that, films prepared are in nanocrystalline range. Polycrystalline nature of zinc oxide films and lattice parameter (C) has been determined which agree with the standard data. From the X-ray diffraction analysis 0.1, 0.15, 0.2 and 0.3 M films show hexagonal structure along with c-axis oriented (0 0 2) plane. Optical transmittance properties of the optimized ZnO thin films were investigated by using UV-Vis spectroscopy. The transmittance of above 70% in the visible region has been observed for precursor concentration and increase the concentration caused by the band gap to become broader. An intensive blue luminescence peak around 495 nm is observed at room temperature.

## Conflict of Interests

The author(s) have not declared any conflict of interests.

## ACKNOWLEDGEMENTS

The authors would like to thank Dr. SanjeeViraja

Chinnappanadar, Department of Physics, Alagappa University, Karaikudi and Department of Botany, V.H.N.S.N.College, Virudhunagar for this valuable help to obtaining their characterization of work.

## REFERENCES

Achour ZB, Ktari T, Ouertani B, Touayar O, Bessais B, Brahim JB (2007). Effect of doping level and spray time on zinc oxide thin films produced by spray pyrolysis for transparent electrodes applications. Sens. Actuators, A 134(2):447-451. http://dx.doi.org/10.1016/j.sna.2006.05.001

Ayouchi R, Leinen D, Martın F, Gabas M, Dalchiele E, Ramos-Barrado JR (2003). Preparation and characterization of transparent ZnO thin films obtained by spray pyrolysis. Thin Solid Films, 426(1):68-77. http://dx.doi.org/10.1016/S0040-6090(02)01331-7

Chu D, Hamada T, Kato K, Masuda Y (2009). Growth and electrical properties of ZnO films prepared by chemical bath deposition method. Phys. Status Solid A, 206(4):718-723. http://dx.doi.org/10.1002/pssa.200824495

Gaikwad RS, Patil GR, Shelar MB, Pawar BN, Mane RS, Han SH, Joo OS (2012). Nanocrystalline ZnO films deposited by spray pyrolysis: Effect of gas flow rate. Int. J. Self Propag. High Temp. Synth. 21(3):178-182. http://dx.doi.org/10.3103/S106138621203003X

Gencyilmaz O, Atay F, Yavuz I (2013). Preparation and Characterization of Transparent, Conductive ZnO Thin Film for Photovoltaic Solar Cells. J. Selcuk Univer. Nat. Appl. Sci. pp. 752-759.

Godbole B, Badera N, Shrivastava S, Jain D, Ganesan V (2011). Growth mechanism of ZnO films deposited by spray pyrolysis technique. Mater. Sci. Appl. 2:643. http://dx.doi.org/10.4236/msa.2011.26088

Hongxia L, Jiyang W, Hang L, Huaijin Z, Xia L (2005). Zinc oxide films prepared by sol–gel method. J. Cryst. Growth, 275:e943-e946. http://dx.doi.org/10.1016/j.jcrysgro.2004.11.098

Joseph B, Gopchandran KG, Manoj PK, Peter K, Thomas PV, Vaidyan VK (1999b). Optical and electrical properties of zinc oxide films prepared by spray pyrolysis Bull. Mater. Sci. 5:921-926. http://dx.doi.org/10.1007/BF02745554

Joseph B, Gopchandran KG, Thomas PV, Koshy P, Vaidyan VK (1999a). A study on the chemical spray deposition of zinc oxide thin films and their structural and electrical properties. Mater. Chem. Phys. 58(1):71-77. http://dx.doi.org/10.1016/S0254-0584(98)00257-0

Kathirvel P, Manoharan D, Mohan SM, Kumar S (2009). Spectral investigations of chemical bath deposited zinc oxide thin films ammonia gas sensor, J. Optoelectr. Biomed. Mater. 1(1):25-33.

Kuo SY, Chen WC, Lai FI, Cheng CP, Kuo HC, Wang SC, Hsieh WF (2006). Effects of doping concentration and annealing temperature on properties of highly-oriented Al-doped ZnO films. J. Cryst. Growth, 287(1):78-84. http://dx.doi.org/10.1016/j.jcrysgro.2005.10.047

Mahalingam T, John VS, Sebastian PJ (2003). Growth and characterization of electrosynthesised zinc oxide thin films. Mater. Res. Bull. 38(2):269-277. http://dx.doi.org/10.1016/S0025-5408(02)01036-X

Mani B, Manjon FJ, Mollar M, Cembrero J, Gomez R (2006). Photoluminescence of thermal- annealed nanocolumnar ZnO thin films grown by electrodeposition. Appl. Surf. Sci. 252 2826-2831. http://dx.doi.org/10.1016/j.apsusc.2005.04.024

Manouni AE, Manjón FJ, Mollar M, Marí B, Gómez R, López MC, Ramos-Barrado JR (2006). Effect of aluminium doping on zinc oxide thin films grown by spray pyrolysis. Superlattices Microstruct. 39(1):185-192. http://dx.doi.org/10.1016/j.spmi.2005.08.041

Natsume Y, Sakata H (2000). Zinc oxide films prepared by sol-gel spin-coating. Thin solid films, 372(1):30-36. http://dx.doi.org/10.1016/S0040-6090(00)01056-7

Shiyi Z, Yuying X, Min G (2009). Magnetic properties of ZnO: Cu thin films prepared by RF magnetron sputtering. J. Semicond. 30(5): 052004. http://dx.doi.org/10.1088/1674-4926/30/5/052004

Van Heerden JL, Swanepoel R (1997). XRD analysis of ZnO thin films prepared by spray pyrolysis. Thin Solid Films, 299(1):72-77. http://dx.doi.org/10.1016/S0040-6090(96)09281-4

Wang J, Du G, Zhang Y, Zhao B, Yang X, Liu D (2004). Luminescence properties of ZnO films annealed in growth ambient and oxygen. J. Cryst. Growth, 263(1):269-272. http://dx.doi.org/10.1016/j.jcrysgro.2003.11.059

Wei XQ, Zhang ZG, Liu M, Chen CS, Sun G, Xue GS, Zhuang HZ, Man BY (2007). Annealing effect on the microstructure and photoluminescence of ZnO thin films. Mater. Chem. Phys. 101(2):285-290. http://dx.doi.org/10.1016/j.matchemphys.2006.05.005

Yoon KH, Cho JY (2000). Photoluminescence characteristics of zinc oxide thin films prepared by spray pyrolysis technique. Mater. Res. Bull. 35(1):39-46.http://dx.doi.org/10.1016/S0025-5408(00)00183-5

Zahedi F, Dariani RS (2012). Effect of precursor concentration on structural and optical properties of ZnO microrods by spray pyrolysis.Thin Solid Films, 520(6):2132-2135. http://dx.doi.org/10.1016/j.tsf.2011.09.006

Saleem M, Fang L, Wakeel A, Rashad M, Kong CY (2012). Simple Preparation and Characterization of Nano-Crystalline Zinc Oxide Thin Films by Sol-Gel Method on Glass Substrate. World J. Conden. Matter Phys. 2:10-15. http://dx.doi.org/10.4236/wjcmp.2012.21002

Chougule MA, Patil SL, Pawar SG, Patil VB (2010). Transparent and conductive ZnO: Al thin films prepared by sol-gel process. Archives Phys. Res. 1(1):100-107.

Cullity BD (1978) Elements of X-ray Diffraction, Addition Wesley, Reading, M.A, P. 102.

Gao XD, Li XM, Yu WD (2004). Preparation, structure and ultraviolet photoluminescence of ZnO films by a novel chemical method. J. Solid State Chem .177(10):3830-3834. http://dx.doi.org/10.1016/j.jssc.2004.07.030

Sagar P, Shishodia PK, Mehra RM, Okada H, Wakahara A, Yoshida A (2007). Photoluminescence and absorption in sol–gel-derived ZnO films. J. Lumin. 126(2):800-806. http://dx.doi.org/10.1016/j.jlumin.2006.12.003

Shan FK, Liu GX, Lee WJ, Shin BC (2006). Stokes shift, blue shift and red shift of ZnO-based thin films deposited by pulsed-laser deposition. J. crystal growth, 291(2):328-333. http://dx.doi.org/10.1016/j.jcrysgro.2006.03.036

# 21

# Studies on structural, surface morphology and optical properties of Zinc sulphide (ZnS) thin films prepared by chemical bath deposition

S. Thirumavalavan[1], K. Mani[2] and S. Sagadevan[3]

[1]Department of Mechanical Engineering, Sathyabama University, Chennai-600 119, India.
[2]Department of Mechanical Engineering, Panimalar Engineering College, Chennai-602103, India.
[3]Crystal Growth Centre, Anna University, Chennai-600 025, India.

Zinc sulphide (ZnS) thin films have been prepared by chemical bath deposition method. X-ray diffraction (XRD) is used to analyze the structure and crystallite size and scanning electron microscopy is used to study the particle size and morphology of ZnS thin film. Optical studies have been carried out using UV-Visible-NIR absorbance spectrum. The band gap value of the film is calculated and it is found to be 3.45 eV. The dielectric properties of ZnS thin films have been studied in the different frequency at different temperatures.

Key words: Zinc sulphide (ZnS) thin films, X-ray diffraction (XRD), scanning electron microscopy (SEM), dielectric studies.

## INTRODUCTION

Zinc Sulphide (ZnS) belongs to II-VI group compound with wide direct band gap value ranging from 3.4 to 3.70 eV. Zinc sulfide has cubic or hexagonal crystal structure or both at the same time. ZnS has significant potential applications such as in antireflection coating for light emitting diode (Katayama et al., 1975) for heterojunction solar cells (Bloss et al., 1988) and other optoelectronic devices such as electro luminescence devices and photovoltaic cells which are used in the field of displays (Beard et al., 2002), blue light emitting diode (Coe et al., 2002) sensors and lasers (Klimov et al., 2000) etc. In recent years nanocrystalline ZnS attracts much consideration because the properties in nano form vary significantly from those of their bulk.

Therefore, effort has been made to control the size, morphology and crystallinity of ZnS thin film in a wide variety of applications (Mach and Mueller, 1991; Varitimos and Tustison, 1987). ZnS thin films have been prepared by various methods such as thermal evaporation (Dimitrova, 2000), spray pyrolysis (Mustafa et al., 2007), sputtering (Shao et al., 2003), chemical vapor deposition (Icimura et al., 1999), successive ionic layer adsorption and reaction (Nomura et al., 1995), and metal organic vapour phase epitaxy (MOVPE) (Roy et al., 2006). In the present investigation, we report the synthesis and characterization of ZnS thin films. The ZnS thin films were characterized by X-ray diffraction, and scanning electron microscopy (SEM), for microstructure

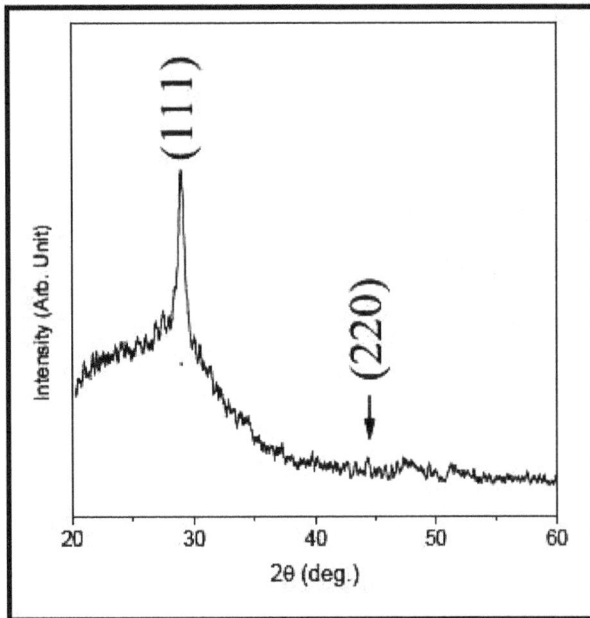

**Figure 1.** XRD spectrum of ZnS thin films.

and morphology respectively, while UV-VIS-NIR analysis and dielectric for optical studies.

### EXPERIMENTAL PROCEDURE

The preparation of ZnS thin films is based on the chemical bath deposition method which was carried out by dissolving zinc acetate and thiourea in double distilled water and at the deposition temperature of 60°C. Then, Ammonia solution was added slowly with constant stirring to form the complex and pH of 10 was achieved. The deposition was done by keeping the substrates vertically inside the chemical bath. Deposition time of five minutes was recorded and the films were rinsed in double distilled water and then dried in air atmosphere. The X-ray diffraction (XRD) pattern of the ZnS thin films was recorded by using a powder X-ray diffractometer (Schimadzu model: XRD 6000 using CuKα (λ=0.154 nm) radiation, with a diffraction angle between 20 and 60°. The crystallite size was determined from the broadenings of corresponding X-ray spectral peaks by using Debye Scherrer's formula. Scanning electron microscopy (SEM) studies were carried out on JEOL, JSM- 67001. The optical absorption spectrum of the ZnS thin films has been taken by using the VARIAN CARY MODEL 5000 spectrophotometer in the wavelength range of 300 to 600 nm. The dielectric properties of the ZnS thin films were analyzed using a HIOKI 3532-50 LCR HITESTER over the frequency range 50Hz-5MHz.

## RESULTS AND DISCUSSION

### X- ray diffraction analysis

Structural identification of ZnS films has been done using with X-ray diffraction in the range of angle $2\theta$ between 20° to 60° as shown in Figure 1. The excellent peaks

(111) and (220) have been obtained. This film shows reflection along (111), (220) peaks corresponding to formation of hexagonal structure of ZnS. The broadened peak shows the nanometer-sized crystallites. The average nano-crystalline size (D) was calculated using the Debye-Scherrer formula:

$$D = \frac{0.9\lambda}{\beta\cos\theta} \qquad (1)$$

where $\lambda$ is the X-ray wavelength (CuKα radiation and equals to 0.154 nm), $\theta$ is the Bragg diffraction angle, and $\beta$ is the FWHM of the XRD peak appearing at the diffraction angle $\theta$. The average crystalline size is calculated from X-ray line broadening using (111) peak and Debye-Scherrer equation to be about 16.4 nm.

### Scanning electron microscope (SEM)

Scanning electron microscope (SEM) was used for the study of surface structure and roughness of the ZnS thin films. Figure 2 shows the SEM images of the ZnS thin films. The spherical crystallites which have the mean particle size of ~ 12 nm are visible through the SEM analysis.

### UV-VIS-NIR spectral analysis

Study of materials by means of optical absorption provides a simple method for explaining some features concerning the band structure of materials. The optical absorption spectrum of ZnS films has been recorded in the wavelength region 300 – 600 nm and it is shown in Figure 3. It is important to note that ZnS films were highly transparent in the visible region. The absorption edge has been obtained at a shorter wavelength. The broadening of the absorption spectrum could be due to the quantum confinement of the ZnS thin films. As it is clear from spectrum, the films have a steep optical absorption feature, indicating good homogeneity in the shape and size of the nano-crystallines and low defect density near the band edge. Generally, the wavelength of the maximum exciton absorption decreases as the particle size decreases, as a result of the quantum confinement of the photo generated electron–hole pairs. The blue shift in the absorption spectrum is mainly attributed to the confinement of charge carriers in the nanoparticles. The dependence of optical absorption coefficient on photon energy helps to study the band structure and type of transition of electrons.

The optical absorption coefficient ($\alpha$) was calculated from transmittance using the following relation:

$$\alpha = \frac{1}{d}\log\left(\frac{1}{T}\right) \qquad (2)$$

**Figure 2.** SEM Image of the ZnS thin films.

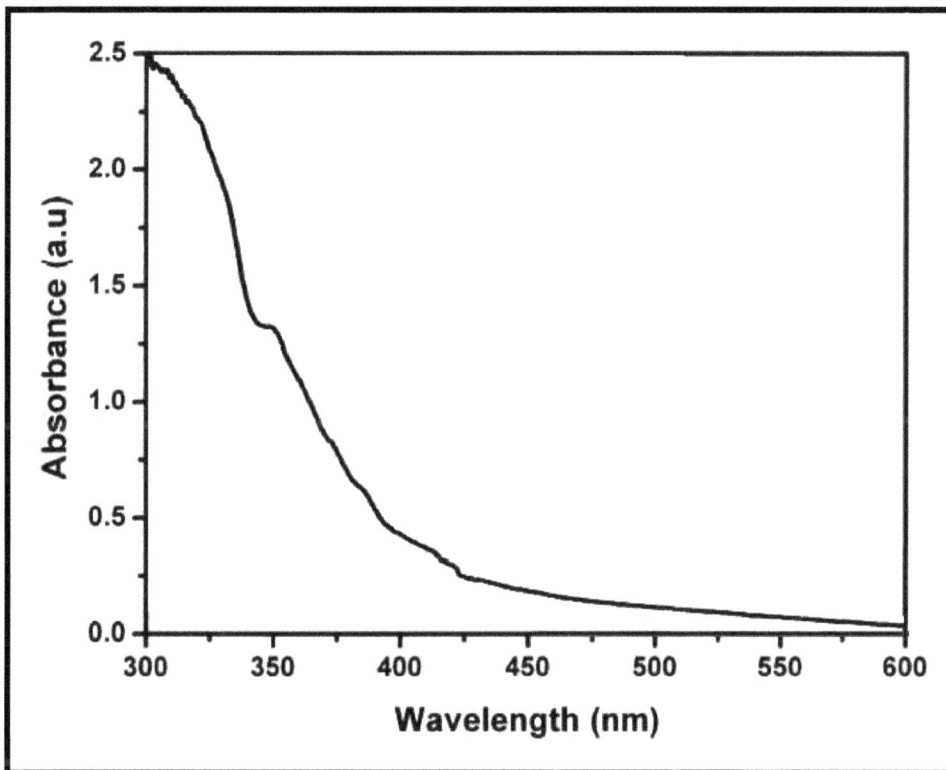

**Figure 3.** UV-Vis absorbance spectrum of ZnS films.

where T is the transmittance and d is the thickness of the film. As a direct band gap material, the film under study has an absorption coefficient (α) obeying the following relation for high photon energies (hv):

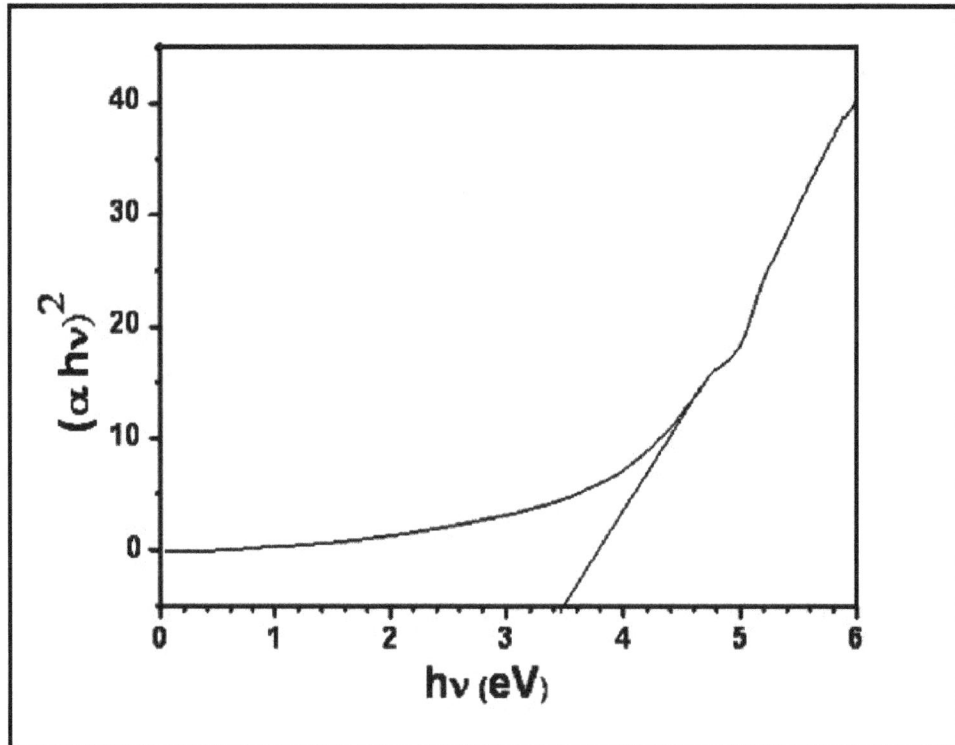

**Figure 4.** Plot of $(\alpha h\nu)^2$ Vs photon energy.

$$\alpha = \frac{A(h\nu - E_g)^{1/2}}{h\nu} \tag{3}$$

where $E_g$ is the band gap of the ZnS films and A is a constant. A plot of variation of $(\alpha h\nu)^2$ versus $h\nu$ is shown in Figure 4. Using Tauc's plot, the energy gap ($E_g$) is found to be 3.50 eV which agrees well with the reported values (Shinde et al., 2011). The energy absorption gap is of direct type and the large band gap clearly indicates the wide transparency of the film. The result could be attributed to the quantum size effects as expected from the nanocrystalline nature of the ZnS thin films.

**Dielectric properties**

The dielectric properties of the ZnS thin films have been measured at different frequency and temperatures. The dielectric constant has been calculated with frequency at different temperatures as shown in Figure 5, whereas the following dielectric losses are depicted in Figure 6. Figure 5 showed that the dielectric constant decreases exponentially with increasing frequency and then reaches almost a constant value in the high frequency region. This also shows that the value of the dielectric constant increases with an increase in the temperatures. The large value of the dielectric constant is owing to the fact that

ZnS thin films acts as a nanodipole under electric fields. The small-sized particles need a large number of particles per unit volume, ensuring increase of the dipole moment per unit volume, and a high dielectric constant (Suresh and Arunseshan, 2014). The decrease in the dielectric constant is due to electronic polarization which is quite less. Dipolar polarization is also expected to decrease with temperature as it is inversely proportional to temperature (Suresh, 2014). The polarizability of the space charge depends on the purity of the nanoparticles. As the temperature increases, the space charge effect towards polarization may have a tendency to increase (Sagadevan, and Sundaram, 2014). In Figure 6, the curves show that the dielectric loss is dependent on the frequency of the applied field, comparable to that of the dielectric constant. The dielectric loss decreases with an increase in the frequency at almost all temperatures, but appears to achieve saturation in the higher frequency range at all the temperatures (Sagadevan and Murugasen, 2014).

**Conclusion**

The Zinc sulphide (ZnS) thin films have been prepared by chemical bath deposition technique. Structural and morphology of the ZnS thin films were investigated by XRD and SEM. The XRD studies show the well crystallized

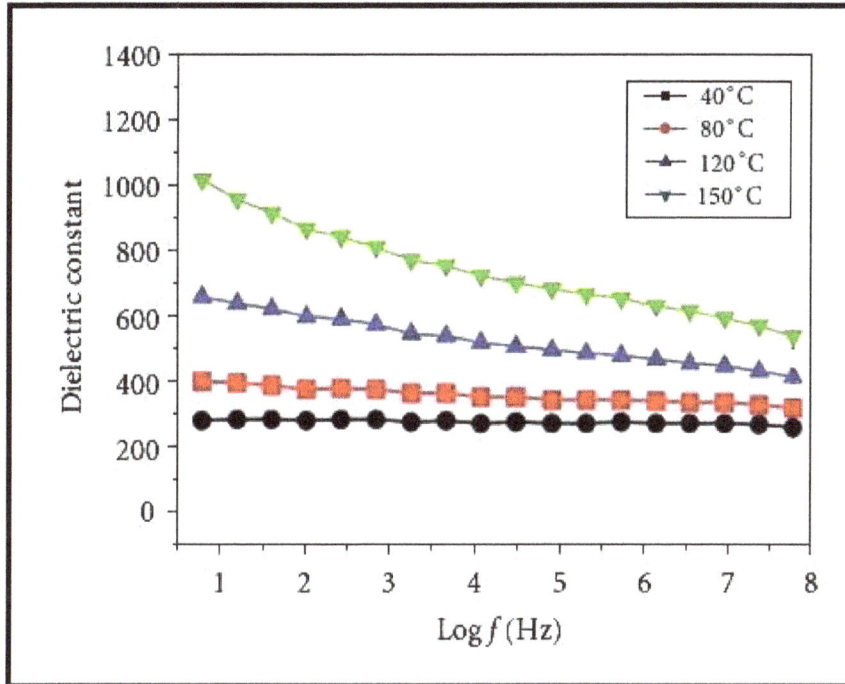

Figure 5. Dielectric constant of ZnS thin films.

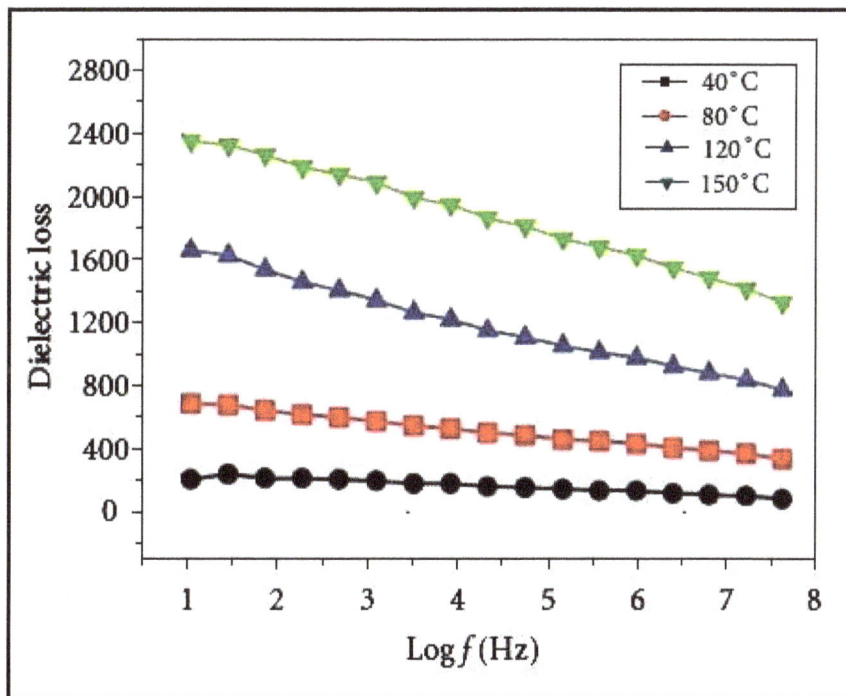

Figure 6. Dielectric loss of ZnS thin films, as a function of frequency.

and cubic structure of ZnS thin films. The UV-Visible absorbance spectrum shows excellent transmission in the entire visible region. The optical band gap is found to be 3.45 eV. The dielectric constant and dielectric loss of the ZnS thin films are calculated in the different frequency and temperatures.

## Conflict of Interest

The authors have not declared any conflict of interest.

## REFERENCES

Beard MC, Turner GM, Schmuttenmaer CA (2002). Size-dependent photoconductivity in CdSe nanoparticles as measured by time-resolved tetrahertz spectroscopy. Nano Lett. 2:983.

Bloss WH, Pfisterer F, Schock HW (1988). Advances in solar energy: An Annual review of research and development 4:275.

Coe S, Woo WK, Bawendi MG, Bulovic V (2002). Electroluminescence from single monolayers of nanocrystals in molecular organic devices. Nature 420:800.

Dimitrova JT (2000). Synthesis and characterization of some ZnS-based thin film phosphors for electroluminescent device applications. Thin Solid Films 365(1):134-138.

Icimura M, Goto F, Ono Y, Arai E (1999). Deposition of CdS and ZnS from aqueous solutions by a new photochemical technique. 198-199, 308-312.

Katayama H, Oda S, Kukimoto H (1975). ZnS blue-light-emitting diodes with an external quantum efficiency of $5 \times 10^{-4}$. Appl. Phys. Lett. 27:657.

Klimov VI, Mikhailovsky AA, Xu S, Malko A, Hollingsworth JA, Leatherdale CA, Eisler HJ, Bavendi MG (2002). Optical gain and stimulated emission in nanocrystal quantum dots. Science 290:314.

Mach R, Mueller GO (1991). Physics and technology of thin film electroluminescent displays. Semicond. Sci. Technol. 6:305.

Mustafa O, Bedir M, Ocak S, Yildirim RG (2007). The role of growth parameters on structural, morphology and optical properties of sprayed ZnS thin films. J. Mater. Sci. Mater. Electron. 18(5):505-512.

Nomura R, Murai T, Toyosaki T, Matsuda H (1995). Single-source MOVPE growth of Zinc sulfide thin films using zinc dithiocarbarnate complexes. Thin Solid Films 271(1-2):4-7.

Roy P, Ota JR, Srivastava SK (2006). Crystalline ZnS thin films by chemical bath deposition method and its characterization. Thin Solid Films 515(4):1912-1917.

Sagadevan S, Murugasen P (2014). Structural and electrical properties of copper sulphide thin films by chemical bath deposition method. Int. J. Chem. Tech. Res. 6(14):5608-5611.

Sagadevan S, Sundaram AS (2014). Dielectric properties of lead sulphide thin films for solar cell applications. Chalcogenide Lett. 11(3):159-165.

Shao L, XChang KH, Hwang HL (2003). Zinc sulfide thin films deposited by RF reactive sputtering for photovoltaic applications. Appl. Surf. Sci. 212-213, 305-310.

Shinde MS, Ahirrao PB, Patil RS (2011). Structural, optical and electrical properties of nanocrystalline zns thin films deposited by novel chemical route. Arch. Appl. Sci. Res. 3(2):311-317.

Suresh S (2014). Studies on the dielectric properties of CdS nanoparticles. Appl. Nanosci. 4:325-329.

Suresh S, Arunseshan C (2014). Dielectric Properties of Cadmium Selenide (CdSe) Nanoparticles synthesized by solvothermal method. Appl. Nanosci. 4:179-184.

Varitimos TE, Tustison RW (1987). Ion beam sputtering of ZnS thin films. Thin Solid Films 151:27.

# 22

# Study on laser etching mechanism of aluminum thin film on polyimide

Liu Xiao-Li, Xiong Yu-Qing, Ren Ni, Yang Jian-Ping, Wang Rui, Wu Gan and Wu Sheng-Hu

Science and Technology on Vacuum Technology and Physics Laboratory, Lanzhou Institute of Physics, Lanzhou 730000, China.

In order to study the laser etching mechanism for aluminum thin film on polyimide substrate, the etching process was simulated by the finite element analysis software ANSYS, and etching profile was predicted. A theoretical model was established by comparing the simulated etching results with calculated ones; it was presumed that the etching process was firstly a thermal dominant one, then a photochemical interaction dominant one, and finally a thermal one again.

Key words: Laser etching, aluminum thin film, polyimide, etching profile.

## INTRODUCTION

Based on the principle of high-power short-pulsed laser interaction with matters, laser etching was widely employed for micromachining of multi-layer thin films of metal and polymer (Liu et al., 2014; Hu et al., 2011). In this case, the absorption of laser energy occurs rapidly and only in a very thin layer of metal film on the surface, and the film is thus instantaneously evaporated and removed while the substrate is unaffected due to the precise "cool" etching process. Understanding the mechanism of laser etching is essential to predict etching result and for a better improvement of the etching quality and precision.

The mechanism of laser etching has been studied extensively, Shin et al. (2007) suggested that laser etching of multi-layer thin films of metal and polymer is a combination of photochemical evaporation and thermal melt expulsion (Srinivasan et al., 1986). Some researches investigate into the properties of laser removal of metal

thin film on polymer samples was carried out theoretically and experimentally in order to understand the mechanism of laser etching (Zhang et al., 2010; Zhou et al., 2005).

In this paper, in order to predict the etching profile of laser etching for aluminum thin film on polyimide substrate and enhance the micro-processing accuracy and quality, a theoretical model, called the Srinivasan–Smrtic–Babu (SSB) model, was adopted, which takes both thermal and photochemical effects of the laser etching into consideration. Parameters for SSB model were obtained by fitting the experimental data from relevant literature (Yoon and Bang, 2005; Shin et al., 2007) and ANSYS software was employed to study the temperature distribution within aluminum thin film on polyimide (PI) substrate during 355 nm laser etching process. The operating parameters include a TEM00 Gaussian distribution laser beam, pulse repetition frequency of 20 kHz, pulse width of 100 ns. By comparing the simulated etching results with predicted ones, the processing of 1aser energy absorption and thermal conduction were analyzed, and a mechanism of metallic film/ polyimide substrate interface separation while metallic film in solid state due to the polyimide material thermal decomposition was also proposed subsequently.

## THEORETICAL MODELS

### A theoretical model for PI removal by laser etching

Photochemical and thermal effects were considered for ablation of polyimide (PI), $H_c$ is the etching depth per laser pulse by the photochemical effect according to the assumption of Beer's law (Li et al., 2006), which can be expressed as follows:

$$H_c = a_{eff}^{-1} \ln \left( \frac{F}{F_{th}} \right) \tag{1}$$

Where $a_{eff}^{-1}$ is the absorption coefficient(cm$^{-1}$), $F$ is the laser fluence per pulse (J/cm$^2$), and $F_{th}$ is the threshold fluence (J/cm$^2$).

When processing PI, the thermal effect is non-ignorable, the single pulse etching depth expressed by $H_T$, from pseudo-zeroth order rate law:

$$-\frac{dc}{dt} = kc \Rightarrow -\int_{c_0}^{c} \frac{dc}{c} = k \int_0^t dt \tag{2}$$

$$H_T = k_0 \cdot e^{-E/RT} \tag{3}$$

Where $k_0$ is the effective frequency factor (µm/pulse), $E$ is the activation energy (kJ/cm$^3$), $R$ is the gas constant (371 J/(cm$^3$·K)), and $T$ is the temperature averaged in some manner over the irradiated region and is controlled by the incident fluence and the photon absorption dynamics of the PI. Temperature within the laser exposed region is both time and position dependent. In the absence of knowledge of the exact functional dependence, it is only possible to proceed phenomenologically by assuming a one dimensional heat flow. The fluence dependence of local temperature can be modeled by an expression similar to that proposed by Yung et al. (2000):

$$T(x) = \frac{a_{eff} F}{\rho C_P} \cdot e^{a_{eff} \cdot x} \tag{4}$$

Where $\rho$ is density(kg/cm$^3$), $C_P$ is heat capacity(J/kg·K), and $x$ is the etching depth(cm).

Calculate the $T_{avg}$ (average temperature) from $x=0$ to $x=d_p$:

$$T_{avg} = \frac{\int_0^{d_p} \frac{a_{eff} F}{\rho C_P} e^{-x a_{eff}} dx}{\frac{1}{a_{eff}}} = \frac{a_{eff} F}{\rho C_P} \cdot \frac{1}{a_{eff}} (1 - e^{-a_{eff} d_p}) \tag{5}$$

It can be simplified as:

$$dp = \frac{1}{a_{eff}} \cdot \ln \frac{F}{F_{th}} \Rightarrow dp \cdot a_{eff} = \ln \frac{F}{F_{th}} \tag{6}$$

By combining Equation (5) and (6), it can be obtained that:

$$T_{avg} = \frac{a_{eff}}{\rho C_P} (F - F_{th}) \tag{7}$$

By putting Equation (7) into Equation (3), $H_T$ (The etching depth by the photothermal effect) can be obtained as:

$$H_T = k_0 \cdot e^{[(-E \cdot \ln F/F_{th})/(a_{eff}(F-F_{th}))]} \tag{8}$$

**Table 1.** Parameters of PI for 355 nm UV laser etching.

| Parameter | Values |
|---|---|
| $F_{th}$(J/cm$^2$) | 0.1 |
| $a_{eff}$(cm$^{-1}$) | 0.2×10$^5$ |
| $k_0$(µm/pulse) | 8.86 |
| $E^*$( kJ/cm$^3$) | 207.6×10$^3$ |

Where $E^* = \frac{E \cdot \rho \cdot C_P}{a_{eff}}$ is the effective activation energy (kJ/cm$^3$), when photochemical and thermal effects were take into consideration, the etching depth per laser pulse, $H_{PI}$ can be expressed as follows:

$$H_{PI} = \frac{1}{a_{eff}} \cdot \ln \frac{F}{F_{th}} + k_0 \cdot e^{[(-E^* \cdot \ln F/F_{th})/(a_{eff}(F-F_{th}))]} \tag{9}$$

By putting parameters in Table 1 into Equation (9), theoretical relationship between etching depth and fluence can be obtained, as shown in Figure 1.

In Figure 1, which shows the theoretical results in logarithmic scale, the etching depth will maintain at a constant value as the laser fluence increases. When the laser fluence is smaller than 1 J/cm$^2$, while photochemical effect is dominant, the etching depth as a logarithmic function increases rapidly, and then stays at a constant value. Reversely, the etching depth by thermal effect increases dramatically in the region of laser fluence greater than 5 J/cm$^2$. Although the etching depth is an exponential function of laser fluence, the logarithmic function in the exponent reduces the etching rate rapidly in the region of fluence greater than 25 J/cm$^2$. This indicates that a laser fluence greater than a certain value contributes little to the etching depth. Thus, it is meaningless to increase the laser fluence above this value.

### A theoretical model for Al removal by laser etching

Considering that thermal effect is dominant during laser etching process of Al, we adopt a theoretical model which takes only thermal effects into account, to predict the etching depth of laser etching of Al films. $H_{AL}$ can be expressed as follows:

$$H_{AL} = H_T = k_0 \cdot e^{[(-E \cdot \ln F/F_{th})/(a_{eff}(F-F_{th}))]} \tag{10}$$

By putting parameters of Al in Table 2 into Equation (10), theoretical relationship between etching depth and laser fluence can be obtained, as shown in Figure 2. Figure 2 shows the theoretical results of etching depth considering only thermal effect in decimal scale. In the region of fluence smaller than 3.2 J/cm$^2$, the etching depth is not measurable; and in the region of fluence greater than 4 J/cm$^2$, the etching depth increases rapidly.

## SIMULATION SOLUTION

Here, the adopted equations for laser-material interaction will be presented. Non-steady-state temperature distribution within thin film and substrate can be obtained by applying heat transfer and phase change model of laser heating. Ignoring gas and liquid dynamic effects (Li et al., 2008), differential equation describing temperature $T(x,y,t)$ of this process is:

$$c(T)\rho \frac{\partial T(x,y,t)}{\partial t} = \nabla(K(T) \cdot \nabla T(x,y,t) + Q(x,y,T,t)) \tag{11}$$

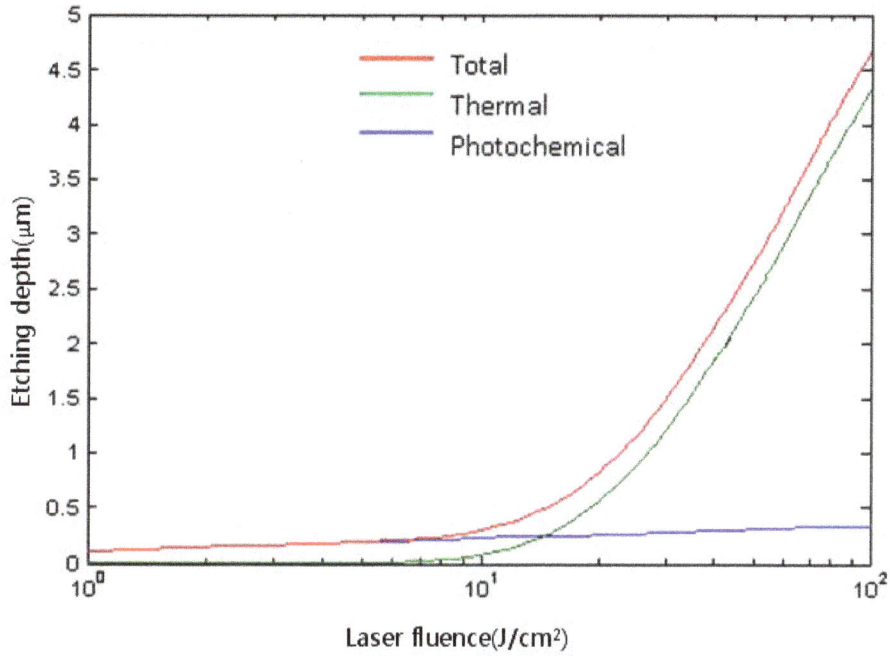

**Figure 1.** Relationship between etching depth and laser fluence for PI etched by 355 nm UV laser.

**Table 2.** Parameters of Al for 355 nm UV laser etching.

| Parameter | Values |
|---|---|
| $F_{th}$(J/cm$^2$) | 0.32 |
| $a_{eff}$(cm$^{-1}$) | $5.917 \times 10^5$ |
| $k_0$(μm/pulse) | 78 |
| $E^*$( kJ/cm$^3$) | $1.13 \times 10^7$ |

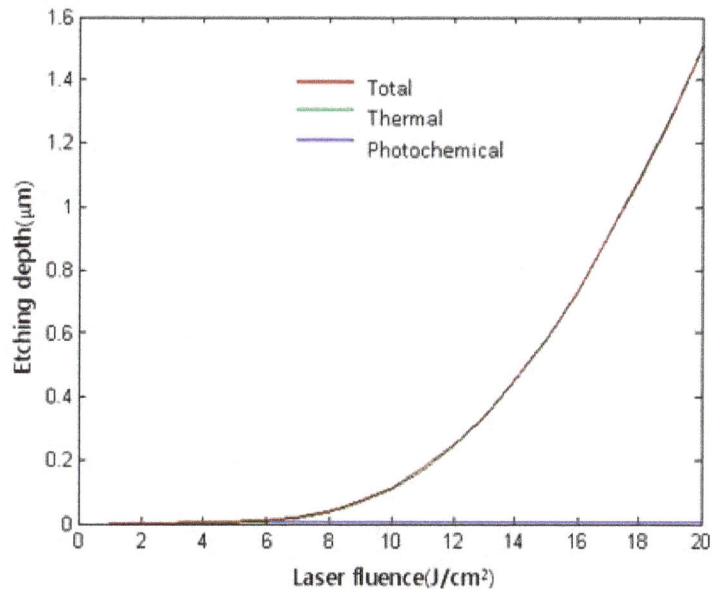

**Figure 2.** Relationship between etching depth and laser fluence for Al etched by 355 nm UV laser.

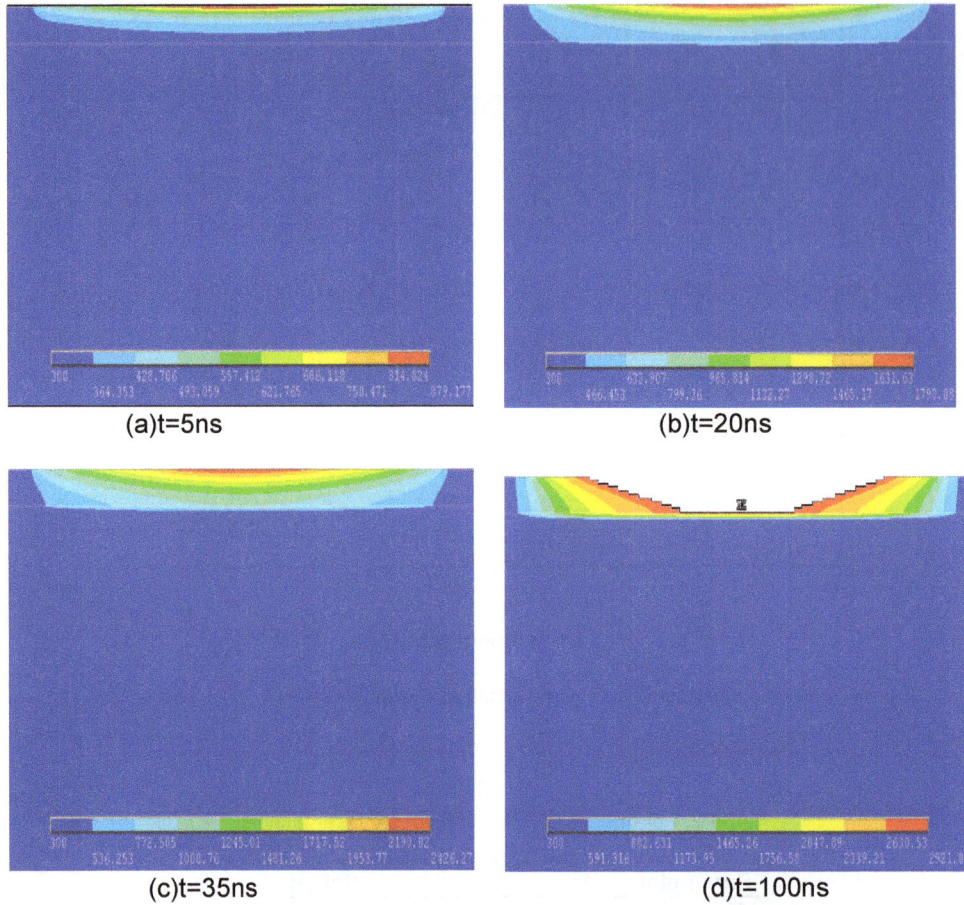

**Figure 3.** Simulated temperature field caused by laser irradiation with different times (Different color represents different temperature).

Where $c$ is specific heat capacity, $\rho$ is material density, $K$ is thermal conductivity, and $Q$ is the heat generation rate. The initial temperature before laser irradiation equals to environmental temperature:

$$T(x, y, 0) = T_{env} \qquad (12)$$

The boundary conditions for temperature of the heated top surface and the other surfaces are different. Because of the high heat flow gradients during laser heating, no heat is lost from the heated side. Therefore, the Neumann boundary condition was used for boundary at $y=0$ (Hu and Yin, 1997). However, as the laser affected depth is much smaller than thickness of thin film, the opposite surface is assumed to stay at environmental temperature according to the Dirichlet boundary condition, expressed as follows:

$$\frac{\partial T(x,y,t)}{\partial z}\Big|_{z=0} = 0 \qquad (13)$$

$$T(x, y, t)\Big|_{z=thickness\ of\ sample} = T_{env} \qquad (14)$$

According to the theory of heat transfer in solid, analytical solution to Equation (11) can be obtained when laser power density is independent upon time. However, as one laser pulse generally offers several times the amount of energy required for one unit mass to rise from room temperature to evaporation temperature of Al within a very short period of time, the large heat flow gradient makes thermal properties of Al thin film temperature-dependent. Therefore, finite element method implemented in ANSYS was used to obtain the approximate numerical solution of the nonlinear model. This paper is based on theoretical and simulative analysis. The process of laser removal of Al thin film on PI samples was studied and the mechanism of the laser etching was analyzed.

**Numerical simulation for process of etching of aluminum thin film on PI**

Before simulation, the constant temperature boundary was assumed ($T_{int}$=300 K). Figure 3 shows the temperature distribution during the laser etching of Al/PI with different irradiation pulse. Figure 3(a) displays

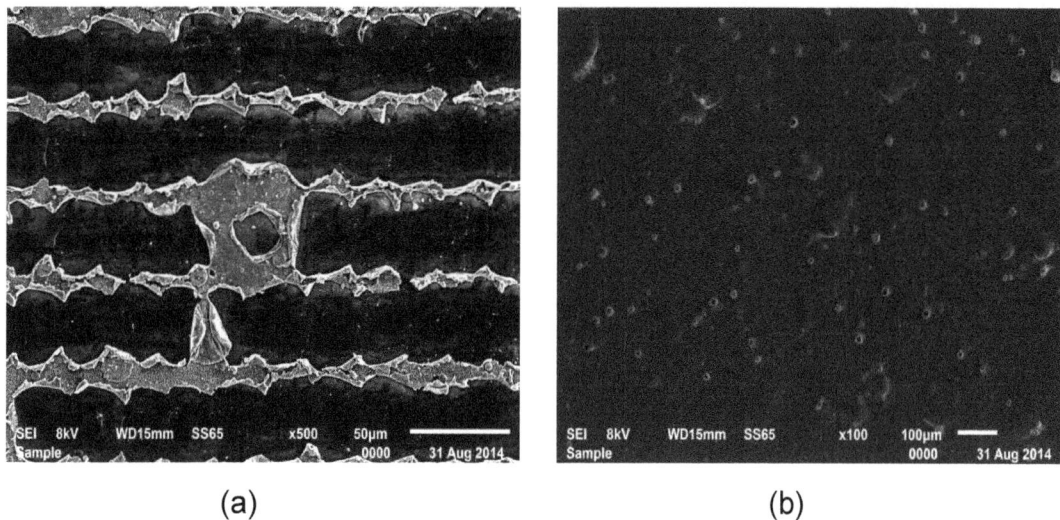

(a)                                                    (b)

**Figure 4.** SEM images of etched morphology with different irradiation pulse.

temperature distribution within Al/PI with a laser output power of 4W and an irradiation pulse of 5 ns. It can be seen that the maximum center temperature on the surface is 814.824 K and the interface temperature retains 300 K, which means that the heat does not transfer to the substrate, and laser etching is a thermal process, for which the excitation energy is instantaneously transformed into heat. Over a period of time, as shown in Figure 3(b), temperature at interface reaches 466.453 K, which is lower than the substrate's decomposing temperature, 500 K. Figure 3(c) shows temperature distribution within Al/PI for a pulse of 35 ns, it can be seen that the maximum temperature at the interface reaches about 772.505 K, which is higher than substrate's decomposing temperature, which can result in substrate decomposition, and the pressure formed cause a state of separation between Al films and PI substrate. Meanwhile, heat conduction was blocked and the total energy is transformed to heat, induce rapid evaporation of Al films. As shown in Figure 3(d), it can be found that with the laser irradiation pulse change from 5 ns to 100 ns, the surface temperature increases gradually and Al thin film was removed rapidly.

Experimental etching of Al thin film on PI was conducted to verify the simulated results. It shown that the etched Al thin film presents different edges with different irradiation pulse, corresponded to different laser fluence. The coarse edge of etched thin film in Figure 4(a) indicate that it was peeled off by pressure formed by decomposition of PI, and Figure 4(b) shows the bubbles formed on surface of PI caused by decomposition clearly.

## CONCLUSION

By comparing of the simulated and calculated etching

results, it can be concluded that the etching process was firstly a thermal interaction dominant one, aluminum thin film absorbs portion of the laser energy and transferred it into heat; then, when the temperature raised to decomposing temperature of polyimide, a photochemical interaction was dominant and polyimide at interface decomposed, a pressure is formed above the interface and peeled aluminum thin film off from polyimide substrate; Finally, temperature of floated aluminum increased and gasified, thermal mechanism is determinant again at this stage. The simulation was verified by experiment, and the simulated results can be used to optimize the etching parameters to obtain a smooth etching edge.

## Conflict of Interest

The authors have not declared any conflict of interest.

## ACKNOWLEDGEMENT

This work is supported by National Natural Science Foundation of China (NSFC. 51135005)

### REFERENCES

Liu TH, Hao ZQ, Gao X, Liu ZH, Lin JQ (2014). Shadowgraph investigation of plasma shock wave evolution from Al target under 355-nm laser ablation. Chinese Phys. B 23(8):085203.
Hu HF, Ji Y, Hu Y, Ding XY, Liu XW, Guo JH, Wang XL, Zhai HC (2011). Thermal analysis of intense femtosecond laser ablation of aluminum. Chinese Phys. B. 20(4):044204.
Srinivasan V, Smrtic MA, Babu SV (1986). Excimer laser etching of polymers. J. App. Phys. 59(11):3861-3867.
Zhang F, Duan J, Zeng XY, Li XY (2010). 355 nm DPSS UV

Laser Micro-Processing For Semiconductor and Electronics Industry, Proc. SPIE. 7584:75840-75850.

Zhou XL, Wen W, Zhang JS, Sun DB (2005). A mathematical model of the removal of gold thin film on polymer surface by laser ablation. Surf. Coat. Tech. 190:260–263.

Yoon KK, Bang SY (2005). Modeling of polymer ablation with excimer lasers. J. Korean Soc. Precision Eng. 22(9):60-68.

Shin BS, Oh JY, Sohn H (2007). Theoretical and experimental investigations into laser ablation of polyimide and copper films with 355-nm Nd: YVO$_4$ laser. J. Mater. Process. Tech. 187-188:260-263.

Li L, Zhang D, Li Z, Guan L, Tan X, Fang R, Hu D, Liu G (2006). The investigation of optical characteristics of metal target in high power laser ablation. Physica B. 383(2):262-266.

Yung WKC, Liu JS, Man HC, Yue TM (2000). 355 nm Nd:YAG laser ablation of polyimide and its thermal effect. J. Mater. Process. Technol. 101:306~311.

Li J, Peterson GP, Cheng P (2008). Dynamic characteristics of transient boiling on a square platinum microheater under millisecond pulsed heating. Int. J. Heat Mass Tran. 51:273–282.

Hu B, Yin HM (1997). On Critical Exponents for the Heat Equation with a Mixed Nonlinear Dirichlet-Neumann Boundary Condition. J. Math. Anal. Appl. 209:683-711.

# 23

# Optical properties of ZnO thin films deposed by RF magnetron

**R. Ondo-Ndong[1,2], H. Z. Moussambi[1], H. Gnanga[1], A. Giani[2] and A. Foucaran[2]**

[1]Laboratoire pluridisciplinaire des sciences, Ecole Normale Supérieure, B.P 17009 Libreville, Gabon.
[2]Institut Electronique du Sud, IES-Unité mixte de Recherche du CNRS n° 5214, Université Montpellier II, Place E. Bataillon, 34095 Montpellier cedex 05- France.

We have grown ZnO thin films on glass substrate by RF magnetron sputtering using metallic zinc target. The influences of some parameters on thin film optical properties were assessed. They exhibited extremely high resistivity of $10^{12}$ $\Omega$.cm, an energy gap of 3.3 eV at room temperature. It was found that a RF power of 50 W, a target to substrate distance of 70 mm, very low gas pressures of 3.35 x $10^{-3}$ Torr of argon and oxygen mixed gas atmosphere gave ZnO thin films with a good homogeneity and a high crystallinity. All the films are transparent in the visible region (400 to 800 nm) with average transmittance above 80%. The optical transmittance and refractive index, calculated from the spectra of optical absorbance, show a significant dependence on the growth parameters. As for the sample grown at 100°C, the average transmittance is about 80% in the visible wavelength range and the refractive index is estimated to be 1.97.

**Key words:** ZnO, RF sputtering magnetron, X-ray diffraction, transmittance, refractive index.

## INTRODUCTION

Zinc oxide is one of the most interesting II–IV compound semiconductors with a wide direct band gap of 3.3 eV (Meng and Dos Santos, 1994; Inukai et al., 1995; Han and Jou, 1995; Craciun et al., 1995; Subramanyam et al., 1999; Sanchez-Juarez et al., 1998; Sourdi et al., 2012; Yang Ming Lu et al., 2007). It has been investigated extensively because of its interesting electrical, optical and piezoelectric properties making suitable for many applications such as transparent conductive films, solar cell window and MEMS waves devices (Craciun et al., 1995). The thermal stresses were determined by using a bending-beam Thorton method (Han and Jou, 1995) while thermally cycling films. ZnO has hexagonal Wurtzite structure and some properties are determined by the crystallite orientation on the substrate. For example, for piezoelectric applications, the crystallite should have the c-axis perpendicular to the substrate. According to the literature, the reactive sputtering technique has received a great interest because of its advantages for film growth, such as easy control for the preferred crystalline orientation, epitaxial growth at relatively low temperature, good interfacial adhesion to the substrate and the high packing density of the grown film. These properties are mainly caused by the kinetic energy of the clusters given by electric field (Molarius et al., 2003; Lin et al., 2008; Kim et al., 1997). This energy enhances the surface migration effect and surface bonding state.

In previous work, we investigated the effect of the substrate temperature and the oxygen-argon mixture gas on the properties of ZnO films. It has been found that the structural properties of ZnO films depend very much on the substrate temperature. Indeed, a ZnO hexagonal wurtzite structure and properties are determined by the orientation of the crystallites on the substrate. FWHM of the (002) X-ray rocking curve must be less than 0.32 for an effective electromechanical coupling (Ondo-Ndong et al., 2003). In continuation of this work, the optical properties of ZnO structures have been investigated based on the deposition parameters.

## EXPERIMENTAL

Zinc oxide films were deposited by RF magnetron sputtering using a zinc target (99, 99%) with diameter of 51 and 6 mm thick. Substrate is p-type silicon with (100) orientation. The substrates were thoroughly cleaned with organic. Magnetron sputtering was carried out in an oxygen and argon mixed gas atmosphere by supplying RF power at a frequency of 13.56 MHz. The RF power was about 50 W. The flow rates of both the argon and oxygen were controlled by using flow meter (ASM, AF 2600). The sputtering pressure was maintained at $3.35.10^{-3}$ torr controlling by a Pirani gauge. Before deposition, the pressure of the sputtering system was under $4.10^{-6}$ torr for more than 12 h and were controlled by using an ion gauge controller (IGC – 16 F).

Thin films were deposited on silicon, substrate under conditions listed in Table 1. These deposition conditions were fixed in order to obtain the well-orientation zinc oxide films. The presputtering occurred for 30 min to clean the target surface. Deposition rates covered the range from 0.35 to 0.53 µm/h. All films were annealed in helium ambient at 650°C for 15 mn. Measurements of transmittance in the range from 300 to 900 nm are made using a UV-Visible CARY spectrometer.

The device has a pulsed xenon lamp, which produces only a flash in each acquisition of a measurement point. A quartz bulb, not glass, let's UV radiation through. The emerging beam is then a cylindrical diverging beam that will cover the entire surface of the mirror M1. A mirror allows the orientation and focus of the useful part of the beam emerging from the input to the network and then diffracted by the latter towards a beam exit slot gap. The assembly constituted by the entrance slit and the first mirror is called collimator. A blade placed on the path L of monochromatic radiation is used to reflect a portion of the intensity of the wave to a photoreceptor which measures the intensity of radiation which will pass through the vessel containing the sample to take into account small fluctuations the intensity of the light source. The rough surface materials with in homogeneities or imply low volume detected signal. Thus, we must make an adjustment, before any measurement: The 100% for power transmission Pyrex substrate as a reference. Piloting, digital capture and processing of data is performed by a microcomputer.

### Theoretical model for complex index

To calculate the optical constants, are often used to model on a volume of isotropic and homogeneous material. In reality, the behavior of thin films obtained from the ideal model overflows due to the inhomogeneity of layers and the dispersion of the refractive.

**Table 1.** ZnO sputtering conditions.

| | |
|---|---|
| Sputtering pressure | $3.35 \times 10^{-3}$ Torr |
| Mixture gas | $A_r + O_2 = 80 - 20\%$ |
| Power RF | 50 W |
| Sputtering time | 6 h |
| Substrate temperature | 100°C |
| Target-substrate distance | 7 cm |

These optical constants are represented by the index of refraction that is, in the general case, depends on the complex wavelength dependence $\tilde{n} = n(\lambda) + ik(\lambda)$. Is the real part of the refractive index. The complex index $\tilde{n} = n(\lambda) + ik(\lambda)$ is very important for the dielectric characterization of materials. $\tilde{n}$ Provides, at infinity, the complex permittivity $\tilde{\varepsilon} = \varepsilon_1 + i\varepsilon_2$ through:

$$\varepsilon_1 = n^2 - k^2$$
$$\varepsilon_2 = 2n \cdot k$$

And also, the relative permittivity $\varepsilon_r$ and electrical conductivity to the required frequency $\sigma$ through:

$$\varepsilon_r = \varepsilon_1$$
$$\sigma = \varepsilon_2 \cdot \omega \cdot \varepsilon_0$$

The measurement of light transmission through a parallel plate dielectric film, in the working range considered, is sufficient to determine the real and imaginary parts of the complex refractive index and thickness. Wales and Lyashenko developed a method using the successive approximations and interpolations for calculating these three quantities (Wales et al., 1967; Lyashenko and Miloslavskii, 1964).

Manifacier et al. (1976) have developed a method, like in the same range of applicability but differs from Lyashenko and Miloslavskii (1964) accuracy by: Firstly, the calculation processing and the data is easier, and secondly it provides an explicit expression for n, k and thickness. This last method we have used to characterize our samples of zinc oxide thin film prepared. Figure 1 shows a thin layer complex refractive index $\tilde{n}$, linked by two transparent media $n_1$ and index $n_0$. With $n_0$ the index of air ($n_0 = 1$) and $n_1$ the index of the substrate.

In the case of normal incidence, the amplitude of the transmitted wave length is given by Wales et al. (1967) and Lyashenko and Miloslavskii (1964):

$$A = \frac{t_1 t_2 \exp\left(-2i\pi nd/\lambda\right)}{1 + r_1 r_2 \exp\left(-4i\pi nd/\lambda\right)} \tag{1}$$

Where $t_1$, $t_2$, $r_1$, $r_2$, n and d are respectively the transmission and reflection coefficients of the front and rear faces of the sample, the refractive index and the thickness of the material.
The transmission of the layer is given by:

$$T = n_1/n_0 \mid A \mid^2 \tag{2}$$

In the case of low absorption along with,

$$k^2 \ll (n - n_0)^2 \text{ et } k^2 \ll (n - n_1)^2 \tag{3}$$

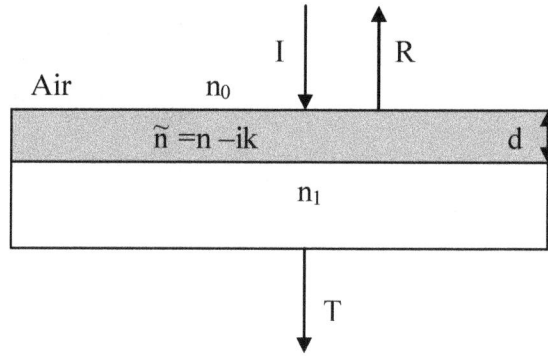

**Figure 1.** Optical transmittance on the sample.

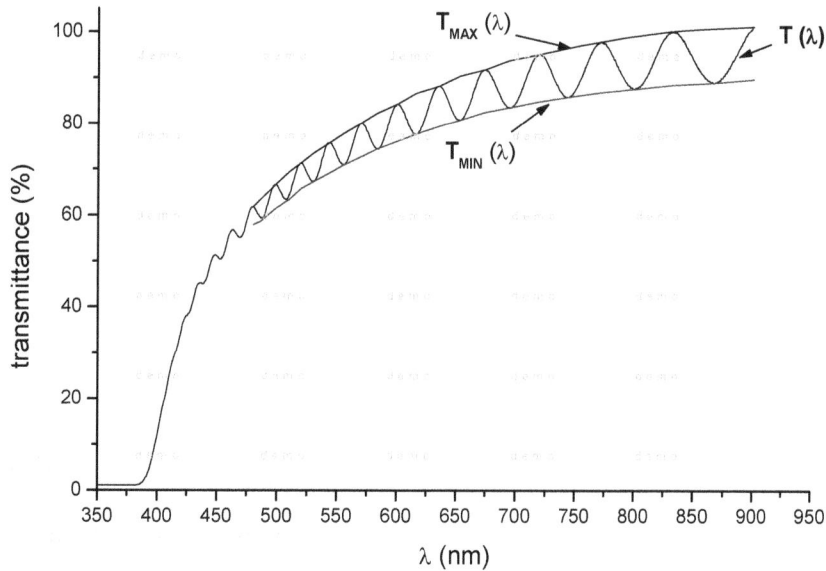

**Figure 2.** Optical transmittance spectra Example on the sample.

$$T = \frac{16\,n_0 n_1 n^2 \alpha}{C_1^2 + C_2^2 + 2C_1 C_2 \alpha \cos 4n\,\pi d/\lambda} \qquad (4)$$

Where $\quad C_1 = (n + n_0)(n_1 + n), \quad C_2 = (n - n_0)(n_1 - n) \qquad (5)$

And $\quad \alpha = \exp(-4\pi k d/\lambda) = \exp(-\beta d) \qquad (6)$

$\beta$ is the absorption coefficient of the thin film, k is the extinction coefficient and the percentage absorption $\alpha$.

Generally, outside the region of the fundamental absorption or free carrier absorption (for higher wave lengths), the dispersions of n and k are large. The maxima and minima of transmission in Equation (4) to occur:

$$\frac{4\pi n d}{\lambda} = m\pi \qquad (7)$$

Where m is the wave number.

In corresponding to a thin layer of transparent semiconductor substrate non-absorbent, $C_2 < 0$ usual cases, the extreme values of the transmission are given by the formula:

$$T_{max} = \frac{16\,n_0 n_1 n^2 \alpha}{(C_1 + \alpha C_2)^2}, \qquad T_{min} = \frac{16\,n_0 n_1 n^2 \alpha}{(C_1 - \alpha C_2)^2} \qquad (8)$$

Combining Equations (8) relationship Lyashenko developed an iterative method for the determination of n and $\alpha$, and using Equations (6) and (7), determining k and d. We propose a major simplification of this method. Indeed, we consider $T_{min}$ and $T_{max}$ as continuous functions of n ($\lambda$) and $\alpha$ ($\lambda$). Indeed, the two envelopes of the measured transmittance form a non-linear system of two equations in two unknowns n ($\lambda$) and $\alpha$ ($\lambda$), which can be solved by iteration. These functions, which are envelopes of the maxima $T_{max}$ ($\lambda$) and the minimum $T_{min}$ ($\lambda$) in the transmission spectrum are shown in Figure 2. $\alpha$ coefficient is given by the ratio of Equations (8).

$$\alpha = \frac{C_1 \left[1 - (T_{max}/T_{min})^{1/2}\right]}{C_2 \left[1 + (T_{max}/T_{min})^{1/2}\right]} \qquad (9)$$

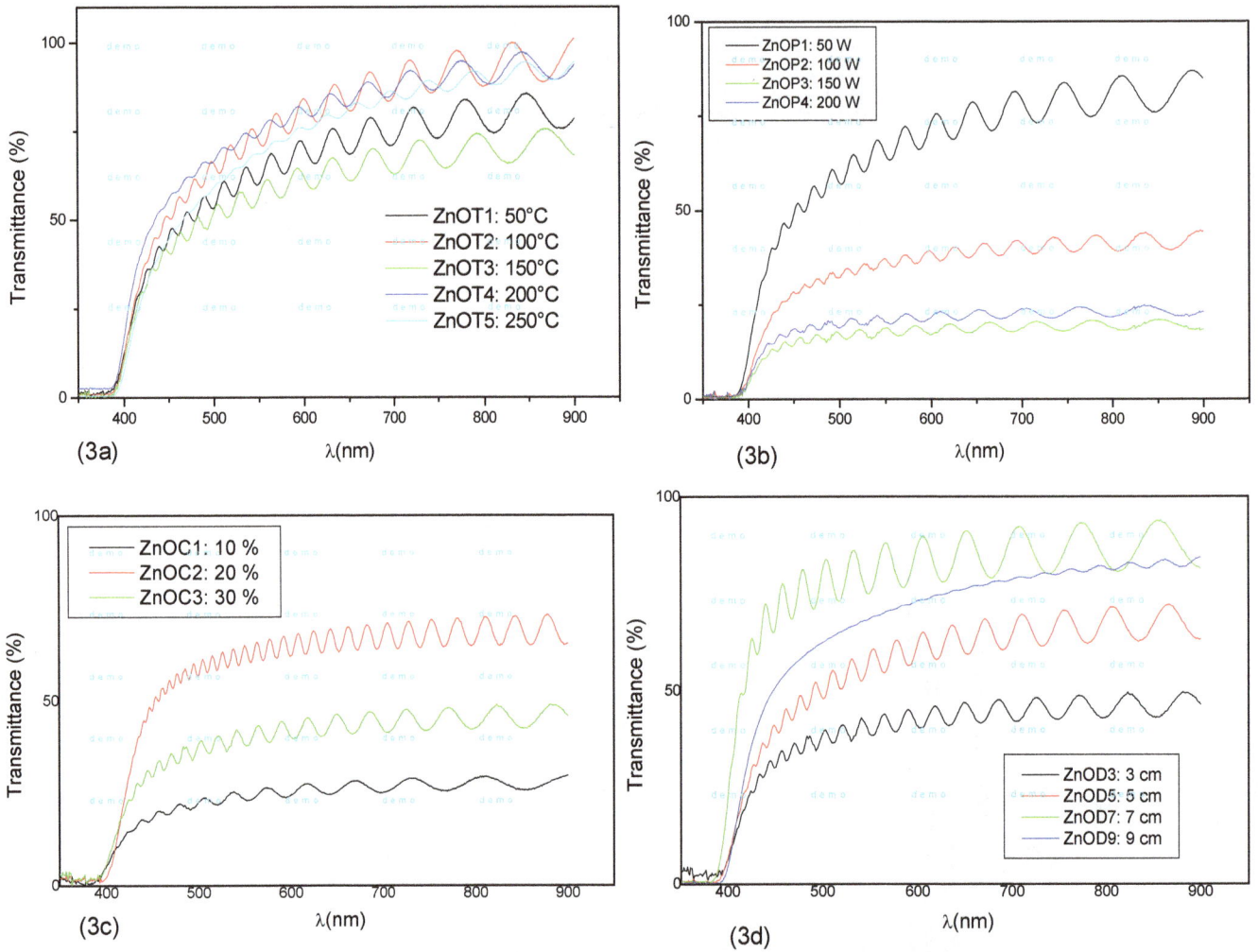

**Figure 3.** Optical transmittance spectra of ZnO thin films: a) at various substrate temperatures, b) at different power, c) based on the rate of oxygen, d) depending on the target-substrate distance.

Then we deduce Equation (8) the relationship of the refractive index of the thin layer.

$$n = \left[ N + \left( N^2 - n_0^2 n_1^2 \right)^{1/2} \right]^{1/2} \qquad (10)$$

With $\quad N = \dfrac{n_0^2 + n_1^2}{2} + 2 n_0 n_1 \dfrac{T_{max} - T_{min}}{T_{max} \, T_{min}} \qquad (11)$

N is a constant.
    The Equation (8) shows that the refractive index n is determined explicitly.
Knowing n can be determined by the above equation $\alpha$. The thickness d of the layer may be calculated by two maxima or minima using the equation below.

$$d = \dfrac{M \lambda_1 \lambda_2}{2 \left[ n(\lambda_1) \lambda_2 - n(\lambda_2) \lambda_1 \right]} \qquad (12)$$

Where in M is the number of oscillations between two extreme points (M =1 between two consecutive minima or M=2 two

consecutive maxima), $\lambda_1$, n $(\lambda_1)$ and $\lambda_2$, n $(\lambda_2)$ levels are matching wave the wavelength and the refractive index.

## RESULTS AND DISCUSSION

### Transmittance

To know the parameter values that seem to be making the best deposits from a structural point of view, we have undertaken, optical characterizations in order to identify the influence of the four parameters of deposits. And to avoid the effect of film thickness on the optical properties, we worked on samples of similar thicknesses of 2.8 to 2.9 µm.

Figure 3 shows the transmission spectra of ZnO films. We observe that the powers in Figure 3a transmission are high (80 to 90%) for all samples. It is also observed that the absorption front at a wavelength for which the transmission is reduced to 50% is set to 375 to 400 nm.

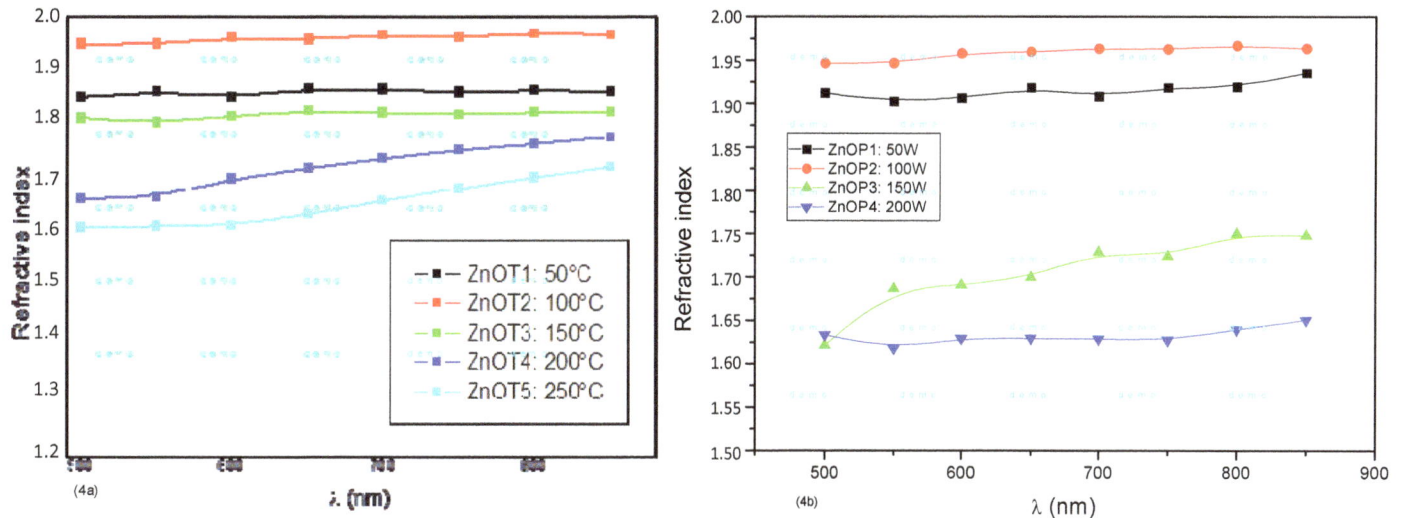

**Figure 4.** Refractive index as a function of wavelength: a) for different substrate temperatures; b) different powers.

The difference of the extreme of all samples was 100°C maximum in Figure 3a. Subramanyam et al. (1999) confirm these results. Indeed, they get a power transmission of the order 86% and observed a decrease in optical transmission with temperature in the range 300-400°C.

In the spectral range considered, we have represented in Figure 3b changing transmission thin zinc oxide layers for different powers. We note that the sample of ZnO prepared to 50 W has a maximum transmission of about 80%. By cons, for the other samples, the transmission spectra are of little use. These observations clearly confirm the results obtained by ray diffraction where we found that the samples are poorly crystallized and the grain size is unusually small compared to that obtained with the sample 50 W. The effect of oxygen on the transmission rate of the ZnO films showed in Figure 3c. A decrease in transmittance was observed to measure the percentage of oxygen in the gas mixture (Ar - $O_2$) increases in the region of short wavelength. The maximum transmission is observed for the oxygen content equal to 20% and the minimum transmission are higher. Moreover, the difference between the transmissions of the extrema ($T_{max}$ - $T_{min}$) is greater for the sample. This reflects a higher refractive index.

We studied the influence of the target-substrate distance watching the optical transmission of the ZnO thin films. Figure 3d shows that the optical properties of zinc oxide are dependent of the target-substrate distance. Based on the experimental conditions, we can say that the target-substrate distance equal to 7 cm is ideal for making our ZnO films. Indeed, the power transmission of this sample was very high (95% $\lambda$ = 600 nm). It is estimated to have homogeneous layer thickness minimum distance. Indeed, at this distance, the

thermalization of the structure is efficient. The discharge (plasma) is maintained with a minimum of particle collision and the efficiency of the pulverization is effective.

**Refractive index n**

Changes in the refractive index as a function of wavelength at different substrate temperatures are shown in Figure 4a. The refractive index has a high dispersion to the layers developed to above 150°C temperatures. Figure 4b shows the variation of the refractive index as a function of wavelength at different powers. We see that the index decreases with power. It varies from 1.97 to 1.6 in from 50 to 200W at 600 nm. Figure 5 shows that a given wavelength, the refractive index increases from 1.87 to 1.97 when the substrate temperature ranges from 50 to 100°C. Above 100°C, the value of the refractive index decreases as the substrate temperature increases to 1.63. To highlight these observations, we have shown in Figure 6 changes in the refractive index as a function of oxygen concentration in the gas mixture at a given wavelength ($\lambda$ = 600 nm). We find that the influence of the gas mixture on the refractive index is significant only when we have an oxygen level of 20%. In addition, we note a decrease in the index with increasing oxygen content in the gas mixture. We attribute this phenomenon, compared with the X-ray crystallographic disorientation of the structure. Indeed, the structural study showed that the optimum oxygen level, to develop well-crystallized films of ZnO was 20%.

Figure 7 shows that the target distance of 7 cm substrate is ideal for obtaining zinc oxide layers of good quality, taking into account, of course, the deposition

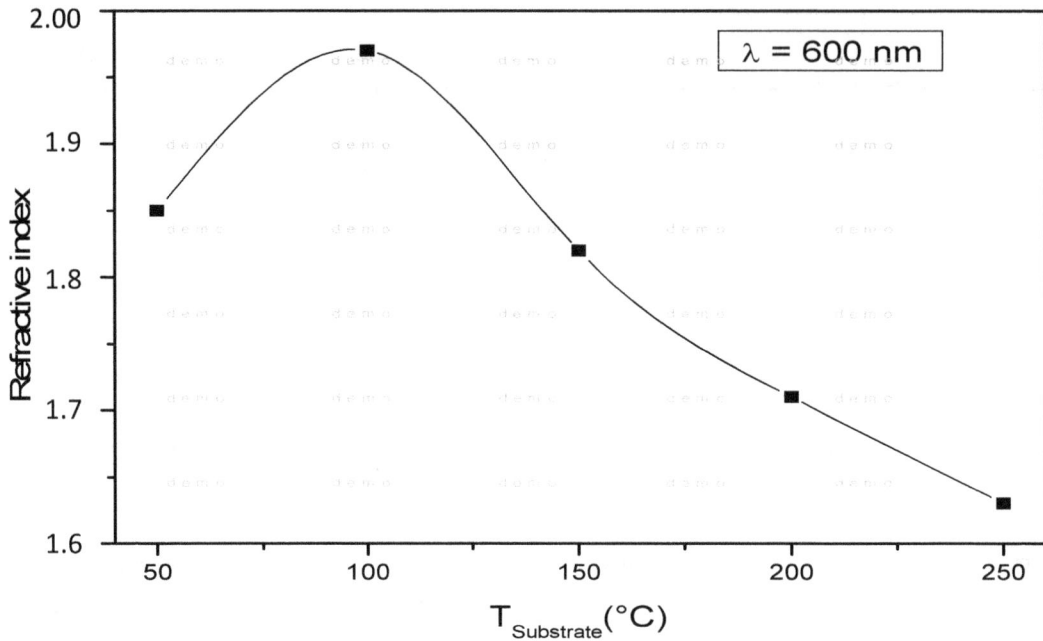

**Figure 5.** Refractive index as a function of the substrate temperature constant λ.

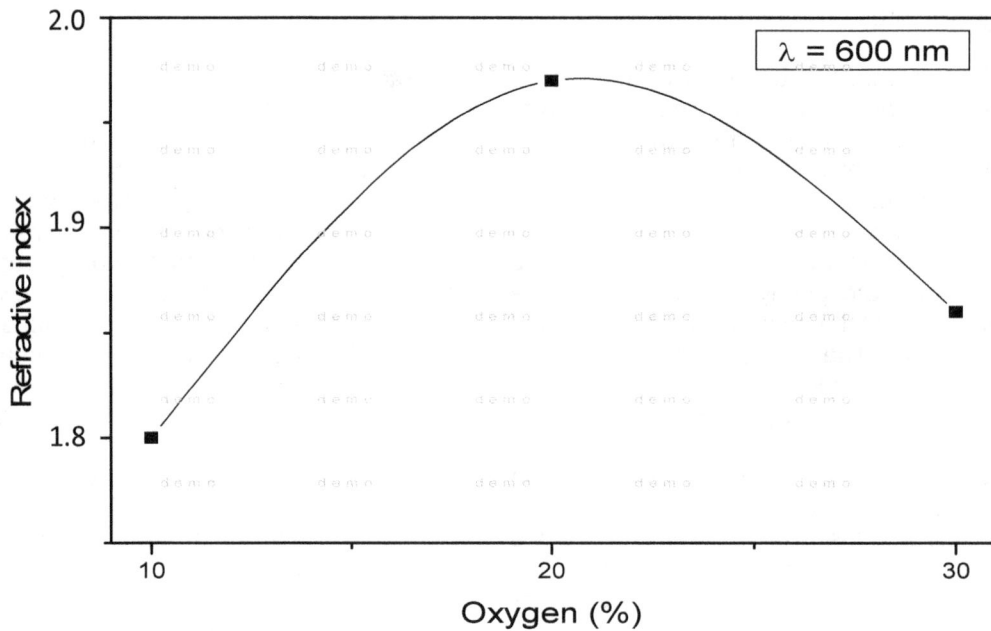

**Figure 6.** Refractive index as a function of oxygen concentration at constant λ.

conditions. Knowing that the optical properties of thin films depend on the thickness, we determined the thickness of our ZnO films by Equation (12) and we have compared to values determined by profilometry. Figure 8 shows the variation of the thickness of the ZnO thin film as a function of the launched power. We find that the thicknesses obtained by optical determination decrease as the power increases. This explains, perhaps, the low transmittance samples drawn over 50 W. In addition, the evolution of the diffraction peak as a function of RF power, shows that the samples prepared at most 50 W exhibit crystallization defects.

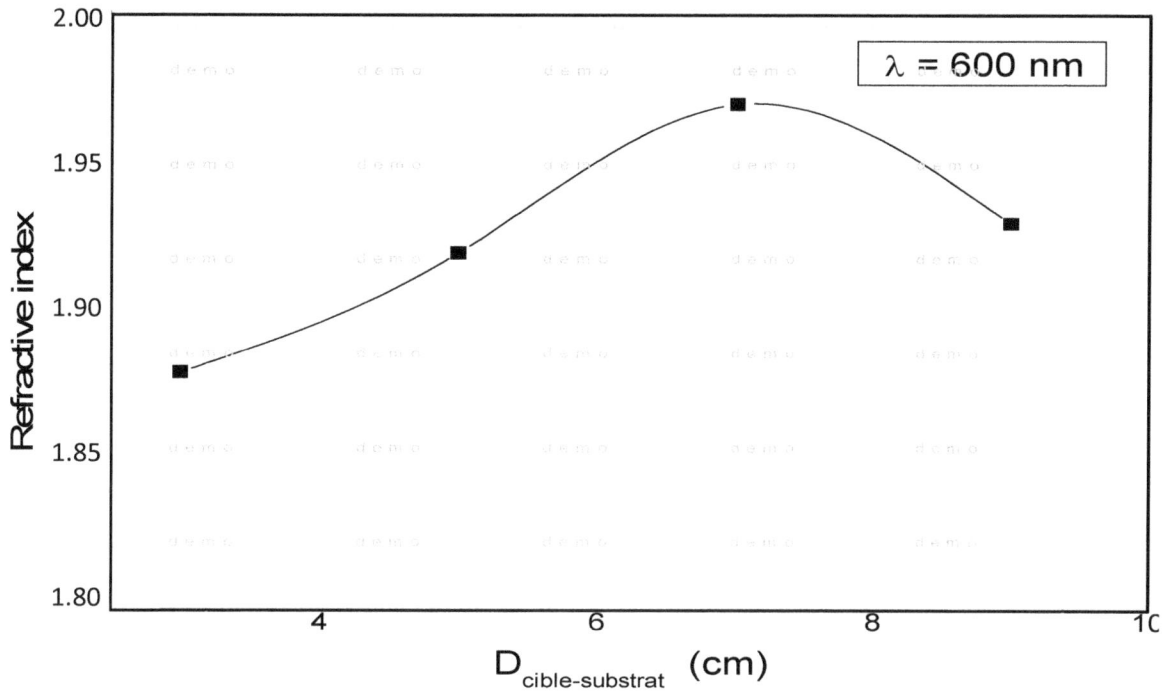

**Figure 7.** Refractive index as a function of the substrate-target distance constant $\lambda$.

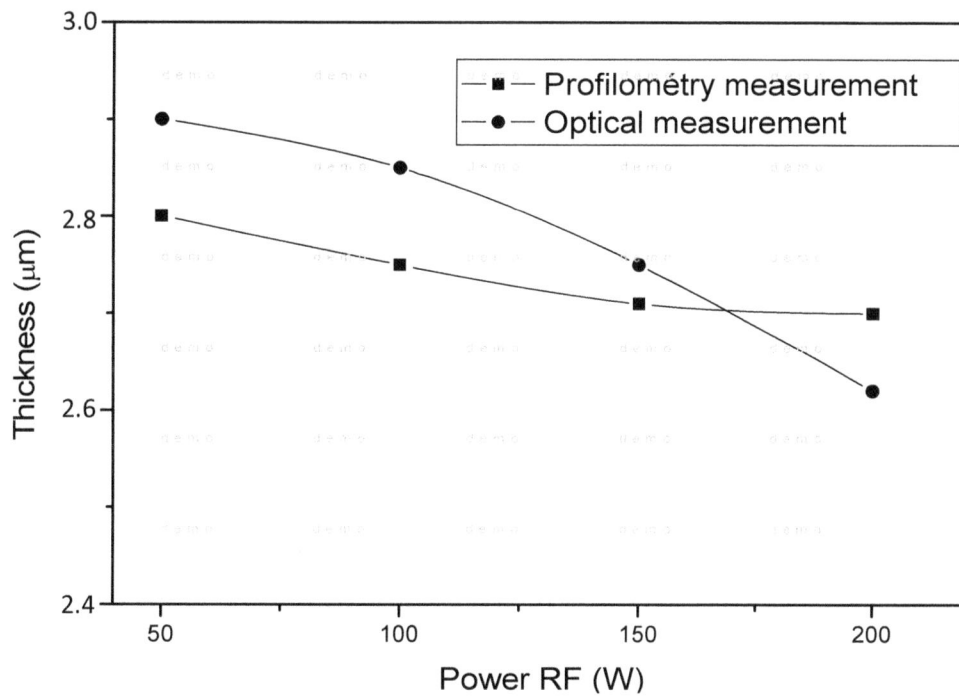

**Figure 8.** Evolution of the thickness of the ZnO film in function of the power RF.

## Determination of the optical gap $E_g$

The X-ray part presents a spectrum of electromagnetic radiation. The wave-particle duality of radiation is expressed by such a relation between the energy of a photon, and the wavelength $\lambda$.

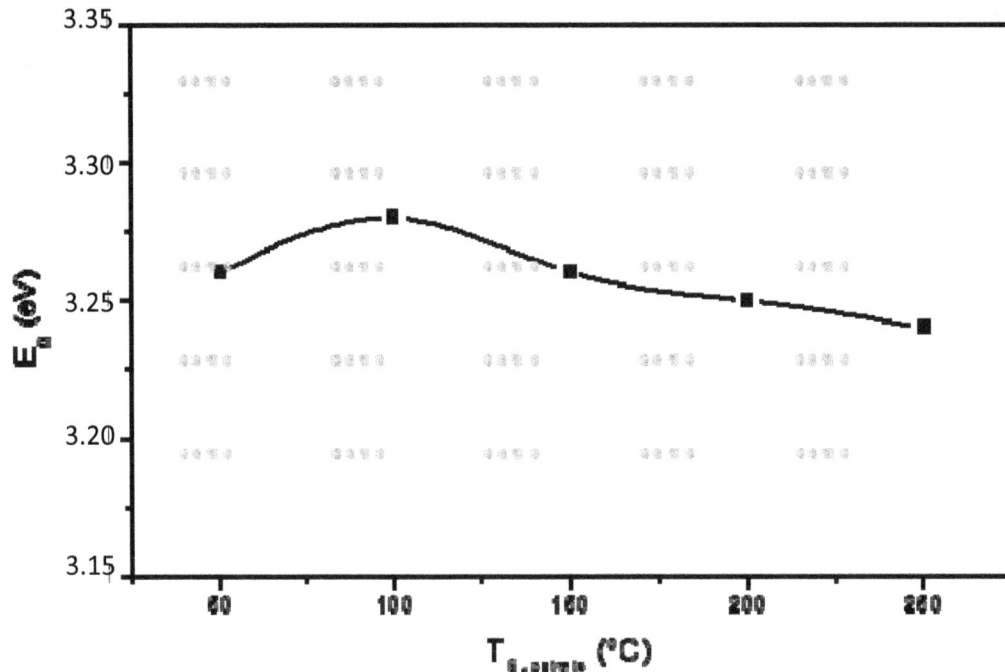

**Figure 9.** Dependence of the gap as a function of substrate temperature.

$$E = h\nu = \frac{hc}{\lambda} \quad \text{or} \quad \lambda = \frac{hc}{E} \tag{13}$$

The simple and well-known formula is:

$$\lambda_{nm} = \frac{1,24}{E(eV)} \tag{14}$$

It is from this relationship that we determine the energy of the band gap of our prepared by RF magnetron sputtering thin layers. In fact, we make a linear extrapolation at the absorption front of our power transmission layer. This straight line intersects the wavelength axis at a value of $\lambda$. For the different samples, we determined the optical gap from Equation (14). Indeed, the level of the linear variation of the absorption front, we drew a curve tangential to the front. This linear extrapolation, which cuts the axis of wavelengths, we can determine the optical gap. Figure 9 shows the evolution of the gap energy as a function of substrate temperature band. We find that the evolution of the energy of the forbidden band as a function of substrate temperature is almost constant. However, the sample prepared at 100°C gives an energy gap greater range compared to those given by the other samples.

## Conclusion

Here, the effect on different experimental parameters on the growth and the properties of thin layers of zinc oxide has been studied. We performed several sets of samples we characterized optically. This systematic study led us to an area of very specific definition of manufacturing parameters for obtaining ZnO films of good quality. Indeed, the numerical parameters of the manufacturing balance sheet are as follows: 100°C for the substrate temperature, 50 W RF power injected into the discharge, 20% to the oxygen content in the gas mixture and 7 cm for the target-substrate distance.

These results, which are consistent with those found in the literature on zinc oxide prepared by RF magnetron sputtering, using a zinc target, will allow us to achieve the intended applications.

## Conflict of Interest

The authors have not declared any conflict of interest.

### REFERENCES

Craciun V, Elders J, Gardeniers JGE, Geretovsky J, Boyd IW (1995). Growth of ZnO thin films on GaAs by pulsed laser deposition. Thin Solid Films 259:1-4. doi:10.1016/0040-6090(94)09479-9

Han MY, Jou JH (1995). Determination of the mechanical properties of rf magnetron sputtered zinc oxide thin films on substrate. Thin Solid Films. 260:58-64. doi:10.1016/0040-6090(94)06459-8

Inukai T, Matsuoka M, Ono K (1995). Characteristics of zinc oxide thin films prepared by r.f. magnetron-mode electron cyclotron resonance sputtering. Thin Solid Films. 257:22-27. doi:10.1016/0040-6090(94)06325-7

Kim YJ, Kim YT, Yang HK, Park JC, Han JI, Lee YE, Kim HJ (1997). Epitaxial growth of zno thin films on r-plane sapphire substrate by radio frequency magnetron sputtering. J. Vacuum Sci. Technol. A. 15:1103-1107. DOI: 10.1116/1.580437

Lin YC, Hong CR, Chuang HA (2008). Fabrication and analysis of ZnO thin film bulk acoustic resonators. Appl. Surf. Sci. 254(13):3780-3786. doi:10.1016/j.apsusc.2007.11.059

Lyashenko SP, Miloslavskii VK (1964). A simple method for the determination of the thickness and optical constants of semiconducting and dielectric layers. Opt. Spectrosc. 16:80-81.

Manifacier JC, Gasiot J, Fillard JP (1976). A simple method for the determination of the optical constants n, k and the thickness of a weakly absorbing thin film. Phys. E: Sci. Instrum. 9(1976):1002. doi:10.1088/0022-3735/9/11/032

Meng LJ, Dos Santos MP (1994). Direct current reactive magnetron sputtered zinc oxide thin films the effect of the sputtering pressure. Thin Solid Films 250:26-32. doi:10.1016/0040-6090(94)90159-7

Molarius J, Kaitila J, Pensala T, Ylilammi M (2003). Piezoelectric ZnO films by r.f. sputtering. J. Mater. Sci.: J. Mater. Electron. 14:431-435. DOI: 10.1023/A:1023929524641

Ondo-Ndong R, Ferblantier G, Al Kalfioui M, Boyer A, Foucaran A (2003). Properties of rf magnetron sputtered zinc oxide thin films. J. Cryst. Growth. 255:130-135. doi:10.1016/S0022-0248(03)01243-0

Sanchez-Juarez A, Tiburcio-Silver A, Ortiz A, Zironi EP, Rickards J (1998). Electrical and optical properties of fluorine-doped ZnO thin films prepared by spray pyrolysis. Thin Solid Films 333:196-202. doi:10.1016/S0040-6090(98)00851-7

Sourdi I, Mamat MH, Abdullah MH, Ishak A, Rusop M (2012). Optical properties of nano-structured zinc oxide thin films deposited by radio-frequency magnetron sputtering at different substrate temperatures. Humanities, Science and Engineering Research (SHUSER), 2012 IEEE Symposium on 24-27 June 2012:607–611. DOI: 10.1109/SHUSER.2012.6268895

Subramanyam TK, Srinivasulu Naidu B, Uthanna S (1999). Structure and optical properties of dc reactive magnetron sputtered zinc oxide films. Cryst. Res. Technol. 34:981–988. DOI: 10.1002/(SICI)1521-4079(199909)34:8<981::AID-CRAT981>3.0.CO;2-G

Wales J, Lovitt GJ, Hill RA (1967). Optical properties of germanium films in the 1–5 µ range. Thin Solid Films. 1:137-150. doi:10.1016/0040-6090(67)90010-7.

Yang Ming L, Shu Yi T, Jeng Jong L, Min Hsiung H (2007). The structural and optical properties of zinc oxide thin films deposited on PET Substrate by r.f. Magnetron Sputtering. Solid State Phenomena.121–123 :971-974. 10.4028/www.scientific.net/SSP.121-123.971.

# Permissions

# List of Contributors

**Sahebali Manafi**
Department of Engineering, Shahrood Branch, Islamic Azad University, Shahrood, Iran

**Mojtaba Jafarian**
Department of Engineering, Shahrood Branch, Islamic Azad University, Shahrood, Iran

**S. Ourabah**
Faculty of Physics, University of Sciences and Technology, Houari Boumediène, BP 31 El Alia, Bab Ezzouar, 16111 Algiers, Algeria

**A. Amokrane**
Faculty of Physics, University of Sciences and Technology, Houari Boumediène, BP 31 El Alia, Bab Ezzouar, 16111 Algiers, Algeria
Preparatory National School for Engineer Studies, Rouiba, Algiers, Algeria

**M. Abdesselam**
Faculty of Physics, University of Sciences and Technology, Houari Boumediène, BP 31 El Alia, Bab Ezzouar, 16111 Algiers, Algeria

**Juu-En Chang**
Department of Environmental Engineering, National Cheng Kung University, Taiwan

**Yi-Kuo Chang**
Department of Safety Health and Environmental Engineering, Central Taiwan University of Science and Technology, 666 Po-Tze Road, Peitun District, Taichung city 406, Taiwan

**Min-Her Leu**
Department of Environmental Engineering, Kun Shan University of Technology, Taiwan

**Ying-Liang Chen**
Sustainable Environment Research Center, National Cheng Kung University, Taiwan

**Jing-Hong Huang**
Department of Environmental Engineering, National Cheng Kung University, Taiwan

**Olushola S. Ayanda**
Department of Chemistry, Faculty of Applied Sciences, Cape Peninsula University of Technology, P. O. Box 1906, Cape Town, South Africa

**Folahan A. Adekola**
Department of Chemistry, University of Ilorin, P. M. B. 1515, Ilorin, Kwara State, Nigeria

**Alafara A. Baba**
Department of Chemistry, University of Ilorin, P. M. B. 1515, Ilorin, Kwara State, Nigeria

**Bhekumusa J. Ximba**
Department of Chemistry, Faculty of Applied Sciences, Cape Peninsula University of Technology, P. O. Box 1906, Cape Town, South Africa

**Olalekan S. Fatoki**
Department of Chemistry, Faculty of Applied Sciences, Cape Peninsula University of Technology, P. O. Box 1906, Cape Town, South Africa

**P. Ganesh Babu**
Department of Physics, Sri Ramanujar Engineering College, Chennai-600 048, India
Department of Ceramic Technology, A. C .Tech Campus, Anna University, Chennai-25, India

**P. Manohar**
Department of Ceramic Technology, A. C .Tech Campus, Anna University, Chennai-25, India

**C. Mahamadi**
Department of Chemistry, Bindura University of Science Education, P. Bag 1020, Bindura, Zimbabwe

**B. Oto**
Department of Physics, Faculty of Science, Yüzüncü Yil University, 65080 Van, Turkey

**A. Gür**
Department of Chemistry, Faculty of Science, Yüzüncü Yil University, 65080 Van, Turkey

**U. Böyük**
Department of Science Education, Education Faculty, Erciyes University, Kayseri, Turkey

**S. Engin**
Department of Physics, Institute of Science and Technology, Erciyes University, Kayseri, Turkey

**H. Kaya**
Department of Science Education, Education Faculty, Erciyes University, Kayseri, Turkey

**N. Mara sli**
Department of Physics, Faculty of Science, Erciyes University, Kayseri, Turkey

**E. Çadirli**
Department of Electronics and Automation, Technical Vocational School of Sciences, Niğde University, Niğde, Turkey

**M. Sahin**
Department of Electronics and Automation, Technical Vocational School of Sciences, Niğde University, Niğde, Turkey

**M. Aliahmad**
Department of Physics, University of Sistan and Baluchestan, Zahedan, Iran

**M. Noori**
Department of Physics, University of Sistan and Baluchestan, Zahedan, Iran

**N. Hatefi Kargan**
Department of Physics, University of Sistan and Baluchestan, Zahedan, Iran

**M. Sargazi**
Faculty of Science, University of Payam-e- Noor, Tehran, Iran

**Naveed Ahmed**
Department of Mathematics, Faculty of Sciences, HITEC University, Taxila Cantt, Pakistan

**Umar Khan**
Department of Mathematics, Faculty of Sciences, HITEC University, Taxila Cantt, Pakistan

**Sheikh Irfanullah Khan**
Department of Mathematics, Faculty of Sciences, HITEC University, Taxila Cantt, Pakistan
COMSATS Institute of Information Technology, University Road, Abbottabad, Pakistan

**Yang Xiao-Jun**
College of Science, China University of Mining and Technology, Xuzhou, Jiangsu, 221008, China

**Zulfiqar Ali Zaidi**
Department of Mathematics, Faculty of Sciences, HITEC University, Taxila Cantt, Pakistan
COMSATS Institute of Information Technology, University Road, Abbottabad, Pakistan

**Syed Tauseef Mohyud-Din**
Department of Mathematics, Faculty of Sciences, HITEC University, Taxila Cantt, Pakistan

**Fatma Meydaneri**
Department of Metallurgy and Materials Engineering, Faculty of Engineering, Karabük University, 78050, Karabük, Turkey

**Buket Saatçi**
Department of Physics, Faculty of Arts and Sciences, Erciyes University, 38039, Kayseri, Turkey

**Ahmet Ülgen**
Department of Chemistry, Faculty of Arts and Sciences, Erciyes University, 38039, Kayseri, Turkey

**Nitin Kumar**
ECE Department, Uttarakhand Technical University, Dehradun, Uttarakhand, India

**S. C. Gupta**
ECE Department, DIT University, Dehradun, Uttarkhand, India

**Olatunde Stephen Olatunji**
Department of Chemistry, Faculty of Applied Sciences, Cape Peninsula University of Technology Bellville Western Cape, South Africa

**Oladele Osibanjo**
Basel Regional Coordination Centre, Faculty of Science, University of Ibadan, Ibadan, Oyo State Nigeria

**S. O. Yakubu**
Department of Mechanical Engineering, Nigerian Defence Academy, Kaduna

**N. N. Garba**
Department of Physics, Ahmadu Bello University, Zaria, Nigeria

**Y. A. Yamusa**
Centre for Energy Research and Training, Ahmadu Bello University, Zaria, Nigeria

**A. Isma'ila**
Department of Physics, Ahmadu Bello University, Zaria, Nigeria

**S. A. Habiba**
Department of Physics, Ahmadu Bello University, Zaria, Nigeria

**Z. N. Garb**
Department of Chemistry, Ahmadu Bello University, Zaria, Nigeria

**Y. Musa**
Centre for Energy Research and Training, Ahmadu Bello University, Zaria, Nigeria

**S. A. Kasim**
Centre for Energy Research and Training, Ahmadu Bello University, Zaria, Nigeria

**Seyed Hossein Hosseini**
Department of Chemistry, Faculty of Science, Islamshahr Branch, Islamic Azad University, Tehran-Iran

**A. Asadnia**
Young Researchers Club, Center Tehran Branch, Islamic Azad University, Tehran-Iran

**Damaris Mbui**
Department of Chemistry, College of Biological and Physical Sciences P. O. Box 00100-30197, University of Nairobi, Kenya

**Duke Omondi Orata**
Department of Chemistry, College of Biological and Physical Sciences P. O. Box 00100-30197, University of Nairobi, Kenya

**Graham Jackson**
University of Cape Town P. O. Private Bag Rondebosch 7701, South Africa

**David Kariuki**
Department of Chemistry, College of Biological and Physical Sciences P. O. Box 00100-30197, University of Nairobi, Kenya

**Alla Srivani**
Department of Physics, T. J. P. S College and Sri Mittapalli college of Engineering, Guntur, Andhra Pradesh, India

**Vedam Ram Murthy**
Department of Physics, T. J. P. S College and Sri Mittapalli college of Engineering, Guntur, Andhra Pradesh, India

**G. Veera Raghavaiah**
Department of Physics, T. J. P. S College and Sri Mittapalli college of Engineering, Guntur, Andhra Pradesh, India

**Seyed Hossein Hosseini**
Department of Chemistry, Faculty of Science, Islamic Azad University, Islamshahr Branch, Tehran, Iran

**Prabakaran Kandasamy**
Research Center in Physics, VHNSN College, Virudhunagar – 626001, Tamilnadu, India

**Amalraj Lourdusamy**
Research Center in Physics, VHNSN College, Virudhunagar – 626001, Tamilnadu, India

**S. Thirumavalavan**
Department of Mechanical Engineering, Sathyabama University, Chennai-600 119, India

**K. Mani**
Department of Mechanical Engineering, Panimalar Engineering College, Chennai-602103, India

**S. Sagadevan**
Crystal Growth Centre, Anna University, Chennai-600 025, India

**Liu Xiao-Li**
Science and Technology on Vacuum Technology and Physics Laboratory, Lanzhou Institute of Physics, Lanzhou 730000, China

**Xiong Yu-Qing**
Science and Technology on Vacuum Technology and Physics Laboratory, Lanzhou Institute of Physics, Lanzhou 730000, China

**Ren Ni**
Science and Technology on Vacuum Technology and Physics Laboratory, Lanzhou Institute of Physics, Lanzhou 730000, China

**Yang Jian-Ping**
Science and Technology on Vacuum Technology and Physics Laboratory, Lanzhou Institute of Physics, Lanzhou 730000, China

**Wang Rui**
Science and Technology on Vacuum Technology and Physics Laboratory, Lanzhou Institute of Physics, Lanzhou 730000, China

**Wu Gan and**
Science and Technology on Vacuum Technology and Physics Laboratory, Lanzhou Institute of Physics, Lanzhou 730000, China

**Wu Sheng-Hu**
Science and Technology on Vacuum Technology and Physics Laboratory, Lanzhou Institute of Physics, Lanzhou 730000, China

**R. Ondo-Ndong**
Laboratoire pluridisciplinaire des sciences, Ecole Normale Supérieure, B.P 17009 Libreville, Gabon
Institut Electronique du Sud, IES-Unité mixte de Recherche du CNRS n° 5214, Université Montpellier II, Place E. Bataillon, 34095 Montpellier cedex 05- France

**H. Z. Moussambi**
Laboratoire pluridisciplinaire des sciences, Ecole Normale Supérieure, B.P 17009 Libreville, Gabon

**H. Gnanga**
Laboratoire pluridisciplinaire des sciences, Ecole Normale Supérieure, B.P 17009 Libreville, Gabon

**A. Giani**
Institut Electronique du Sud, IES-Unité mixte de
Recherche du CNRS n° 5214, Université Montpellier II,
Place E. Bataillon, 34095 Montpellier cedex 05- France

**A. Foucaran**
Institut Electronique du Sud, IES-Unité mixte de
Recherche du CNRS n° 5214, Université Montpellier II,
Place E. Bataillon, 34095 Montpellier cedex 05- France